Recent Advances in Renewable Energy

(*Volume 3*)

(Renewable Energy Engineering: Solar, Wind, Biomass, Hydrogen and Geothermal Energy Systems)

Edited by

Emmanuel D. Rogdakis & Irene P. Koronaki

Thermal Engineering Section, School of Mechanical Engineering,
National Technical University of Athens, 9th Heroon Polytechneiou St.
15780 Zografou, Athens, Greece

Recent Advances in Renewable Energy

Volume # 3

Renewable Energy Engineering: Solar, Wind, Biomass, Hydrogen, and Geothermal Energy Systems

Editors: Emmanuel D. Rogdakis & Irene P. Koronaki

ISSN (Online): 2543-2397

ISSN (Print): 2543-2389

ISBN (Online): 978-1-68108-719-1

ISBN (Print): 978-1-68108-720-7

General:

1. Any dispute or claim arising out of or in connection with this License Agreement or the Work (including non-contractual disputes or claims) will be governed by and construed in accordance with the laws of the U.A.E. as applied in the Emirate of Dubai. Each party agrees that the courts of the Emirate of Dubai shall have exclusive jurisdiction to settle any dispute or claim arising out of or in connection with this License Agreement or the Work (including non-contractual disputes or claims).
2. Your rights under this License Agreement will automatically terminate without notice and without the need for a court order if at any point you breach any terms of this License Agreement. In no event will any delay or failure by Bentham Science Publishers in enforcing your compliance with this License Agreement constitute a waiver of any of its rights.
3. You acknowledge that you have read this License Agreement, and agree to be bound by its terms and conditions. To the extent that any other terms and conditions presented on any website of Bentham Science Publishers conflict with, or are inconsistent with, the terms and conditions set out in this License Agreement, you acknowledge that the terms and conditions set out in this License Agreement shall prevail.

Bentham Science Publishers Ltd.
Executive Suite Y - 2
PO Box 7917, Saif Zone
Sharjah, U.A.E.
Email: subscriptions@benthamscience.org

BENTHAM SCIENCE

CONTENTS

PREFACE

This book examines the recent advances from the theoretical as well as applied perspectives addressing the major issues associated with renewable energy systems. Each chapter is self-contained and tackles the fundamental issues and latest developments of a particular sub-topic. This book gives the opportunity to even non-specialized readers to understand the complexity of each topic, and to access the most up-to-date literature. Moreover, it enables specialized readers to broaden their understanding of complex renewable energy topics and it provides a comprehensive overview of the cutting-edge developments of the issues covered by the book.

This book covers important themes including CSPs, thermal energy storage systems, bioenergy applications, hydrogen production storage systems and normal shallow geothermal systems as well as the measurement techniques that are used for these systems. "Recent advances in Renewable Energy Systems" is a reference book for professional engineers from power, refrigeration companies as well as engineering students because it connects the theory to applications in such a way that can be easily understood.

It is intended for researchers and postgraduates with an interest in energy, climate change and environmental economics, and also policymakers and energy companies.

Chapter 1 introduces CSP as mature technology according to its widespread deployment. CSP is the only type of renewable energy that allows long term energy storage over sufficiently long periods of time and at large scales to completely eliminate the intermittent nature of the solar resource. However, cost reduction is essential in order to compete with alternative sources. Existing technologies are described in detail highlighting improvement opportunities. Increasing the overall solar-to-electric energy conversion efficiency by developing new power conversion pathways and looking for alternative markets for CSP such as desalination, process heat, enhanced oil recovery or hybridization appear to be the best options for the future.

In **Chapter 2**, Thermal Energy Storage (TES) systems are presented, supported by renewable energy sources (mainly solar energy as an effective means of achieving the aforementioned goal). This study reviews the available TES systems. In this context, sensible TES systems which utilize liquid and solid storage media and their applications are presented. Furthermore, the usage of ice and other solid-liquid phase change materials in latent heat storage systems is investigated in terms of materials, applications and future trends. Finally, the utilization of thermochemical reactions in TES systems is presented.

In **Chapter 3**, new and innovative ideas about the adoption of solar energy systems in buildings are presented. Simple and low cost solar collectors which can produce heating in low and medium temperatures levels are analyzed. Emphasis is given in the utilization of Phase Change Materials, as well as in the utilization of solar assisted heat pumps. Moreover, innovative passive heating systems, as Trombe wall are presented with details.

Chapter 4 presents the use of solid, liquid and gaseous biomass as an energy source, with particular emphasis being given to biogas production by anaerobic digestion.

Chapter 5 presents the evolution of wind energy throughout the last 30 years, along with its prospects for covering a considerable percentage of the future global electrical demand. Furthermore, available information concerning the major wind energy markets has been analysed and revealed in general the existing trends of wind energy for the next years to come. In this context, technology and financial aspects along with environmental issues arising from wind power projects' implementation are investigated in order to provide all the necessary data for acquiring an integrated view of the wind energy future.

Chapter 7 provides basic operational and construction guidelines concerning all possible geothermal plants (closed or open loop) along with representing drawings. The chapter begins with basic geological information regarding the ability of the subsoil to absorb or supply heat via the integrated heat exchanger. Subsequently, a summary on the operation of heat pumps in general and the advantages of geothermal pumps is provided. Then, the two main categories of geothermal loops are analyzed: closed loop horizontal or vertical geothermal exchangers and open loop system installations. Finally, the proposed methodology to follow when such a system should be designed is discussed, containing cost elements issues of such systems.

Chapter 8 presents the essential continuous monitoring of outdoor natural qualities for optimizing operation, preventing small and large scale damages and adapting design according local conditions. Common requirement for such applications is saving data for a long period of time and/or the ability to send them on line. This is why most of the instrumentation presented in this chapter is based on transducers: flow transducers, temperature transducers solar radiation transducers, humidity transducers *etc*.

Emmanuel D. Rogdakis & Irene P. Koronaki
Thermal Engineering Section
School of Mechanical Engineering
National Technical University of Athens
9th Heroon Polytechneiou St.
15780 Zografou, Athens,
Greece

List of Contributors

A.A. Nikoglou	Engineering Measurements Laboratory, School of Mechanical Engineering, National Technical University of Athens – NTUA, 9 Iroon Polytechniou, Zografos Campus, Athens 15780, Greece
Christos Tzivanidis	Solar Energy Laboratory, Thermal department, Heroon Polytechniou 9, 15780 Athens, Greece
Dimitrios Grimekis	Laboratory of Steam Boilers and Thermal Plants, National Technical University of Athens, 9th Heroon Polytechniou St., Athens, 15780, Greece
Efstratios G. Papoutsis	Laboratory of Applied Thermodynamics, Zografou, 15780, Greece
Evangelos Bellos	Solar Energy Laboratory, Thermal department, Heroon Polytechniou 9, 15780 Athens, Greece
Emmanouil D. Rogdakis	Laboratory of Applied Thermodynamics, National Technical University of Athens, Greece
E.P. Hinis	Engineering Measurements Laboratory, School of Mechanical Engineering, National Technical University of Athens – NTUA, 9 Iroon Polytechniou, Zografos Campus, Athens 15780, Greece
Irene P. Koronaki	Laboratory of Applied Thermodynamics, Zografou, 15780, Greece
John K. Kaldellis	Soft Energy Applications and Environmental Protection Laboratory, Piraeus University of Applied Sciences, P.O. Box 41046, Athens, 12201, Greece
J. Apostolou	Soft Energy Applications and Environmental Protection Laboratory, Piraeus University of Applied Sciences, P.O. Box 41046, Athens, 12201, Greece
Konstantinos Braimakis	Laboratory of Steam Boilers and Thermal Plants, National Technical University of Athens, 9th Heroon Polytechniou St., Athens, 15780, Greece
Michael T. Nitsas	Laboratory of Applied Thermodynamics, Zografou, 15780, Greece
Michalis G. Vrachopoulos	Energy and Environmental Research Laboratory, Psachna, 34400, Euboea, Greece
Maria K. Koukou	Energy and Environmental Research Laboratory, Psachna, 34400, Euboea, Greece
Nikolaos P. Tsolakoglou	Energy and Environmental Research Laboratory, Psachna, 34400, Euboea, Greece
P.K. Rouni	Engineering Measurements Laboratory, School of Mechanical Engineering, National Technical University of Athens – NTUA, 9 Iroon Polytechniou, Zografos Campus, Athens 15780, Greece
Panagiotis I. Bitsikas	Laboratory of Applied Thermodynamics, National Technical University of Athens, Greece
Petros Vlakakis	Laboratory of Steam Boilers and Thermal Plants, National Technical University of Athens, 9th Heroon Polytechniou St., Athens, 15780, Greece
Sonia Fereres	ThermoFluids Laboratory, Abengoa Research, Campus Palmas Altas, Energia Solar 1, Seville, 41014, Spain
Sotirios Karellas	Laboratory of Steam Boilers and Thermal Plants, National Technical University of Athens, 9th Heroon Polytechniou St., Athens, 15780, Greece

Trifon Roumpedakis Laboratory of Steam Boilers and Thermal Plants, National Technical University of Athens, 9th Heroon Polytechniou St., Athens, 15780, Greece

CHAPTER 1

Solar Thermal Energy Systems

Sonia Fereres*

ThermoFluids Laboratory, Abengoa Research, Campus Palmas Altas, Energia Solar 1, Seville, 41014, Spain

Abstract: To date, concentrated solar power is the only type of renewable energy that allows for long term energy storage over sufficiently long periods of time and at large scales to completely eliminate the intermittent nature of the solar resource. It is a proven and mature technology as shown by its widespread deployment. However, cost reduction is essential in order to compete with alternative sources. Existing technologies are described in detail highlighting improvement opportunities. Increasing the overall solar-to-electric energy conversion efficiency by developing new power conversion pathways and looking for alternative markets for Concentrated Solar Power such as desalination, process heat, enhanced oil recovery or hybridization appear to be the best options for the future.

Keywords: Alternative solar thermal markets, Concentrated solar power, Heat transfer fluids, Solar receivers, Thermal energy storage.

INTRODUCTION

Overview of Solar Thermal Energy Systems

Globally, the current production of energy from non-renewable fossil fuel sources is unsustainable both from the point of view of their depletion and their impact on global warming and climate change. The energy problem is of such complexity, that there is not a unique technological solution or a single source of renewable energy that would drastically alter the present situation in the short run. Sustainable and low-carbon technologies will play a crucial role in the energy mix in the years to come. Solar thermal energy is one of those technologies.

Solar thermal energy encompasses a large number of technologies that harness solar energy to produce heat. This thermal energy, depending on its temperature range and on the scale of production, can be utilized directly as heat for residential

* **Corresponding author Sonia Fereres:** ThermoFluids Laboratory, Abengoa Research, Campus Palmas Altas, Energia Solar 1, Seville, 41014, Spain; Tel: +(34) 954936185, +34 626170992; Email: sonia.fereres@abengoa.com

Emmanuel D. Rogdakis & Irene P. Koronaki (Eds.)

or industrial applications. The solar heat can be further converted into electricity or into other energy vectors such as fuels.

This chapter will focus on the generation of electricity from solar thermal energy. The market and current deployment of solar thermal electric energy (STE) or Concentrated Solar Power (CSP) systems will be discussed, highlighting the space where Concentrated solar power plants can be profitable and compete with other renewable and non-renewable energy sources. The different Concentrated solar power technologies will be described, discussing in detail the main elements in a Concentrated solar power system and where the current technological challenges reside. Recent research and developments in the different areas will be analysed and, finally, the current trends and opportunities towards achieving further efficiency improvements and cost reductions will be discussed.

Solar Energy Conversion to Electricity: Concentrated Solar Energy

Concentrated Solar Power (CSP) or Solar Thermal Energy Electricity (STE) generates electricity by concentrating solar radiation to heat a material (typically a fluid). The heated material, referred to as Heat Transfer Fluid (HTF) or Heat Transfer Medium (HTM), is then used to generate steam (or to heat a different working fluid such as air) to drive a turbine-generator set in the power block. As a result of this solar-to-heat conversion step, thermal energy storage is easily incorporated, making it the main competitive advantage of Concentrated solar power systems. Thermal storage allows dispatchable renewable energy electricity generation at any time of the day, at night or during periods of low solar insolation. The excess heat produced during peak insolation periods can be used to heat up the heat transfer fluids and storage materials. This heat can be discharged later when the sun is not shining, decoupling heat generation from electricity production.

Concentrated solar power also has the advantage of being able to produce electricity at a utility scale. Furthermore, because the power block unit in Concentrated solar power is similar to that of conventional fossil fuel thermal power systems (*e.g.* steam cycles, an ORC, *etc.*), the system is easy to operate and allows easy hybridization particularly with natural gas combined cycle plants. Recent years have seen a rapid growth in STE installed capacity, price, and in technological and performance improvements. However, for Concentrated solar power to be cost-competitive with fossil fuels and with photovoltaics (PV), additional developments are still needed to further reduce its costs. Fossil fuel generation such as natural gas combined cycle (NGCC) power plants might still need to be appropriately penalized for carbon dioxide emissions in order to provide a fair comparison to STE electric generation.

In the past 5-10 years, the generation of solar thermal electricity through Concentrated solar power plants has grown robustly yet much slower than anticipated in the past [1]. The first commercial CSP plants were deployed in California, USA, in the 1980s. In the late 2000s, the economic crisis limited the resurgence of CSP plant development in Spain, which had expanded thanks to government subsidies. Construction of CSP in the USA was slow until 2013 due to competition of cheap conventional and unconventional gas sources (*i.e.* shale gas) and constantly decreasing PV prices. However, the capability of CSP plants to supply electricity on demand through their built-in storage will continue to gain importance until other intermittent renewable energy forms such as photovoltaics (PV) and wind power increase their shares of global electricity and essentially until large scale battery technology finally takes off.

There are currently over 10 GW worldwide installed (or under construction/ development) solar thermal power plants [2]. This number includes all different types of technology (parabolic trough, linear Fresnel reflector, power tower, and dish/engine systems). Of the 10 GW, 44% correspond to operational facilities, 14% are under construction, and 42% are under development. A current map of solar thermal projects with the power generated by country or region and by power plant status can be seen in Fig. (**1**), even though at the moment it is uncertain if all projects will achieve the required permits, financing, and power purchase agreements [1].

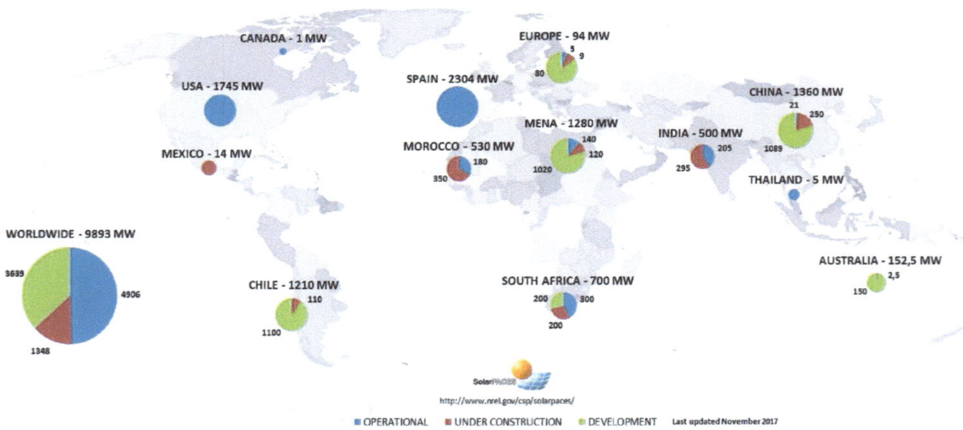

Fig. (1). CSP Projects around the world [2].

Unfortunately, even though CSP allows for large scale generation, it is also very sensitive to scale [3]. CSP plants need scales of megawatts or larger to maximize efficiency and minimize costs, requiring very large capital investments and financial risks that not everybody can take on. It is a technologically viable solution for areas with large solar resource and land availability, but it is not currently cost-competitive without subsidies. This is the reason for important cost reduction investment efforts worldwide. For example, the United States Department of Energy launched a very ambitious research funding program in 2010 named Sunshot Initiative to reduce the leveled cost of solar generated electricity by 75% in a decade [4] without employing subsidies. To understand where to focus these cost reduction efforts, Sunshot Initiative proposed a simplified cost breakdown of CSP electricity that can be seen in Fig. (**2**) comparing typical costs from 2010 with the projected necessary reductions to achieve competitiveness in 2020. The solar field and thermal energy storage are the areas with largest cost reduction potential (above 75%), followed by the receiver, heat transfer fluid, and power plant.

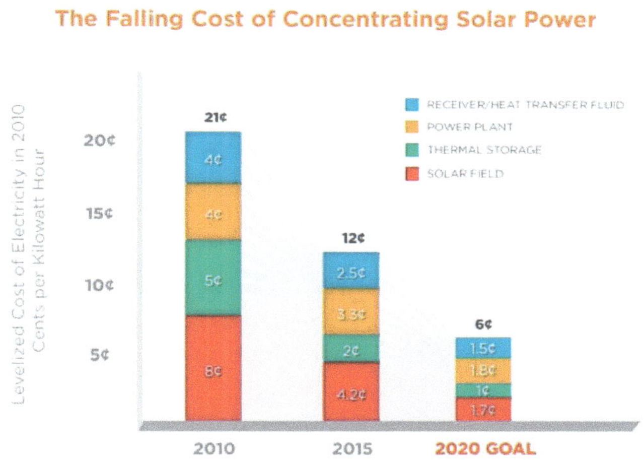

Fig. (2). United States Department of Energy - SunShot Program cost reduction targets for CSP systems [4].

Similarly, the International Energy Agency (IEA) SolarPACES Strategic Plan from 2012 to 2016 [2] also has concentrated in achieving a significant cost reduction for new plants, aiming to guarantee a high performance over the power plant life time. Cost reduction is suggested to be mainly attained through simplifications, mass production and evolution in the manufacturing processes but without forgetting the need for new developments of advanced materials, processes, and concepts.

Other limitations of CSP technology are the availability of high voltage transmission connections to major electricity consuming centers in the vicinity of such large installations. CSP plants also require vast land use and large water supply in locations where water is a scarce resource such as desserts. Most of the water use in a CSP plant is for cooling purposes, whereas a comparatively minimal amount of water is required to clean the solar collector reflective surfaces. A contemporary wet-cooled parabolic trough CSP plant requires 2.9–3.2 liters per kilowatt-hour (kWh) of water for cooling purposes [3]. Consequently, the development of dry cooling technologies is growing rapidly [4]. Unfortunately, dry cooling technologies currently only represent a very small fraction of the installed CSP plants [5].

In terms of technological limitations, the main challenge remains increasing the thermodynamic cycle upper working temperatures. A further increase of operational temperatures can have major impact on energy conversion efficiency and cost reductions, however higher efficient power cycles that can be coupled to the solar field are yet to be fully demonstrated at the commercial scale. Moreover, current STE designs are approaching container material thermal limits. Additionally, there are no ideal candidates for new higher temperature heat transfer media. New molten salt mixtures that can be used at higher temperatures typically also have higher corrosion rates with common container materials and/or have higher freezing points, merely shifting (but not increasing) the operating temperature range. Novel HTM will involve rethinking current STE power plants and will consequently increase the risk (and financing challenges) of the future installations.

In spite of the aforementioned shortcomings, the CSP installations have continued to grow. The CSP deployment in the US in the past few years can be seen in Table **1**. It includes both trough and tower technology, at scales comparable to current fossil power plants (> 100 MW, most approaching 300 MW) and not all of them include thermal storage.

Table 1. Recent CSP installations in the United States (Source: SunShot- DoE [4]).

Project	**Solana**	**Genesis**	**Mojave**	**Ivanpah**	**Crescent Dunes**
Technology	Trough	Trough	Trough	Tower	Tower
Storage	2-Tank Molten Salt (6 hr)	None	None	None	2-Tank Molten Salt
Size (MW)	280	250	280	392	110
Company	Abengoa	NextEra	Abengoa	BrightSource	SolarReserve
Utility	APS	PG&E	PG&E	SCE + PG&E	NVE
Location (State)	Arizona	California	California	California	Nevada
Commissioning Date	October 2013	March 2014	Late 2014	February 2014	Late 2014

The business case for concentrated solar power will depend on the market. In spite of the ever-growing PV market, there is still room for electricity generation early in the morning or in the evening, when PV can no longer be a useful source. According to the SunShot initiative [4], the ideal solar power plant of the future for the US market will be different from these current large-scale utility size installations. It is believed to consist of smaller, more modular systems (in the 1-50 MW size range), with a low-cost, mass-manufactured solar field, a more efficient power block, coupled to low cost thermal energy storage, more capital-efficient (needing smaller capital investments per plant or including new financing mechanisms such as yieldco′s and green bonds) and it should be co-optimized with PV.

From the technical point of view, there is also still room for improvement. Recent studies [3] claim that there is still a substantial opportunity to increase the overall solar-to-electricity conversion efficiency in a concentrated solar power plant. It implies rethinking the power conversion scheme. In a typical concentrated solar power mirror array and thermal receiver there can be losses around 42% [3]. Independently of the efficiency of the collector system, in the end, due to the thermodynamics of the Rankine cycle, only 40% of the captured and concentrated thermal energy can be converted to electricity. Gross thermal-to-electric conversion efficiencies are typically 35%–45% [4]. If the solar field has an overall efficiency around 42%, after the power plant needs are met, that means that only about 16% of the incident radiation is successfully transformed into net electric output.

The technologies deployed in concentrated solar power plants to generate electricity also show significant potential for supplying specialized demands such as process heat for industry; co-generation of heating, cooling and power; and water desalination. They could also produce concentrating solar fuels (CSF, such as hydrogen and other energy carriers) – an important area for further research and development [1]. Solar-generated hydrogen can help decarbonize the transport and other end-use sectors by mixing hydrogen with natural gas in pipelines and distribution grids, and by producing cleaner liquid fuels.

In summary, to this date, concentrated solar power is the only type of renewable energy that allows for long term energy storage over sufficiently long periods of time and at large scales to completely eliminate the intermittent nature of the solar resource. However, cost reduction is fundamental in order to compete without subsidies. Increasing the overall energy conversion efficiency by developing new power conversion pathways and looking for alternative markets for concentrated solar power such as desalination, process heat, enhanced oil recovery or hybridization seem like the best options for the future.

CONCENTRATED SOLAR POWER TECHNOLOGIES

As mentioned above, concentrated solar power employs a solar collector (*e.g.* mirrors) to reflect, concentrate, and focus the solar radiation to heat a heat transfer fluid (HTF) that is circulating through a receiver/absorber. The HTF transfers the absorbed heat to a heat engine (turbine) to convert the thermal energy into mechanical energy, with the possibility of having or not an intermediate thermal energy storage system. A turbine-generator set transforms the mechanical energy into electricity in the power block. A schematic describing this general concentrated solar power concept and components is shown in Fig. (**3**).

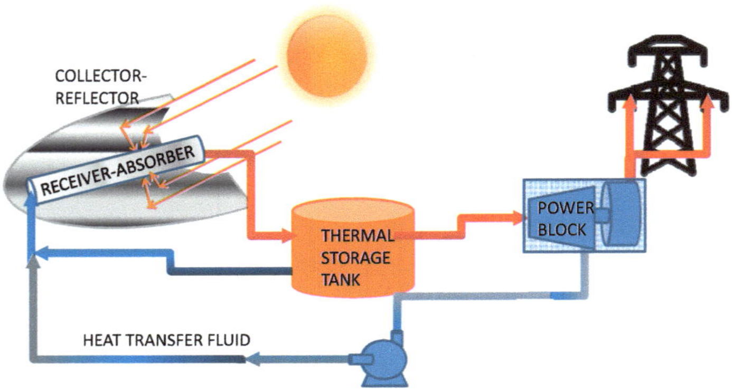

Fig. (3). General schematic of a concentrated solar power system with its main components: collector/reflector, receiver/absorber, heat transfer fluid, thermal energy storage, and power block.

Depending on the way the solar radiation is focused, *i.e.* the type of collector-receiver, concentrated solar power systems can be line-focus (linear Fresnel, parabolic trough) or point-focus (dish-Stirling, central receiver power tower, and beam down). The main concentrated solar power technologies can be seen in Fig. (**4**). Though all these technologies use different methods to focus the incident solar radiation, their operating principle is the same and they share similar elements such as HTF, thermal energy storage systems, and power blocks.

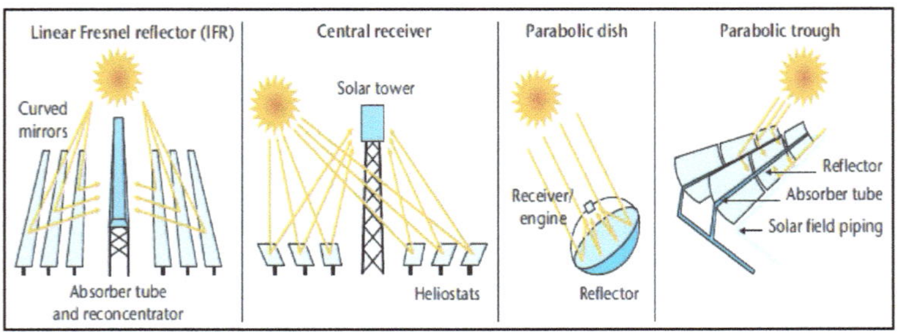

Fig. (4). Main CSP technologies from Ref [1].

Data regarding the actual concentrated solar power plant projects that are currently operational or under construction worldwide can be readily found at the National Renewable Laboratory (NREL) website [5]. It is data compiled by the SolarPACES (Solar Power and Chemical Energy Systems) organization. It gives a clear picture of the concentrated solar power technology maturity, installed capacity, project location and development trends. For example, as Fig. (**5**) shows, parabolic trough projects outnumber any other technology (over 4 GW operational in 77 plants with an average plant size of approximately 50 MW). Power tower technology only has a total of 567 GW operational plants; however, the plant size distribution is very uneven: there are 6 plants with turbine gross capacity around 2 MW each (typically technology demonstrators), 3 plants with an average of 17 MW and 2 more plants exceeding 100 MW. Looking at power tower plants that are currently under construction, all of them are projects where each concentrated solar power plant produces 50 MW or more. It is also interesting to highlight that there are no current operational dish-engine plants. It is also expected to see a rise in the development of LFR plants in the coming years, since they are well suited for other uses besides electricity production and can be easily coupled with new hybrid technology such as thermophotovoltaics (TPV).

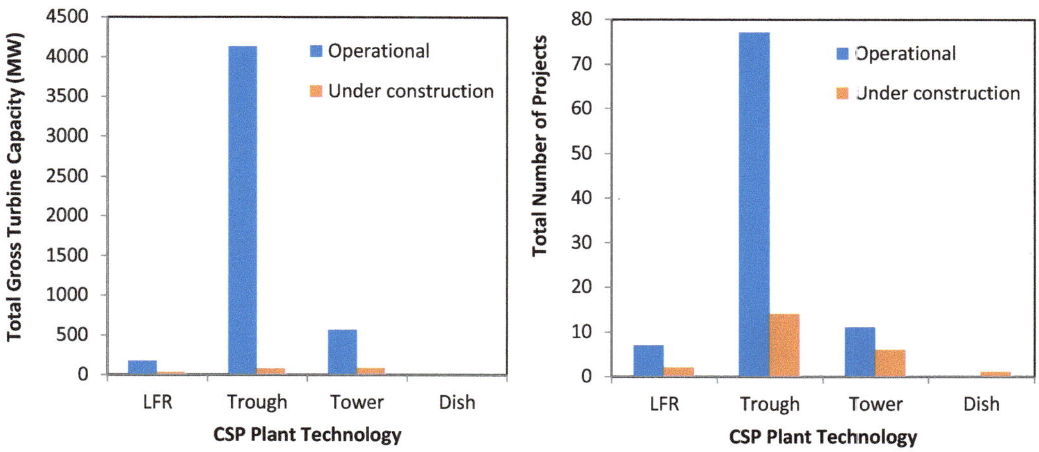

Fig. (5). Current concentrated solar power installed gross turbine capacity and number of projects by technology type and status (data compiled from [5]).

Table **2** summarizes the main performance metrics for the different types of concentrated solar power technologies. The main parameter describing the degree of solar radiation concentration is the concentration ratio, defined as the ratio between the collector aperture area and the receiver aperture area. The highest concentration ratios are achieved with parabolic dish technology (well above 1000) and the lowest are typically found in Linear Fresnel (LFR) (approximately 30), although they can improve slightly if a secondary receiver is installed. Although parabolic dishes have the highest concentration ratios and highest solar-to-electric conversion efficiency, there are no current commercial concentrated solar power plants with this technology mainly because it is not cost-effective.

According to the International Energy Agency [1, 6] the highest outlook for improvements can be found in power tower technology, essentially related to the development of advanced power cycles and new heat transfer fluids capable of operating at higher temperatures. Parabolic trough is the most mature technology and there seems to be limited space for further developments, however there are still many trough projects under construction (adding a total of almost 1 GW according to SolarPACES/NREL [5]).

Table 2. Comparison of CSP technologies [1, 6 - 8].

	Parabolic Trough	**Power Tower**	**LFR**	**Dish-engine**
Annual solar-to-electric efficiency (%)	13-15	14-18	9-13	22-24
Peak efficiency (%)	25	35	18	32
LCOE (ÚSD/kWh)	0.16-0.40	0.13-0.30	0.14-0.45	-
Maturity	High	Medium	Medium	Low
Outlook for improvements	Limited	Very significant	Significant	Through mass production
Land occupancy	Large	Medium	Medium	Small
Typical plant capacity	50 MW	50-100 MW	<50 MW	3-30 kW
Operating Temperature (ºC)	290-390	290-565	270-390	550-750
Concentration ratio (suns)	80	1000	30	1500

The main technologies, their characteristic, challenges and opportunities for improvements are described in detail below.

Parabolic Trough Systems

Parabolic trough systems are the most mature of the concentrated solar power technologies. They have been used for utility scale power generation since the 1980s. Trough system components have been tested and optimized over many design generations. Operation and maintenance costs are also well understood and can be easily planned. Over three decades of commercial operation experience reduce the system risk, making parabolic trough the easiest technology to currently finance.

Long parabolic mirrors focus the incident sunlight onto receiver tubes located at the focal line of the mirrors (Fig. **6**). The mirror and receiver tube structure tracks the sun throughout the day, rotating along its axis. A heat transfer fluid (HTF) is heated as it flows through the receiver tubes. This HTF is commonly synthetic oil but recent developments also use molten salts (Archimede Solar Plant in Italy [9]) or even air at ambient pressure (Airlight Energy Ait Baha Plant [5]). The heated HTF transfers the thermal energy in a heat exchanger to steam, which is then used to generate electricity through a conventional Rankine cycle. Other systems have explored Direct Steam Generation (DSG), producing water steam in the receiver tubes directly and eliminating the need for a HTF-to-steam heat exchanger.

Fig. (6). Parabolic trough collectors and aerial view from Abengoa´s Mojave Solar plant in California, with a gross capacity of 280 MW covering nearly 2 square miles with 2,200 mirrored parabolic trough collectors and 1.5 million square meters of reflective area [10].

The main trough components are: the frames and support structures, the reflector surfaces, receiver tubes, sun-tracking drive and control system, and heat transfer fluid system. Main engineering efforts focus on reducing the structural support system aiming towards lighter, modular and scalable designs for easy, inexpensive assembly and deployment.

Parabolic trough technology is inherently modular and can be easily up-scaled. The basic plant layout consists of a number of collectors connected to form a single unit (or loop) that is connected to the heat transfer fluid system circuit and repeated over the solar field. The heat transfer fluid travels from the loops to and from a steam generator to drive the steam turbine to generate power. Many of these systems incorporate 2-tank sensible heat molten salt storage systems, for which a HTF-salt- steam heat exchanger is also needed.

The main drawbacks of this technology are:

a. Large land requirements for a vast extension of collector rows and loops.
b. Upper temperatures of the HTF limited to 400°C (synthetic oil), reducing the thermodynamic efficiency of the system
c. Thermal losses (convection, re-radiation)
d. Degradation of receiver tube performance after operation

There is a decreasing number of R&D activities in this type of technology, since it is relatively mature and well extended commercially. Many incremental developments are under way to explore new reduced cost collector designs, performance improvements such as new absorber coatings, or mitigation strategies to avoid receiver tube performance loss over the lifetime of the plant.

Linear Fresnel Systems

In linear Fresnel (LFR) systems, a series of flat or slightly curved mirrors arranged closed to the ground focus the incident radiation on an elevated receiver tube assembly, but in contrast to parabolic trough installations, the structure is stationary. The primary reflectors (mirrors) track the sunlight while the receiver assembly is fixed. The fixed receiver consists of one or more receiver tubes arranged in a flat or low profile array and an optional secondary reflector. This flat or nearly flat reflector profile leads to higher concentration ratios without significantly increasing wind loads. This leads to much lower capital and operating costs, but also much lower conversion efficiencies mainly due to greater optical losses.

Traditionally, LFR have been used for low- and medium-temperature heat generation for a wide variety of applications such as building heating and cooling, industrial process heat supply, or water treatment. Advanced state-of-the-art LFR are targeting higher temperature applications for both heat and electricity production [11].

A substantial limitation with traditional LFR is that large space is required between reflectors to avoid shading, which leads to large ground utilization requirements for a given collector area. In order to overcome this space requirement, a compact linear Fresnel (CLFR) [12] has been proposed, where two separate receivers share the reflectors located between them. A reflector will track one receiver or the other depending on which one provides less shading/blockage loss (Fig. 7). The CLFR system increases the optical efficiency compared to a conventional LFR, but it adds complexity to the reflector tracking mechanism [11].

Except for a few exceptions, current operational LFR plants use direct steam generation (DSG), *i.e.* water/steam is the heat transfer fluid that circulates through the receiver tubes. The only reported operational LFR plant using a different HTF is in Rende, Calabria (Italy) using thermal oil instead [5]. New higher temperature designs are also exploring the use of molten salts as heat transfer fluid in LFR plants. A molten salt LFR is conceptually feasible and at least two different demonstration prototypes have been built since 2014 [13, 14] but commercial plants have yet to be constructed and freeze recovery systems must be developed and tested.

Fig. (7). AREVA Solar's Compact Linear Fresnel Reflector (CLFR) solar collector [13].

The main challenges in LFR developments are, similar to other concentrated solar power systems, related to adapting the technology for higher temperature operation to increase efficiency. In this sense, material developments for higher temperature absorber coatings are needed for LFR and other technologies. However, LFR have larger difficulties in maintaining the optical efficiency while increasing the concentration ratio compared to other concentrated solar power systems.

LFR systems are particularly well suited for the combination of concentrated solar power, solar heating/cooling, and photovoltaic (PV) or concentrated PV in integrated installations due to the design flexibility of the fixed receiver structure [11]. Parabolic troughs have significantly higher design challenges for these types of STE-PV systems due to the lack of a secondary reflector and the whole movement of the structure. The limited aperture size in the tower receivers is also not economically viable for such STE-PV hybrid systems. Consequently, more progress in LFR is to be expected in this area in the following years.

Dish Systems

A parabolic dish tracks the sun during the day moving along two axes, concentrating the radiation at a focal point situated above the center of the dish. At the focal point, there is an independent engine/generator (such as a Stirling machine or a micro-turbine). Parabolic dishes are limited in size (typically 5-25 kWe) and are usually used as stand-alone electric generators (such as replacements of diesel generators), since thousands would be needed to produce a large-scale concentrated solar power plant. Dishes offer the highest solar-t--electric conversion performance of any concentrated solar power system (25--% annual solar-to-electrical conversion efficiency).

Several features such as their compact size, low maintenance cost, and the absence of cooling water put parabolic dishes in competition with concentrating photovoltaics (CPV) modules. The modularity of concentrated solar powerdish-Stirling systems allows them to be deployed individually for remote applications, or grouped together for small-grid or end-of-line utility applications. Mass production, cost reduction and improving their compatibility with thermal storage and hybridization to increase dispatchability will allow dishes to compete both with PV and with larger solar thermal systems.

In the current market, there seems to be a limited number of dish-Stirling system suppliers. While several companies have developed dish/Stirling systems over the years, many have gone into bankruptcy. Currently only one company appears poised to deploy the technology immediately on a commercial basis (Ripasso Energy AB [15], Fig. **8**) after Infinia Corporation (with its 3 kW, 6.7 m diameter " PowerDish") filed for bankruptcy in 2013. Another major player, Stirling Energy Systems (SES)- Tessera, with a 25 kW 11.6 m diameter dish product called "SunCatcher", also filed for bankruptcy in 2011.

As far as large scale concentrated solar powerdish-Stirling projects, two pilot plants have been tested. The first operational concentrated solar power dish-Stilring facility was the Maricopa Solar Project of 1.5 MW in Arizona, USA was completed in 2010, formed by 60 SunCatcher dishes by SES and owned by Tessera-Solar. It was decommissioned in 2011 and sold to CondiSys Solar Technology of China. Another concentrated solar power dish-Stirling facility in current operation is the ECube Energy Dish pilot plant of 1 MW in China.

Some of the advantages of the dish-Stirling technology include low maintenance costs; easy installation since the components are shipped to the site, can be assembled at ground level, no special skills are required, and needs no mirror adjustment. Due to their modularity, dish-engine systems can be also used as stand-alone applications such as water pumping in remote areas. Although their installation and maintenance requirements are much smaller than in other concentrated solar power systems, it still might be a challenge for remote applications. It is most likely that the entry in stand-alone markets will occur after the utility and intermediate scale applications (small grids) become more mature, since they can take advantage of the economies of scale of adding multiple units.

Fig. (8). Ripasso Energy´s Dish Stirling Hybrid modules are rated at 33 kW each and have a solar-to-electric conversion of 32% [15].

Dish-Stirling systems have four major components: 1) concentrator, 2) tracking system, 3) cavity receiver, and 4) Stirling engine. The receiver has two functions: a) absorb as much solar radiation coming from the concentrator, and b) transfer this energy as heat to the engine working gas. The Stirling engine working gas is usually hydrogen or helium at temperatures around 650-700°C and pressures as high as 20 MPa. The receiver in a dish-Stirling system can be either directly irradiated (small tubes placed in the receiver form the absorber´s surface through which the engine´s working gas is circulated and heated as the radiation is concentrated on tubes) or indirectly irradiated (a liquid-metal intermediate heat transfer fluid in a heat pipe or in pool boiler is used to transfer the concentrated radiation from the absorber surface to the engine head, where the surface of tubes with engine working gas are located). Generally speaking, the receivers are capable of absorbing high levels of solar radiation (on the order of 75 W/cm^2).

Most R&D efforts are currently aiming at integrating thermal energy storage (TES) within the receiver [16] and /or hybridizing the technology with natural gas burners. With the help of these advances, the solar dish-Stirling system will be able to increase its operating hours and improve its competitiveness.

Dish/engine concentrated solar power systems can be hybridized with fossil fuels to provide dispatch-able power. Typically, an additional gas burner is incorporated into the receiver. While parabolic dish-Brayton systems are inherently easy to hybridize adding a combustor downstream from the receiver, dish-Stirling systems present greater challenges for this task. This is mainly due to the requirements for high temperature isothermal external heat addition in a Stirling cycle. Additionally, geometrical constraints make it difficult to easily integrate both sources of heat (solar & fossil fuel combustion). Although hybrid heat pipe receivers have been under development for the last 20 years with well-known prototypes from Sandia National Labs and DLR, there are yet no commercial hybrid dish-Stirling systems.

Also, due to the high cost of Stirling engines, recent research is looking into using Brayton micro-turbines [17] instead of Stirling engines. However, the efficiency of these Brayton micro-turbines is around 25-33% in contrast to 42% for a standard Stirling engine.

Central Receiver Systems

Central receiver or power towers use an array of mirrors (heliostats) to focus the incident radiation at the receiver located at the top of a tower through a two-axis tracking system. A HTF or HTM flows through the receiver, collecting the heat and transferring it typically to generate steam in a conventional Rankine cycle.

A centralized collector design allows using HTF working at higher temperatures and incorporating more compact and efficient thermal energy storage (TES) systems. Power tower systems have the largest concentration factors, after dish-Stirling, using the added benefit of the economies of scale to achieve further cost reductions. Nevertheless, there is currently no consensus on what is the optimum heliostat (mirror) size for a central receiver tower should be: some developers are increasing the field size and others are suggesting modular smaller tower configurations. Thus, heliostats with a surface area ranging from $1 m^2$ to $160 m^2$ are being used. In any case, heliostats have to maintain enough distance from one and other to avoid shading or blocking from each other. Optimizing heliostat location on the slopes on the terrain may help increase the heliostat density and decrease shading or blocking issues.

In the past 5 years, central receiver systems have emerged as a major option due to the possibility of increasing the system efficiency and further reducing costs with new solar field and receiver designs and incorporating TES (Fig. **9**).

After Abengoa built two direct steam generation (DSG) tower plants near Seville Spain (PS10 and PS20 with a capacity of 11 MW and 20 MW respectively [10]), larger plants have been built in other locations. Also with DSG technology but with a different tower and overall system designs, Bright Source Energy built the largest concentrated solar power installation at the moment, Ivanpah, with 377 MW (net) in total through three distinct towers without storage [1, 5].

Fig. (9). PS10, the first commercial central receiver tower CSP plant by Abengoa producing 11 MW near Seville, Spain ([10]).

The most recent tower developments no longer use water as a HTF. The "Gemasolar" plant built by Torresol Energy was the first tower to use molten salts as both the HTF and the TES ([18]), eliminating the need for a heat exchanger in the system and leading to further cost reductions and efficiency improvements. Gemasolar has a capacity of 19.9 MW and 15 hours of storage. Following that concept, Solar Reserve has built the "Tonopah" plant in Crescent Dunes (Fig. **10**), Nevada the largest single tower with a capacity of 110 MW and 10 hours of storage, using molten salts as both the HTF and the TES material. There are currently three other molten salt towers under construction (at the time of writing) producing: 50 MW with 2.5 hours of storage from the "Supcon Solar Project" in China, 110 MW with 17.5 hours of storage from "Atacama-1" in Chile (Abengoa), and 150 MW with 8 hours of storage from "NOOR III" plant (ACWA) in Morocco.

All the central receiver plants to this date employ a steam Rankine conventional thermodynamic cycle in the power block, which penalizes the overall solar-t--electricity conversion efficiency with its ~40% thermodynamic efficiency. This is expected to change in the future as advanced power cycles offering higher efficiencies become more mature and are ready for commercial demonstration.

Fig. (10). Tonopah Crescent Dunes molten salt power tower plant from Solar Reserve [19].

The key technological challenges in central receiver plants are guiding R&D activities to develop materials, receiver technology, heat transfer fluids, storage, and advanced power cycles that can allow higher operating temperatures in order to increase the overall plant efficiency. All these components are interrelated and will be discussed in detail in the following sections.

SOLAR COLLECTORS AND RECEIVERS

Solar Collector Field Improvement Opportunities

The solar collector uses reflector facets such as mirrors to collect, concentrate and redirect the solar radiation to the solar receiver. The solar collector, whether it is for parabolic trough, power tower or dish systems, represents an attractive target for plant cost reduction potential as collector materials, manufacturing and installation account for the main capital investment in a concentrated solar power plant.

The efficiency of active surfaces in both solar collectors and receivers depends primarily on their optical properties (*e.g.* reflection, absorptance, transmission and re-emission of light). These properties are determined by the incident light wavelength, intensity and angle of incidence, the component materials, and existence of surface texturing and/or coatings. These surfaces must also face harsh environmental conditions and large temperature gradients over a long period of time (*i.e.* the lifetime of a concentrated solar power plant is typically around 20 years) while maintaining high optical properties.

The main desirable characteristics of reflector surfaces are often competing: high reflectance, durability, low cost, low operation and maintenance (O&M), light weight and resistant to abrasion, oxidation or corrosion.

There are certain common improvement opportunities in concentrated solar power collector technology [20 - 22], regardless of the plant type:

1. Optical efficiency improvements in reflectors: increasing the optical performance and the durability of reflectors by using low-cost materials is an active area of research that will also reduce solar collector costs. Reflector films might replace glass, making troughs lighter and cheaper. Traditional mirrors are made of glass and reflective coatings on the back of the glass. High reflectance materials such as silver and aluminum are commonly used as reflective substrates; however, new lower cost oxides are also being explored. There are new viable alternative materials to glass as superstrates for concentrated solar power collector surfaces such as coated aluminum reflectors, and polymeric film substrate reflectors.
2. Reflector anti-soiling coatings and cleaning techniques: soiling and dust collects on collector surfaces over time, decreasing the specular reflectance of the collectors and, consequently, reducing the solar field efficiency. The accumulation of dust or sand is particularly problematic in arid or desert regions, which are typically where concentrated solar power plants are located. Reflector cleaning accounts for a significant portion of the solar field O&M costs. Lower maintenance and /or new cheaper cleaning techniques should be developed, as keeping the reflector surfaces clean play a major role in plant efficiency. Traditional cleaning with water is labor intensive and requires large amounts of water in areas where typically water is scarce. Novel cleaning methods can be passive (*e.g.* anti-soiling and self-cleaning coatings) and active (*e.g.* new cleaning techniques that do not require water).
3. Sun-tracking, drives and control systems:
   - In general, more reliable wireless methods for collector control and communications and self-aligning technology are among the new developments that are needed for future control systems [23].
   - Specifically, for power tower technology: Low cost alternative heliostat drives and control algorithms are being developed to substitute current expensive azimuth drives. Large fields with slanted slopes require especially accurate controls. Also, small heliostats require finer controls for accurate

positioning and wiring since they will need a larger number of elements compared to larger heliostats for a given solar field.

4. Support structure: Wind-loads dominate the structural design criteria for both trough collector and heliostat supports. Pedestal structures have a larger impact on large heliostats as they experience higher wind loads for which they require more structural steel per surface area to survive worst wind loads case scenario. Both analytical modeling and expensive experimental validation of new designs are currently needed. The development of novel non-steel support structures could also lead to further cost reductions.
5. Collector design and manufacturing: low cost, mass-produced and highly automated manufacturing of collector mirrors can significantly reduce the overall system cost. New designs that require less collector material quantity are also being explored as cost-reduction opportunities. Moreover, easy and automatic assembly and installation procedures can reduce the overhead cost during construction and decrease the time to market of new technologies.

Advances in Solar Receivers

The receiver section in a concentrated solar power plant is where the collector/reflector surfaces concentrate the incident solar radiation. Its purpose is to absorb as much of this concentrated radiation and transfer it to the heat transfer medium of choice, while keeping optical and thermal losses to a minimum. Depending on the system design it may have additional reflective surfaces.

Essentially receiver designs must maximize solar irradiance absorptance and minimize heat losses. The heat transfer medium can be heated directly (*e.g.* exposed solid particles) or indirectly (*e.g.* HTF circulating through a tubular receiver) [24]. The main design limitations for receivers typically are: 1) the maximum heat flux that they can absorb and transfer to the heat transfer fluid without overheating the receiver walls or the HTF within them, 2) the maximum temperature gradient the receiver walls can withstand, and 3) reliability over thousands of daily thermal cycles.

Key opportunities for improvement in CSP receiver design include:

- Increase the thermal conversion efficiency by reaching higher temperatures in the receiver.
    - Develop new solar selective absorbers and coatings.
    - Reduce thermal losses and spillage.
    - Increase receiver lifetime by decreasing mechanisms that promote performance degradation.
    - Improve receiver performance characterization and modeling tools.

The different types of concentrated solar power receivers are summarized below, highlighting the main challenges each specific technology has up to this date.

Parabolic Trough Receivers/Absorbers

The parabolic trough receiver consists of an absorber with a solar selective coating and a glass envelope surrounding the absorber tube (Fig. **11**). The glass envelope acts as an anti-reflective and protective coating against dust, oxidation, and abrasion. The enclosure between the absorber tube and the glass envelope is evacuated to reduce the convective heat losses in the receiver. Metal bellows are located at each end of the tube to accommodate thermal expansion with seals that ensure a good vacuum is kept inside the glass enclosure. Certain metallic compounds with affinity for gas molecules called "getters" are installed in the evacuated annular space in order to absorb hydrogen and/or other gases that might permeate into the vacuum over time.

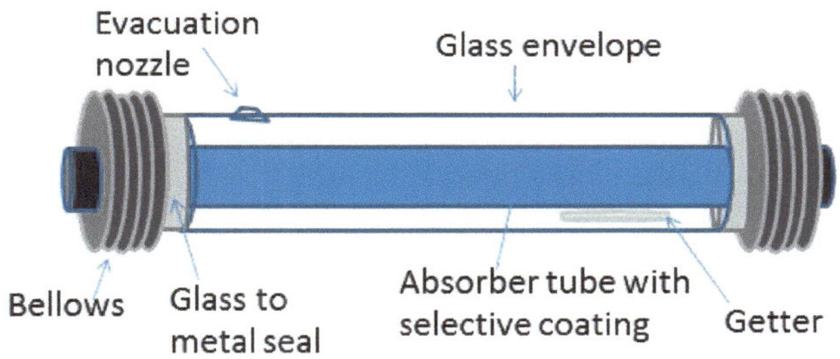

Fig. (11). Schematic of a parabolic trough receiver.

Key problems in the parabolic trough receiver are vacuum losses and receiver lifetime (degradation) [25]. In the past few years, researchers have found that parabolic trough efficiency is greatly affected by heat losses in the receiver tubes and that, over time, these heat losses seem to increase (*i.e.* "hot tube phenomena"). The heat loss is directly related to the vacuum quality inside the glass envelope of the receiver tube. Degradation of the vacuum can be due to a) outgassing of materials in the system, b) gas permeation through walls and c) air leaks penetrating into the vacuum system. It is commonly accepted that the main gas that could permeate through the vacuum walls is hydrogen resulting from the thermal degradation of the thermal oil used as heat transfer fluid [26, 27]. A recent study [25] using residual gas analysis confirms that, although hydrogen is the main gas produced (99%), the getters are capable of absorbing the hydrogen reasonably well. However, other gases such as nitrogen are not removed so efficiently by the getter and, over time and temperature cycles, nitrogen can saturate the getter, negatively affecting the receiver vacuum quality. In other words, contrary to what is commonly accepted (*i.e.* hydrogen is responsible for vacuum loss in parabolic trough receivers), new findings suggest that other gases might be highly responsible for the long term heat loss increase (and performance decrease).

Linear Fresnel Receivers

In contrast to parabolic trough receivers, those used in linear Fresnel (LFR) systems are not evacuated (although they still have a glass enclosure) and are typically arranged forming a flat or low-profile tube array. The lack of vacuum leads to larger convective heat losses and limits the possibility of solar selective surfaces to those that do not thermally degrade in air environments.

LFR can have one or an array of receiver tubes (absorbers). An array of absorber tubes will increase the collector intercept area. The receiver assembly also typically has some sort of insulation to decrease thermal losses. Different receiver cavity designs can be found; a common trapezoidal shape cavity is shown in Fig. (**12**).

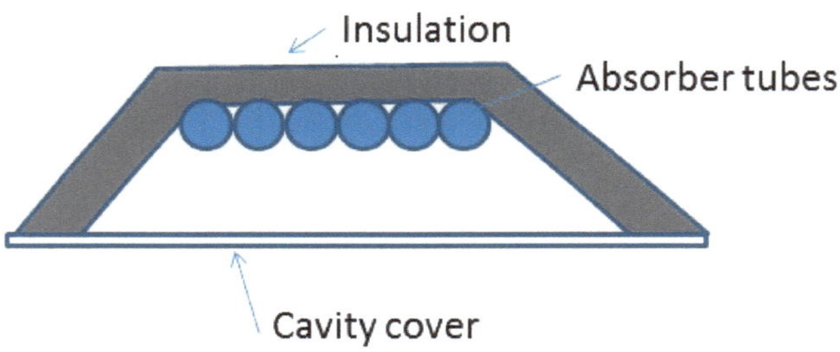

Fig. (12). Schematic of the cross-section of a trapezoidal LFR cavity receiver.

As LFR technology moves towards higher temperature applications, new material developments will be needed and evacuated receiver tubes will be further implemented [11]. As mentioned before, some of these material developments are common to other technologies such as novel absorber coatings. Given that LFR receivers have higher convective heat losses compared to other receiver technologies, new LFR receiver designs are investigating alternatives to conventional receiver cavities and cavity cover materials. For instance, new aerogels can be simultaneously insulating and optically transparent and they are being explored as a cavity covers instead of conventional plastic films in order to reduce the convective heat losses [8].

Power Tower Receivers

Receivers for power tower systems can be classified according to:

a. the type of HTM that is heated in the receivers (liquid, gas, or solid particles)
b. the geometry (tubular, volumetric, liquid film/particle curtain)

The upper temperature limit of the receivers is generally determined by the HTF of choice (further details can be seen in the following section). Power tower receiver technology is relatively mature for liquids, reaching efficiencies close to 90% with the appropriate coatings and designs [24].

Tubular Receivers

Conventional tubular receivers consist of an array of metallic tubes (typically stainless steel or other alloys) where the concentrated radiation is focused heating the HTF indirectly as it passes through the tubes. The tubes can be arranged inside a cavity (*e.g.* cavity receiver) or around the exterior of a cylinder (*e.g.* external receiver). Cavity receivers generally have lower radiation losses (spillage) but higher convective losses than external receivers. Tube size and wall thickness should be designed to maximize heat transfer and minimize pumping power requirements.

The original tubular receivers were designed to work with water to produce steam during the HTF heating process. Examples of water/steam tubular receivers are demonstrator Solar One from the 1980s and the first commercial plant PS10 in Seville, Spain [10] was commissioned in 2007. This technology is mature but its main difficulty remains integrating the TES system, as direct storage of steam is not currently cost effective. The need for an intermediate heat exchanger between the HTF and TES has presently shifted the interest to other high temperature systems such as molten salt tubular receivers.

The first liquid tubular receiver prototypes using molten salts were tested in the 1990s in Solar Two [28]. They have been widely studied in many R&D facilities such as Sandia National Laboratories [29], at Themis, and at the Plataforma Solar de Almería [24]. The first commercial one was tested at Gemasolar power plant near Ecija, Spain [18] in 2011 and it is now the most common power tower receiver type in the market, because of the possibility of using the molten salts directly as thermal storage, reducing system complexity and cost, and increasing plant efficiency. As the need for higher receiver fluxes and temperatures increases, new HTF are being explored such as liquid metals. Their higher thermal conductivity should reduce the large temperature difference between the front irradiated and the back receiver tube walls and the associated thermal stresses. The main challenges to overcome with liquid tubular receivers are related to the HTF characteristics: material compatibility with HTF is difficult especially under high temperature gradients, high heat fluxes and extreme thermal cycling conditions; plus, the HTF can solidify in the receiver cold spots leading to plugs and blockages.

Tubular receivers are also used with gaseous HTF. Prototypes of high temperature tubular pipe receivers to heat up pressurized air have been developed and tested recently for solar-hybrid systems [30, 31]. Although operation of such designs has been successful to this date, there are still many challenges that have to be overcome. It is inherently difficult to effectively transfer the heat from the irradiated receiver tubes to the gas due to the low density and poor heat transfer characteristics of air. The high operating temperatures (between 800-1000°C) also increase the radiative and convective heat losses. Unfortunately, these receivers are subjected to rapid transient thermomechanical loads which negatively affect the receiver structure and fatigue life.

Recent trends pushing toward developing CSP tower plants with supercritical CO_2 as both the HTF and power block working fluid [32] are proposing to use similar high temperature tubular receivers. Small tubes will allow the high pressure requirements for the supercritical state. Analysis and demonstration of s-CO_2 tubular receivers is expected to be the focus of R&D in the next few years.

Volumetric Air Receivers

Volumetric air receivers heat up air by circulating it through a porous structure (*e.g.* porous ceramic or honeycomb). Depending on the porous structure material, metallic, ceramic or silicon carbide, the air can heat up to 800-1000°C, 1200°C or 1500°C, respectively [24].

Like any other receiver, volumetric receivers should minimize radiative heat losses that can be mitigated through creative multilayer semitransparent window designs. Other challenges in volumetric air receivers include flow instabilities and non-uniform heating that can lead to localized temperature peaks overheating the receiver material which can fail if the thermal loads are too high.

Pressurized gas receivers have been recently reevaluated thinking of thermochemical reactor needs. The main challenge regarding these receivers is designing suitable windows that have the desired optical properties, that are durable, and that withstand the high pressures.

A particular type of volumetric cavity receivers have been designed to be used with submicron sized carbon particles suspended in pressurized air [33]. This solid-gas suspension is heated to be directly utilized as the working fluid in a Brayton cycle. These black solid particles increase the absorption of solar radiation throughout the entire volume due to their large surface area (*i.e.* small size). Furthermore, the beauty of the small particle receiver concept is that the carbon particles are oxidized thanks to the large heat fluxes and the resulting HTF flow is particle free, ready to be expanded in the power block.

Solid Particle Receivers

The need for receivers and HTFs capable of withstanding higher temperatures to increase cycle efficiency has recently leaded to the rediscovery of falling particle receiver designs. Solid particles as HTM can reach temperatures in excess of 1000ºC, they are inexpensive and they can be directly stored, eliminating the need for a HTF-TES heat exchanger. In contrast to direct steam generation systems, the HTM cannot be used as the power block working fluid, so a heat exchanger is needed. On the other hand, since the particles directly absorb the concentrated radiation, some of the heat flux limitations found in tubular receivers do not exist in these type of designs.

Particle receivers can have different solid circulation concepts: moving bed, falling film, fluidized beds, spouted beds, and up flow beds, among others [34]. Although solid particle receivers have been studied since the 1980s, it has not been until recently that the first prototypes have been developed (*i.e.* high temperature falling particle receiver using at Sandia National Laboratories). These on-sun tests have not been as successful as planned to this date: receiver efficiencies of only 50% have been achieved and the particles only increased their temperature by 250ºC [35]. Recirculation of particles and other methods to increase the particle residence time during exposure to radiation are currently being explored [24]. For instance, imposing an electro-magnetic field to ferromagnetic particles is being investigated as means to increase particle residence time, reduce flow instabilities, and, hence, increase heat transfer absorption rates [36]. Further research is also needed to develop viable and efficient connecting subsystems such as particle transport systems (conveyers, lifts, elevators, *etc.*), particle storage systems, and particle-fluid heat exchangers.

The latest developments are also using the solid particle receivers for thermochemical processes such as solar-driven ceria reduction [37]. These initial results seem to indicate that direct particle irradiation for such processes with a falling particle receiver is not achieving performance targets and other alternative designs should be explored. There is no doubt that, as progress in thermochemical cycles advances, more variations of these receiver designs will be explored in the coming years.

Receiver Coatings

Incorporating selective absorption coatings to improve the receiver surface absorptivity (α) while minimizing its emissivity (ε) will further increase receiver efficiency and reduce losses. Ideal coatings should be chemically and structurally stable at high temperatures in air, durable, low cost, withstand long term thermal cycling and easy to apply.

There are essentially six major groups of solar selective coatings, which are summarized in Table **3**.

Pigments, like Pyromark 2500 do not have the best combination of absorptance and emittance or long term thermal stability but they are inexpensive and easy to apply, so they are extensively used and annually reapplied. Of the simplest structures, cermets currently have the best combination of adequate tunable properties and cost/ ease of fabrication; this is why they have been successfully commercialized. However, the new photonic crystals seem to have the greatest potential to surpass present efficiency limits due to their ability to control over the spectral and angular selectivity with nanostructure features on the scale of wavelengths. As a matter of fact, the photonic band gap in photonic crystals can suppress emission completely over a range of wavelengths that can be chosen to correspond to the optimum calculated reflectivity [40].

Table 3. Types of solar selective coatings: properties and examples (data extracted from [38 - 40]).

Type of Solar Selective Coating	Main Characteristics	Challenges	Examples	Typical Optical Properties	
				α	ε
Intrinsic selective materials	Homogenous materials with certain spectral selectivity simple & easy to construct	Not ideal performance (↑α but maybe ↑ε)	tungsten copper sulfide (CuS2) zirconium diboride (ZrB2) pyromark 2500	0.96	0.8
Semiconductor-metal tandems	Semiconductor with a metallic back Good a/e ratios (12-15)	structural and thermal mismatch between semiconductor & metal need to prevent interlayer diffusion	Semiconductors: silicon, germanium Backing metals: silver, aluminum	0.79-0.89	0.016-0.073

(Table 3) cont.....

Semiconductor-metal tandems	Semiconductor with a metallic back Good a/e ratios (12-15)	not stable at high T processing challenges to deposit on absorber tubes need anti-reflective coating	Semiconductors: silicon, germanium Backing metals: silver, aluminum	0.79-0.89	0.016-0.073
Multilayer absorbers	stacks of thin alternate layers of dielectrics (absorb in the visible) and semi-transparent metals (reflect in the infrared)	complex high fabrication cost physical & chemical reactions between layers oxygen & water diffusion through layers prone to damage/failure under cyclic thermal loads	dielectrics and high-band gap semiconductors: SiO_2, TiO_2, MgF_2, Al_2O_3, Si_3N_4 metals: tungsten, titanium, aluminum	0.97	0.05
Metal-dielectric composites (Cermets)	Dielectrics (absorb in the visible) with fine metal particles (reflect in infrared) Tunable performance	Intermediate properties to dielectrics and metals	Cr–Cr2O3 Ni–Al2O3	0.94	0.07
Textured surfaces	Porous, granular or needle-like metal structures	Oxidation in air	α-Si nanocones	>0.93	
Photonic crystals & nanostructured absorbers	easy to fabricate & scalable very high absorptance low radiative losses	Need further research		0.92- 0.99	<0.1

HEAT TRANSFER FLUIDS FOR CSP

Heat Transfer Fluids (HTF) are a key element in a CSP plant, transferring the heat collected from the solar resource in the receiver to the power block for electricity generation. If the HTF is the same as the working fluid in the power block, electricity is generated by direct expansion through a turbine. Alternatively, the HTF will transfer the heat to a separate working fluid (often steam or an organic fluid) to generate steam through a standard thermodynamic cycle (steam cycle or an ORC). In concentrated solar power plants integrating thermal energy storage (TES), the HTF will also transfer the heat collected to and from the storage system.

The choice of HTF has a significant effect on the concentrated solar power plant performance and cost. The HTF´s temperature limits determine the plant operating temperature range and, therefore, its thermodynamic efficiency.

Ideal HTF Properties

HTFs must be carefully designed to transfer heat easily, operate in the required temperature range, circulate well in a confined space, and be compatible with containment materials [34].

The ideal HTF the will have the following characteristics:

- High conductivity to enhance the heat transfer
- Low viscosity to reduce fluid pressure drop and required pumping power
- Low working pressure to limit tube wall thickness, minimizing wall thermal gradients and stresses.
- High specific heat capacity will allow direct thermal storage
- Broad operating temperature range (ΔT)
- Low melting temperature to avoid freezing problems in field and the need for expensive heat tracing elements in ducts.
- High thermal stability (low thermal decomposition temperatures) for high thermodynamic efficiency
- Low cost and availability
- Good corrosion behavior and compatibility with container materials
- Non-hazardous: low toxicity and low flammability

All these desired properties depend on the type of HTF. For example, upper and lower temperature limits are a major challenge for liquid HTFs but do not apply for pressurized gases or two-phase HTFs such as water/steam.

For liquid HTFs, the minimum operating temperature is defined by the HTF´s melting point. Lowering operating temperatures to or below ambient standard temperatures can reduce costs by simplifying the HTF systems and eliminating the need for heat tracing elements.

Reaching higher operating temperatures is essential to future cost reductions, as higher temperatures lead to higher power conversion efficiencies. The value of the maximum operating temperature (T_{max}) is essentially determined by the thermal stability limit and thermal decomposition of the HTF, but it could also depend on a high vapor pressure or a high corrosion rate of the containing construction materials. System designs should take into account not only the average fluid working temperatures but also maximum film temperatures, as high temperature peaks can appear in certain receiver locations that can lead to the initiation of the HTF decomposition reactions that can spread.

The thermal stability limit of the HTF can be determined through thermogravimetric analysis (TGA), were the weight of material samples is monitored during specific temperature ramp. This upper temperature limit (T_{max}) is usually defined as the temperature at which the sample has lost 3% of its initial weight at a heating rate of 10K/min and is used at a laboratory scale to compare substances.

Cost is an important consideration for choosing a commercially viable HTF. Many substances with interesting physical properties are sometimes much more expensive, thus engineers must reach a compromise between acceptable physical properties, raw material costs or the cost of compatible construction materials.

Types of HTF: Current and Future Applications

Traditional commercial HTFs are steam, synthetic oil, and molten salts. Research and development efforts are currently targeting higher operating temperature fluids such as liquid metals (sodium), molten glass, pressurized gases, supercritical fluids (such as supercritical carbon dioxide), and solid particle suspensions as heat transfer media. These advanced high temperature HTM will enable high temperature receivers and high efficiency power cycles.

Thermal Oil

concentrated solar power trough plants initially started using synthetic oil as HTF to avoid the high pressure requirements and liquid-vapor phase change of water. Most parabolic trough plants still use synthetic oil as HTF. Synthetic oils are generally preferred over traditional mineral oils because they have better thermo-physical properties (lower viscosity and higher thermal conductivity) and they are less flammable. The most common synthetic oil is a eutectic mixture of biphenyl and diphenyl oxide (DB/DPO) usually referred to by its commercial names, *e.g.* Therminol-VP1, Dowtherm A, and Diphyl. It has the highest thermal stability of organic HTF and it has low viscosity, performing exceptionally well as a HTF. As an organic fluid, it is flammable and has been the cause of some fire incidents in concentrated solar power plants. It solidifies around 12-13ºC and it can be used up to 400ºC without presenting significant thermal decomposition. This upper temperature is the main limitation of the current oil-trough technology.

Since thermal oil performs well as a HTF and its behavior is well known, every now and then researchers test different formulas in order to increase its maximum operating temperature. Researchers have tried various coatings and/or additives to hinder decomposition reactions that appear around 400ºC. Even nanoparticles have been added for increased thermal conductivity [42]. However, these improvements have not yet been tested commercially probably because the risk of using an unknown fluid is higher than the small enhancements that have been reported.

Hydrogen formation is thought to increase due to aging of DB/DPO in parabolic trough installations over prolonged periods of time (above 10 years). This is a relevant issue today, especially considering how many commercial plants are operating with thermal oil. The formation of hydrogen degrades the vacuum inside the trough tube enclosure that is required for an adequate heat transfer. The gas will act as insulation, decreasing the thermal conductivity and, therefore, reducing the solar collector efficiency. Much has been written in the literature about hydrogen formation in thermal oil [43, 27]. The hydrogen production reaction seems catalyzed by both impurities in the oil and the presence of oxide layers on the surface of the steel receiver walls. Troughs have installed hydrogen " getters" in order to vent the gaseous compounds produced, but it is unclear if they can absorb the amounts that are thought to be produced during the long term use of thermal oil. Unfortunately, not much long term operating data has been reported with hydrogen production rates, so the severity of the problem is still unknown. Recently the German Aerospace Center (DLR) has evaluated other promising synthetic oils that can be used up to 425ºC with reduced formation of hydrogen and other gaseous compounds [27].

Molten Salts

Molten salt mixtures are very desirable HTF for concentrated solar power, presenting a good combination of properties: they are thermally stable at high temperatures; they have high energy density, low vapor pressure, high specific heat capacity, and they are non-flammable, abundant and inexpensive. They also present relatively large operating temperature intervals. Their main drawback is that they typically have high melting points. The melting temperature limits the minimum HTF operating temperature in order to prevent the salts from freezing and causing blockages in ducts and heat exchangers. Heat tracing elements are used to maintain the salt circulating above freezing temperatures, leading to high parasitic power consumptions and complicating the solar plant installation and maintenance requirements.

Table **4** summarizes some properties of the most common molten salt mixtures classified according to their main anion group (nitrates, fluorides, carbonates, and chlorides).

Table 4. Properties of selected molten salts for HTF and TES applications.

Salts	Salt Components	Composition mol% (wt%)	T_{melt} (°C)	T_{max} (°C)	ΔHf (J/g)	Cp (J/gK)	ρ (g/cm³)
Nitrates	$NaNO_3$	100	306	520	181	1.66*	1.85
	KNO_3	100	335		99		
	Solar salt: $NaNO_3$-KNO_3	66:34 (60:40)	238	565	113	1.55*	1.84
	Eutectic $NaNO_3$-KNO_3	50:50 (46:54)	221	~550	101	1.52*	1.84
	Hitec salt: $NaNO_3$-$NaNO_2$-KNO_3	7:49:44 (7:40:53)	141	450-540	60	1.54*	1.79
	Hitec XL salt: $Ca(NO_3)_2$-$NaNO_3$-KNO_3	30:21:49	133	460-500		1.43*	1.91
	Eutectic $LiNO_3$-$NaNO_3$-KNO_3	30:18:52	120	435-540		1.64*	1.78
	KNO_3-$LiNO_3$-NH_4	19.3:29.4:51.3	92				
	KNO_3-NH_4-$AgNO_3$	20.7:60.2:19.1	52				
	Saltstream 500 ® (Halotechnics)	-	65	500	<< 5	1.22*	
Fluorides	LiF-NaF-KF (FLiNaK)	46.5:11.5:42 (29:12:59)	454	>700	400	1.89**	2.02
	LiF-BeF_2	67:33 (53:47)	460				
Chlorides	KCl-$MgCl_2$	68:32 (62:38)	435	>700	355	1.15**	1.92
	KCl-LiCl	41:59	353	>700	236	1.20**	1.65
	Eutectic BaCl2-NaC- -$CaCl_2$-LiCl	34.6:12.5:40:12.9	378				
	Saltstream 700e® (Halotechnics)	-	205	>700	57	0.76**	2.24

(Table 4) cont.....

Salts	Salt Components	Composition mol% (wt%)	T_{melt} (°C)	T_{max} (°C)	ΔHf (J/g)	Cp (J/gK)	ρ (g/cm³)
Carbonates	Eutectic $Li2CO_3$-K_2CO_3	62: 38	488				
	Eutectic Li_2CO_3-K_2CO3-Na_2CO_3	43.5:25: 31.5	397	788	221	1.85**	1.98

Table adapted from Refs [8, 41, 44, 45, 46]. *values at 400°C** values at 700°C

The most frequently used molten salt is a sodium-potassium nitrate off-eutectic mixture (60 wt.% NaNO3-40 wt.%KNO3) commonly referred to as "solar salt". Solar salt has been extensively studied and tested [44 - 45, 47], and it is commercially used in many concentrated solar power plants primarily as the storage material but also as a HTF in some cases (Archimede parabolic trough solar power plant in Italy [9], Gemasolar central receiver tower power plant in Spain [18]), eliminating the need for the HTF-TES heat exchanger and, thus, increasing the system efficiency. Solar salt has a melting point around 221°C and it has a slightly higher percentage of $NaNO_3$ than the corresponding $NaNO_3$-KNO_3 eutectic mixture to reduce its cost.

The thermal stability limit of solar salt is around 565°C. As mentioned before, increasing the thermal stability limit of solar salt would allow for more efficient, higher temperature power conversion cycles. The chemical stability of solar salt and other nitrate mixtures has been studied in detail by many authors [47 - 49]. The primary decomposition reaction of solar salt is nitrite formation in the melt with oxygen release. The secondary decomposition reaction is alkali metal oxide formation in the melt with nitrogen or nitrogen oxide release and it is far more complex and less well understood.

The decomposition temperature of solar salt varies with heating rates and with the surrounding atmosphere; hence, it is somewhat system specific. Faster heating rates lead to higher decomposition temperatures, but generally a rate of 10 K/min is used to compare materials at a laboratory scale.

In common 2-tank molten salt indirect storage systems that use oil as the HTF, the salt tanks use pure nitrogen as cover gas. The main reason is to reduce the flammability risk in case of a gas leak in the salt-oil heat exchangers. However, if the salt is used as both the HTF and TES medium (*i.e.* a molten salt tower like Gemasolar with storage [18]) the salt is generally surrounded by dry air. Recent literature results show that the thermal stability of solar salt increases with the partial pressure of oxygen [44, 49]. In particular, Sandia National Laboratory is evaluating extending the operating temperature of solar salt to 650°C by stabilizing with pure oxygen environments [26], which might be interesting from the scientific and performance points of view but presents multiple practical challenges and safety implications.

For years, researchers have tried to develop new molten salt materials capable of operating at higher temperatures and with lower melting temperatures. Salt mixtures have the advantage of having a lower melting temperature than their corresponding pure salt components. Multicomponent salts (ternary, quaternary or even higher order component nitrate, carbonate, and chloride mixtures) have been and are still currently being evaluated for this purpose. The main challenge is that, typically, when the melting point of a salt is lower, the thermal stability also decreases, leading to an overall operational temperature range ($\Delta T = T_{max} - T_{min}$) similar to that of solar salt, as shown in Fig. (**13**) and Table **4**.

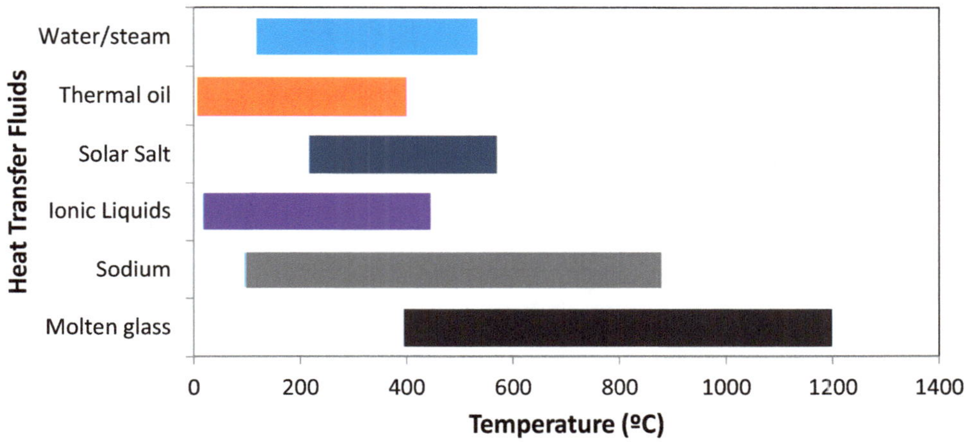

Fig. (13). Operating temperature range of main liquid HTFs (adapted from [8, 41]).

Lower melting temperature nitrates and nitrite-nitrates such as Hitec or Hitec XL salts have been proposed as a single working fluid and TES media for advanced troughs [26], enabling direct TES storage. A single HTF-TES medium reduces plant costs by eliminating the heat exchanger between both components. Additionally, fluids with lower melting temperatures will present fewer freezing issues in the field, lowering the parasitic consumption associated with heat tracing elements installed to avoid duct blockages. Unfortunately, Hitec and Hitec XL salts still melt at temperatures well above ambient (around 130°C), therefore the need for heat tracing elements cannot be fully eliminated. From the practical standpoint of view, the impact of reducing the electricity consumption to keep the ducts 100°C higher or lower is minimal compared to the installation cost of the heat tracing system. Thus, an ideal HTF would melt below ambient temperature and not require heat tracing elements at all. Moreover, these multicomponent nitrates typically have lower thermal stability such that their upper working temperatures can be below that of solar salt. This will negatively impact the concentrated solar power plant efficiency by lowering the upper operational temperature.

For central receiver towers using molten salt as HTF, the focus shifts to finding salts with higher maximum temperatures than solar salt. The objective is to have the same fluid working as both HTF and TES to reduce heat exchange systems and design complexity. In addition, higher temperatures can raise the plant thermal efficiency and are needed to exploit advanced ultra-supercritical steam and supercritical CO_2 cycles. For these types of plants, carbonates and chloride mixtures seem more appropriate. Carbonates typically have higher thermal stability (especially in pure CO_2 environments) but have very high melting points (400ºC or higher), increasing the risk of salt freezing issues and parasitic electric consumption to keep the HTF in the molten state. Chlorides also present higher working temperatures, around or above 700ºC, but they also have large corrosion issues at high temperatures, requiring expensive container materials and coatings with increased maintenance costs.

Improving the operating temperature range of molten salts is still an active area of research. However, experimentally finding and determining the properties of molten salt multicomponent mixtures is very expensive and time consuming. Consequently, researchers have switched to computational high throughput screening to design some of these advanced molten salts. Recently Raade *et al.* [41] reported and commercialized a few novel salt formulations with low melting points and promising operating temperature ranges, such as a nitrate mixture called salt stream 500 or a mainly chloride mixture called salt stream 700 [46]. This advanced nitrate mixture is composed of sodium, potassium, lithium, cesium and calcium nitrates. It has a very low melting point (65ºC), which can be attributed to the cesium addition, and an upper temperature limit of 500ºC. However, the cost of these advanced salt mixtures is still significantly higher than the standard solar salt and they have yet to be implemented in a commercial concentrated solar power facility.

Container material corrosion is another important consideration. Interaction between the salts and the metallic containers can lead to typical corrosion mechanisms (such as corrosion-pitting issues) but also, due to the high thermal loads, stress corrosion cracking of commonly used steels has been found to be important [47]. Even though nitrate salts show negligible corrosion with stainless steel, higher temperature salts such as chlorides and fluorides will require containment elements made of more expensive materials such as Inconel and Hastelloys. Even if nickel based alloys show good resistance to corrosion, they are nearly four times more expensive than iron base steels. Further research to determine experimental corrosion rates under working conditions are still needed.

Water/steam as a HTF

As a HTF, water has good physical properties that are well characterized and known; it is inexpensive, non-toxic, and non-flammable. Its main limitations are the high temperatures and pressures needed for good performance, which result in thick tube walls throughout the system. It is used as a two-phase system, since at the high temperatures required for concentrated solar power water undergoes a liquid-vapor phase transition. Due to the large differences in heat transfer characteristics between water and steam, these fluids are generally separated in different sections in concentrated solar power receivers, *i.e.* the water heating and evaporation (steam production) is usually separated from the superheated vapor section. The upper temperature and pressure limit of water is its critical point (T=374ºC, P=221 bar).

The main power block system that has been used in concentrated solar power plants is a traditional subcritical steam Rankine cycle. Consequently, Direct Steam Generation (DSG) concentrated solar power systems have been proposed to simplify the design and use a single working fluid for both the solar field and power plant. Using the same fluid as HTF and as the working fluid in the power block eliminates the need for an additional heat exchanger, reducing thermal losses. DSG systems are also not constrained by the upper temperatures the HTF can achieve. However, they typically operate with superheated steam temperatures around 530ºC.

DSG has been used in both parabolic trough and central receiver tower concentrated solar power plants. It was initially demonstrated in real working conditions during the late 1990s in the Direct Solar Steam (DISS) facility at the Plataforma Solar de Almeria (PSA) with promising results for trough [50]. Since then DSG has been demonstrated in towers (PS10 and PS20 plants from Abengoa in Sevilla, Spain [10], and most recently Ivanpah towers from Bright Source Energy in Nevada, USA [5]).

In spite of its promising results, DSG systems are currently limited by the lack of cost efficient TES systems using live steam. Current steam accumulators in commercial concentrated solar power plants only provide a couple hours of storage which is used as a buffer during transients and clouds. Directly accumulating the large amount of live steam required for longer storage, *i.e.* around 6 hours to cover the night, is presently not feasible from an economical point of view for large scale concentrated solar power plants. Indirect TES systems based on sensible heat storage (*i.e.* high temperature concrete [51]) and/or based on latent heat storage [52] have also been evaluated in detail to couple to DSG plants, but still further developments are needed for them to be cost competitive with 2-tank molten salt storage.

Other challenges of DSG systems are associated with the large pressure requirements of the HTF, increasing piping and pumping costs.

Liquid Metals

Liquid metals are very efficient heat transfer materials that could lead to the next generation of future high temperature concentrated solar power central receiver plants [53]. They have two significant advantages over traditional HTFs: 1) they can achieve very large heat transfer coefficients, about an order of magnitude higher than standard nitrate-based molten salts, and 2) they are stable at very high temperatures, offering a very wide operational temperature range as a result of low melting points and high boiling temperatures.

The liquid metals that have been evaluated are mainly sodium (Na), sodium-potassium (NaK) [54], lead-bismuth eutectic alloys (LBE) [53] and more recently molten tin (Sn) [55]. They have excellent thermal conductivity, low vapor pressure, and relatively low viscosity. Their temperature range of use is larger than that of any other liquid HTF (see Table **5**) and they represent the only HTF alternative with an upper temperature in the 700-1000°C range beside gases and solid particles. However, they are an important safety hazard and there is limited experience using them in concentrated solar power. Alkali metals react with both air and water, increasing the fire risk and lead metals are extremely toxic, requiring additional safety measures and precautions.

Table 5. Operating temperatures of selected liquid metals used as HTF.

Liquid Metals	**T_{melt} (°C)**	**T_{max} (°C)**
Sodium	98	883
NaK	-13	785
Lead Bismuth Eutectic	125	1553
Tin	232	2686

Of all the liquid metals evaluated, liquid sodium presents the most promising properties, including low cost, high availability, compatibility with common materials (*e.g.* steel). It is well-known as a HTF from the nuclear industry where it has been traditionally used as a reactor coolant but it is a serious fire risk due to its reactivity with water.

In the 1980s, three different receivers were tested with liquid sodium, two at the Plataforma Solar de Almeria, Spain (PSA) and one at the Sandia Central Receiver

Test Facility (CRTF) in Albuquerque, New Mexico (USA). All three projects were discontinued; those at PSA were specifically decommissioned due to a large fire started during a maintenance procedure that caused extensive damage to the plant [54]. Parabolic-dish receivers have also been tested using liquid Na as a HTF in the late 1990s, due to the low economic competitiveness of the dish-Stirling technology these developments have been stalled [53].

Recently, the need for higher temperature HTFs to couple with higher efficiency power cycles (*e.g.* supercritical CO_2) has lead scientists to revisit the use of liquid metals for concentrated solar power as the only promising low-cost and high heat transfer performance alternative. At the temperature range required, the only other possible HTFs are air and other gases, which have much worse heat transfer properties.

However, there are key challenges that need to be addressed for the practical application of liquid metals in concentrated solar power plants. In addition to safety concerns, further development of compatible structural materials is needed; adequate instrumentation and control systems need to be fabricated, and appropriate indirect TES systems must be designed, since liquid metals are not great heat storage materials (*e.g.* high cost, low specific heat capacity).

Other Liquid HTF

Several other interesting fluids have been proposed as HTF for concentrated solar power plants such as Ionic liquids (ILs) and molten glass oxides. The former tackles the lower end of the temperature spectrum whereas the latter is one of the highest operating temperature range liquid HTF proposed to date (Fig. **13**).

Ionic Liquids

Ionic liquids by definition are ionic compounds which are liquid below 100°C. They can be considered a type of molten salt but instead of containing inorganic cations like molten salts, they are based on organic cations (imidazolium, ammonium, pyrrolidinium, *etc.*) with inorganic (Cl-, AlCl4-, PF6-, BF4-, NTf2-, DCA-, *etc.*) or organic anions (CH_3COO-, CH_3SO_3-, *etc.*). Some of the common ILs evaluated for high temperature applications can be seen in Table **6**. Many ILs present melting points below 0°C and can be liquid up to 400°C [56 - 58]. Due to their low melting temperature, low vapor pressure, and high thermal stability, ILs have been considered recently as very promising potential HTF for lower temperature concentrated solar power plants. They have also been investigated as a base fluid for nanoparticle suspensions [56].

Table 6. Selected ionic liquids and main properties (commercial data from [57]).

Ionic Liquid		Melting T (°C)	Decomposition T (°C)	Density (g/cm³)	Viscosity (cP) at 25°C
1-hexyl-3-methylimidazolium tetrafluoroborate	[C6mim][BF4]	-75	403	1.17	108.25
1-butyl-3-methylimidalzolium bis(trifluoromethylsulfonyl)imide	[C4mim][Tf2N]	-4	439	1.43	61.14
1-butyl-2,3-dimethylimidazolium bis(trifluoromethylsulfonyl)imide	[C4mmim][Tf2N]		459	1.44	115.2
1-butyl-3-methylimidazolium hexafluorophosphate	[C4mim][PF6]	6,5 – 10	> 350	1.37	284.49

The density, the specific heat capacity, and the thermal conductivity of ILs are comparable to those of synthetic oils. However, their viscosity is typically an order of magnitude higher, increasing plant pumping power requirements. Other main challenges that still have to be addressed in order for ILs to be used as HTF for concentrated solar power are their high cost but, most importantly, their practical thermal stability and aging characteristics.

Recent work by Fox *et al.* [58] testing the suitability of several ILs at high temperatures over extended periods of time showed that, although ILs are generally thought of as relatively stable fluids, there is limited testing at practical operating conditions and, in reality, most of these fluids are not well suited for applications at moderate to high temperatures. All the ILs tested in this study ([C4mmim] [Tf2N] among others) showed visible aging through a change in color and composition after immediate exposure to heat (200°C). The onset temperatures of decomposition measured through thermogravimetric analysis (TGA) did not significantly change after the extended aging analysis (200°C during 15 weeks). Unfortunately, over the aging process their viscosity significantly increased to the point that it could impact their performance in the field. A few of the ILs had even completely polymerized or solidified after 2000 hours at 200°C and some others had formed a solid-liquid slurry type fluid. These results indicate, although they had promising features, after working extended periods of time at moderately low temperatures, those ILs were not suitable for concentrated solar power plants.

Molten Glass

On the higher end of the operating temperature spectrum, a low melting temperature molten glass has recently been proposed as a potential HTF and TES material for very high temperature concentrated solar power plants [59]. This novel oxide glass material is thermally stable between the temperatures of 400 and 1200°C, a much higher temperature range than any of the molten salt mixtures currently available (Fig. **13**, Table **4**). It is earth abundant, low cost, and has low vapor pressure at high temperatures. The main challenge resides in producing a glass oxide mixture with a low melting point, low viscosity and good heat transfer characteristics. A promising candidate has been reported (Haloglass™ RX [46]) with an acceptable viscosity and thermal stability; it is also compatible with common stainless steels at 400°C but refractory materials are needed for the high temperature components in the plant. However, at the moment, molten glass oxides have been proposed as a more suitable standalone TES material to be coupled to other HTF due to its poor heat transfer capabilities and high viscosity [60].

Gases

Air and other gases (CO_2, H_2 and He) have also been evaluated as HTF for concentrated solar power plants. Pressurized gases have many advantages over traditional HTF such as availability, reduced safety and environmental concerns, easy operation and maintenance, and high operating temperatures. Most importantly, a gaseous HTF could potentially be used also as the working fluid in the power block turbines, eliminating the need for additional heat exchangers (and the associated heat losses). Direct expansion of solar-heated gases allows using higher efficient Brayton cycles, which could be bottomed with conventional Rankine cycles for a combined cycle efficiency above 50%.

However, some of the drawbacks of gases are their poor heat transfer properties. Poor heat transfer results in challenges in receiver designs. Also, the combination of poor heat transfer and low density make coupling to cost-effective thermal storage more difficult, requiring large surface areas for effective heat exchange. Consequently, gases need large pressures for adequate efficiencies, increasing the system installation costs with thick wall structural elements and high pumping power needs.

The use of air as a HTF in central receiver towers has been demonstrated since the 1980s. There are few pre-commercial scale concentrated solar power plants using air as a HTF. In 2009 the DLR commissioned a 1.5 MWe central receiver tower in its research facility in Jülich, Germany (Jülich solar tower) using atmospheric air as a HTF [5]. The air is heated up to 700°C and then used to generate steam. Jülich solar tower was a first stepping stone for testing air receivers and compatible storage, where the HTF works at atmospheric pressure and is not directly expanded in the power cycle, using a conventional steam Rankine cycle. A few years later, the first MW-scale hybrid solar-gas plant called Solugas was a commissioned in 2012 at Abengoa's Solucar R&D complex near Seville, Spain [31]. Solugas used a tubular receiver to heat pressurized air up to 800°C which was then directly fed into the combustion chamber of a modified 4.5 MW commercial Mercury™ 50 gas turbine. Solugas has afterwards been retrofitted to test novel and more efficient volumetric air receivers aimed at reaching 1000°C, which can be used for both hybridized Air-Brayton concentrated solar power plants and for the next generation of thermochemical solar plants.

Inert gases such as CO_2 are other alternative HTF that are recently being reevaluated. Pressurized CO_2 presents advantages from the environmentally and safety perspective. It is non-flammable, non-toxic, and widely available at low cost. Helium has also been evaluated for HTFs in concentrated solar power plants [34]. Its main advantages are that it is inert and its specific heat capacity is much higher than that of air. But, as CO_2, it must be used in a closed cycle (*i.e.* the fluid recirculates from the solar receiver to the turbine, then it is condensed and sent back to the receiver) and leaks are a significant problem.

Supercritical Fluids

The recent surge in the development of supercritical power cycles has made researchers investigate supercritical fluids such as s-CO_2 and s-H_2O as potential HTF that could be directly expanded in these novel turbines [8]. Direct expansion of these fluids could theoretically reduce system costs and increase efficiency by eliminating the need for heat exchangers between the HTF from the solar field and the working fluid in the power block. Unfortunately, storage of supercritical fluids is not a viable option [24, 61] consequently a heat exchanger for the TES system is still needed.

Supercritical fluids can operate at high temperatures like gases but have better heat transfer characteristics, similar to those of liquids. Depending on the working fluid, the operating pressures can be also high. Both s-CO_2 and s-H_2O are non-toxic, non-flammable, and thermodynamically stable at high temperatures. These supercritical fluids are being proposed as the next generation of environmentally friendly heat transfer fluids. However, using supercritical fluids as heat transfer media is a relatively new area of research and practical engineering property databases are very much needed [61].

Supercritical steam (s-H_2O) has been used in conventional power plants for some time now, but it is a novel fluid for concentrated solar power. The critical point of water is 374°C and 220 bar. The high pressures and high temperatures that will be required to use s-H_2O as a HTF in a concentrated solar power receiver will be a tremendous technological challenge for receiver design, piping, heat exchanger, and other components.

The critical point of CO_2 (31°C, 73.8 bar) is close to ambient temperature, but the operating pressure is still relatively high. Further studies to test compatible container materials and their corrosion behavior under practical working conditions are needed.

A more detailed description of the use of supercritical fluids as potential new working fluids for turbines is included in the section dedicated to advanced power cycles.

Particle Suspensions

Another option capable of reaching temperatures in excess of 1000°C and very high heat fluxes in the receiver is using solid particles as the heat transfer medium. These HTM present an interesting new solution for concentrated solar power systems that use advanced higher temperature cycles (as alternatives to other high temperature HTF such as liquid metals) and for thermochemical cycles and processes where the particles are the reactive medium. Moreover, the heated solid particles can be easily directly stored, eliminating the need for intermediate heat exchangers between the heat transfer media and storage.

Ideal solid particles should have the following properties [24]: high solar absorptance, low thermal emittance, low cost, availability, high packing density, high heat capacity, resistance to sintering and agglomeration, corrosion resistance in air or in other media, and resistance to mechanical and thermal shock. The materials that have been proposed for particle suspensions include sand, ceramics (aluminum silica), and certain ferromagnetic particles (magnetite, Alnico, Bismuth ferrite) [36].

Controlling the flow of falling particles is proving to be a challenge and the systems that have been tested so far unfortunately lead to high heat losses and not very high efficiencies. Different methods are currently being explored to control particle flow and increase particle residence time such as imposing an electro-magnetic field.

As an alternative to solid particles as HTM, dense particle suspensions (DPS) are also being investigated [62]. Dense particle suspensions consist of μ-m particles fluidized in a low velocity gas. DPS could potentially combine the ease of handling gases and the high temperatures achievable with gases and solid particles with the good heat transfer properties of liquids.

Nanoparticle suspensions, otherwise known as nanofluids, have also been increasingly studied for solar thermal applications, opening a highly popular area of research in the past few years [8, 63, 64]. Adding nanoparticles to common

HTF can increase some thermo-physical properties (thermal conductivity, heat transfer coefficient but also viscosity) and optical properties of the base fluid. However, the degree of enhancement achievable is uncertain as it depends on many factors (particle size, type, concentration, morphology) [8]. Nanoparticles have been added to a variety of base fluids: synthetic and mineral oil [42], ionic liquids [56], molten salts [65], and water/steam [64]. Further research is needed to understand the particle-particle interaction, the aggregation and agglomeration characteristics, the long term stability of the dispersion, and sedimentation of these fluids before they are industrially used as a HTF.

INCORPORATING THERMAL/THERMOCHEMICAL ENERGY STORAGE

Concentrated solar power is an attractive renewable energy only if thermal energy storage (TES) is incorporated, allowing electricity supply on demand, independently of the solar resource. As more intermittent renewable energy resources are connected to the grid, power fluctuations must be carefully controlled to avoid surges and crashes. There are essentially three main options to balance these power fluctuations in electricity generation: 1) ramp up or down electric production from conventional fossil fuel plants, 2) store electricity (*i.e.* batteries), and/or 3) store thermal energy. At the moment, gas turbines and batteries cannot provide rapid high-power responses and supply energy for long hours. Although there are some other grid-level technologies that are economically competitive with TES such as pumped hydro or compressed air [8], TES has proven to be an inexpensive, large scale, efficient, and reliable method to smooth out short transients and to extend electricity production well passed daylight hours.

Concentrated solar power plants have an intrinsic large thermal inertia providing a "buffering" capacity that allows uniform electricity production during transients (*e.g.* clouds passing) [1]. On the other hand, not all concentrated solar power plants currently in operation include TES systems to provide uninterrupted electricity production [5], but slowly TES systems are being routinely added to modern plants. Unfortunately, rapidly decreasing photovoltaic prices have made concentrated solar power plants without TES nearly irrelevant [1].

The concept of thermal energy storage is simple (Fig. **14**): during the day the excess heat collected by the solar field is diverted to storage and, when production is required, the stored heat is released into the power cycle and the plant continues to produce electricity [1].

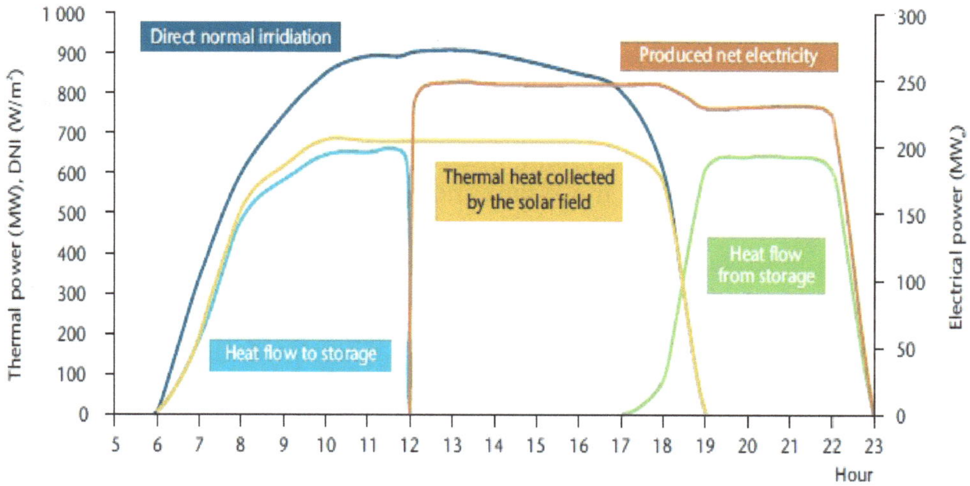

Fig. (14). Example from a 250-MW CSP plant: thermal storage can decouple and shift the electricity generation from daylight to evening hours (from IEA Technology Roadmap: Solar Thermal Electricity [1]).

The performance of a TES system is characterized by its storage capacity (how much heat it can store), its heat transfer rate during storage charge/discharge (storage power) and the storage duration or time (capacity divided by power or hours the plant can run at rated capacity from the TES only). A large storage capacity is desirable, preferably using high energy density materials to reduce the system size and costs. The TES material´s thermal conductivity as well as the system design are the main parameters defining the storage charge /discharge rates. Storage duration requirements are some times more a political or commercial decision rather than a technological one.

Additionally, the TES system must be compatible with the concentrated solar power plant components it exchanges heat with, *i.e.* the heat transfer fluid (HTF) and the power block, working at an appropriate operating temperature range and providing the required temperature drops during the system charge and discharge.

TES System Concepts

In general terms, a TES system can be classified into two distinct types: 1) direct storage, where the heat transfer fluid that is heated in the solar receiver is the same storage material; and 2) indirect storage, where a separate storage medium is employed for the thermal energy storage. It is desirable to have a direct storage system in order to eliminate the need for a heat exchanger between the HTF and TES, reducing the thermal losses inherently associated with any heat transfer process, increasing the system efficiency and lowering the cost. However, in certain situations indirect storage systems are more cost-effective (*i.e.* 2-tank indirect molten salt TES for a superheated steam plant), can provide higher energy density (*i.e.* latent heat TES), or it might be the only viable solution (*i.e.* systems with liquid metals as HTF).

The current commercial TES system integrated in both parabolic trough and central receiver tower plants is the 2-tank molten salt storage using solar salt (60 wt% $NaNO_3$, 40 wt% KNO_3). The 2-tank molten salt system is widely used, there are examples in indirect storage configurations (*e.g.* Andasol Project [5]) and in direct storage (*e.g.* Solar Two Project [5], Gemasolar plant [18]). This TES concept is mature, relatively inexpensive, and well-known (low risk).

A single tank thermocline provides a cost reduction of approximately 35% when compared to the two-tank molten salt TES system [66]. The cost reduction comes from a reduction in tank and TES materials. The single tank can be marginally larger than one tank from the two-tank concept [67]. In a thermocline the hot and cold fluids are separated in the same tank through thermal stratification. A low-cost filler material fills most of the thermocline tank volume in a packed bed configuration, acting as the primary thermal storage medium and substituting the need for a higher volume of molten salts to obtain the required temperature stratification. A thermal insulation baffle can be installed in between both fluids to keep them separated and insulated from each other. However, thermoclines increase system complexity and introduce new problems such as thermal ratcheting of the tank wall. There is currently a commercial installation of this type in a linear Fresnel plant in Spain [5].

Storage Mechanisms and Materials

Ideally, a TES material should have the following desirable characteristics:

- Low cost and wide availability
- High specific heat capacity
- High thermal conductivity, to provide fast system charge/discharge rates
- Non-toxic, non-flammable, environmentally friendly
- Chemical and thermal stability at high temperatures
- Compatibility with construction /container materials, high resistance to corrosion and oxidation

The materials used for TES can be classified into three different categories according to the storage mechanism, in order of increasing energy density: 1) sensible heat TES, that store thermal energy in elements at different temperatures; 2) latent heat TES, where storage is based on the phase change of the material; and 3) chemical or thermochemical TES, which takes advantage of reversible chemical reactions to charge and discharge the thermal storage system.

Sensible heat TES are the most mature technology and the main type of TES commercially implemented but they have the lowest energy density of all three categories, increasing the system size requirements. Latent heat systems have higher energy density but they also typically have poor heat transfer characteristics, requiring thermal conductivity enhancers in most cases. Thermochemical TES have the largest energy density of all three types of TES but they are less developed at the moment. The present R&D effort in thermochemical TES is enormous.

Sensible Heat TES

In sensible heat storage systems, the energy is stored/released by raising/lowering the temperature of the storage material, which can be a solid or a liquid. The energy stored (Q) depends on the amount of material (mass, m), the specific heat capacity of the storage material (Cp) and the temperature difference in the system.

All commercial TES are currently sensible heat storage systems, using materials such as molten salts, concrete or graphite. Other potential materials, their main properties and some cost information can be found in several reviews on TES [7, 68, 69].

Latent Heat TES

Latent heat thermal storage (LHTES) materials, also known as Phase Change Materials (PCM) present higher energy density than simple sensible heat TES materials. Therefore, latent heat TES systems will have smaller storage volumes relative to sensible heat storage. The amount of energy stored is determined by the PCM´s latent heat or enthalpy of phase change (*i.e.* latent heat of fusion in a liquid-solid transition):

Phase change occurs roughly at a constant temperature allowing a minimum temperature difference within the storage media. PCM can be inorganic (salt hydrates, salts, metals, and alloys) and organic (paraffin, fatty acids, alcohols, glycols). However, a barrier to large scale utilization of LHTES is the poor thermal transport properties presented by typical PCMs (thermal conductivity is usually around 1 W/mK or less). Other key challenges concerning presently studied PCM include insufficient storage energy density, limited performance in terms of operating temperature range, inadequate melting point range for specific applications, difficulty controlling and reducing the expansion volume during phase transition, and thermal stability, decomposition, corrosion and durability issues.

Techniques employed to increase the thermal conductivity of PCM include: 1) mixing high conductivity particles or fibrous materials (carbon fibers) into the PCM, 2) encapsulating the PCM at micro or nano-scales within the heat transfer media, 3) creating a composite material with the PCM embedded in a highly conductive, porous matrix such as expanded graphite or metal foams 4) incorporating heat pipes or 5) using extended surfaces to create sandwich-like structures.

PCMs should have melting points in an adequate temperature range depending on their application in addition to other desirable properties such as congruent melting, minimum volume change during phase transition, chemical stability, and reliable convertibility after repeated thermal cycling. For direct steam generation (DSG) saturation temperatures range between 120 and 320°C, thus some suitable PCM that melt in that range are tin, lead and nitrate molten salts. Tin and lead are very costly but they have excellent thermal conductivity compared to the molten salts.

There are excellent reviews on potential PCMs for TES [69 - 72] and the reader is redirected to them for further details.

Thermochemical TES

In thermochemical TES, the energy is stored through a reversible chemical reaction (AB + heat <=> A+B). During the system charge, the endothermic reaction absorbs heat; during the discharge, the heat is released through the exothermic reaction.

Thermochemical TES materials show high potential due to their high energy density and their absence of thermal losses. However, thermochemical TES is at an earlier stage of development compared to both sensible and latent heat storage. Extensive developments are needed until they can be successfully and cost-effectively implemented in CSP plants. Technological challenges include finding materials with appropriate reaction rates, conversion efficiencies, and showing chemical reversibility over the lifetime and cycle requirements of the plant.

For a list of potential suitable materials and reactions for thermochemical storage in concentrated solar power plants, the reader is referred to several reviews on the subject [74, 75].

Future Outlook on TES

R&D in TES systems is mainly focused on developing higher temperature, cost-effective storage solutions with enhanced energy density and increased thermal stability. Some of the approaches engineers and scientists have tried are summarized below.

Given the widespread application and cost-effectiveness of conventional 2-tank molten salt storage systems, developers are always trying to further improve these TES. Some approaches are: a) additives to increase the specific heat capacity of traditional solar salt; and b) using non-nitrate salts that are capable of operating at higher temperatures. A detailed discussion on alternative molten salts is included in the HTF section (page 31).

For latent heat TES, most of the effort has been focused on enhancing the thermal conductivity of such materials through metallic fins, encapsulating the materials, embedding the materials in carbon or metallic fibers, adding highly conductive nanoparticles, embedding heat pipes, *etc* [73]. Encapsulating the PCMs is also required to separate the storage material from the HTF and avoid contamination. However, encapsulation of high temperature PCMs such as molten salts has proven to be challenging [7]. Some of the difficulties are related to PCM/shell material compatibility and corrosion problems, others deal with adequate designs to allow PCM volume expansion during phase change, and thermal gradients affecting the structural integrity of the capsules to withstand a large number of charging/discharging cycles at high temperatures. Interesting prototypes have been developed at the University of South Florida [76]. There is also a pre-commercial prototype of encapsulated molten salts in a packed bed configuration developed by Terrafore Technologies LLC [77], but it has not been implemented in a full scale concentrated solar power plant to this date.

Using multiple PCMs with different melting temperatures in a “cascade” [7, 68] has also been explored as a system improvement to cover larger temperature ranges and provide a better temperature match between the TES and HTF, increasing efficiencies.

Compatibility with containment materials is also an area of concern for all types of TES materials. Most sensible and latent heat TES materials are molten salts (nitrates, chlorides, carbonates, hydroxides, *etc.*), which are typically corrosive with standard metallic containers. Many of the recent developments are tackling the different corrosion mechanisms between common molten salts used in TES and potential containment materials [7]. It is a challenge to find suitable containment materials with acceptable mechanical properties and corrosion resistance over the temperature range of use.

Finally, the main barrier for commercialization of these higher energy density TES solutions is cost. Once appropriate materials have been found with acceptable properties, stable at the temperature range required, compatible with container materials, among other multiple considerations, an in depth cost analysis is required. To this date, unfortunately the two-tank molten salt TES system is sufficiently inexpensive that many of these novel developments have not resulted in cost-effective alternatives yet.

POWER CYCLES FOR CSP PLANTS

In order to obtain electricity from the sun, a concentrated solar power plant uses a thermodynamic cycle to transform the collected heat into mechanical energy and then into electric energy in the power block. The overall plant efficiency must take into account both the solar-to-heat conversion step in the collector-receiver and the heat-to-electricity conversion step from the thermodynamic cycle. Current receiver technology has reached significantly high efficiencies, on the order of 0.9. Therefore, it seems like further improvements in the receiver will only lead to a few efficiency percentage points increase at the most. However, thermodynamic efficiencies are typically only around 50%, leaving much more room for improvement that is presently being explored by the R&D community. Moreover, recent techno-economic analyses indicate that there is a significant cost reduction potential through the improvement of the power block thermal efficiency [3, 23, 78].

The maximum possible thermodynamic cycle efficiency is determined in theory by Carnot: $\eta_{CARNOT}=1-T_C/T_H$, where T_C and T_H are the sink and source temperatures respectively. The source or T_H in a concentrated solar power plant is the maximum plant operating temperature and it is typically limited by the HTF and receiver technology. The sink represents the heat rejected during the cooling step in the thermodynamic cycle and it is limited in practice by ambient temperature. In reality, practical cycles have efficiencies lower than the theoretical Carnot cycle due to system irreversibilities and losses. A common estimate for engineering efficiency limits is to reduce the Carnot efficiency by 2/3 (or sometimes 3/4). Based on the Carnot efficiency, it is clear that if the cold side is constant and equal to ambient temperature, increasing the plant operating temperature will lead to higher efficiencies.

The main power cycles that have been employed up to date in solar facilities are subcritical steam Rankine cycles with low temperature (<400ºC) steam turbines traditionally yielding efficiencies in the 35-42% range. Additionally, Rankine cycles coupled to conventional parabolic trough concentrated solar power plants using thermal oil as a HTF have low turbine inlet temperatures due to the oil´s maximum operating temperature, 400ºC. When molten salts are used in a parabolic trough [9], the solar field can reach higher temperatures, in excess of 400ºC, which potentially enables higher turbine inlet temperatures and, consequently, higher efficiencies. For example, ENEA´s Archimedes molten salt trough plant in Italy is obtaining turbine efficiencies above 39% with wet cooling [5]. Molten salt central receiver towers are limited to 565ºC (or 600ºC if a certain degree of salt thermal degradation is allowed) producing superheated steam around 540ºC and 130 bar, and yielding efficiencies around 42% with wet cooling.

Brayton cycles use a gaseous working fluid throughout the entire cycle and are able to operate up to higher temperatures than a conventional Rankine cycle. As the Carnot efficiency shows, higher temperatures lead to increased thermal efficiencies. When used in conjunction with high temperature tubular and cavity type receivers, pressurized air (or other gases) can be heated to temperatures between 800-1000ºC. Brayton air cycles have been used in concentrated solar power plants mainly in hybrid solar-gas configurations (*e.g.* Solugas [31]). The "solarization" of gas turbines, *i.e.* introducing the concentrated solar radiation in the heat addition step of the cycle, has been extensively studied, with interesting effects on turbine performance [79] and evaluating the integration of thermal energy storage in such systems. However, the main barrier for commercialization at the moment is the lack of a commercial turbine in the scale of 50 MW or higher that has been retrofitted for this use.

Closed helium-Brayton cycles have also been explored for concentrated solar power-power block applications [78, 80]. Helium is a good working fluid choice because it has large specific heat capacity, low density, and permits excellent heat transfer with a relatively low pressure drop, using small, compact turbomachinery. Turbine inlet temperatures on the order of 750-850ºC are predicted to yield thermal efficiencies around 50% with advanced closed Helium-Brayton cycles with multiple reheat stages.

Current and novel cycles can be further optimized if they are coupled in a combined cycle (CC). In a CC, the heat rejected from a high temperature (topping) cycle is used to drive a lower temperature (bottoming) cycle. Combined cycles have been used in traditional power cycles for decades [78] and can reach efficiencies close to 60% or higher. For example, an air-Brayton cycle for a central receiver tower could have a thermal efficiency of approximately 42%. However, if it is used in a CC where the topping cycle is the air-Brayton and the bottoming cycle is a standard subcritical steam-Rankine cycle, the total efficiency can reach values above 52% without major technological hurdles.

Advanced Power Cycles

New higher efficiency power cycles are a growing area of research in order to further decrease costs [23, 61, 78]. The main novel cycles use supercritical fluids, *i.e.* fluids working above their critical point.

Supercritical fluids have excellent heat transfer characteristics, similar to those of liquids, but can operate at much higher temperatures than liquids. Depending on the working fluid, the operating pressures can be also high. Large operating pressures should be avoided because they lead to thicker piping wall requirements and more expensive installations. The critical temperatures and pressures of some common working fluids can be seen in Fig. (**16**) for reference.

Closed cycles using supercritical fluids are generally simpler, more compact, more efficient, and have shorter construction times than conventional Rankine steam cycles. The main advantage of a supercritical cycle is that the working fluid does not pass a distinct phase change during the heat addition step, contributing to achieve higher efficiencies.

Challenges in the design and operation of turbines using supercritical fluids arise from the strong fluctuations these fluids show in their thermophysical properties around the critical point. The turbines might be more sensitive to fluctuations in properties when the heat input comes from a transient source such as the sun. Therefore, supercritical fluid power cycles should be used in conjunction with thermal energy storage to buffer out transients.

Unfortunately, thermal storage of supercritical fluids is not a viable option [61]. As a consequence, a separate fluid/solid material is needed for thermal energy storage and efficient heat exchangers to transfer the heat from the storage media to the turbine working fluid are needed.

Other novel power cycles aim at the low temperature range (*i.e.* supercritical Organic Rankine Cycles or the water-ammonia based Kalina cycle) and are being proposed as bottoming cycles to couple with topping s-CO_2 or air-Brayton cycles. The main characteristics of these cycles and potential challenges are discussed below.

Low Temperature Cycles

Both Organic Rankine Cycles (ORC) and more recently supercritical Organic Rankine Cycles (SORC) are being evaluated for solar thermal applications in the lower temperature range. Only recently have researchers explored the use of ORC at higher temperatures (source at or above 300ºC).

Organic Rankine Cycles (ORCs) have typically been studied for low-temperature or waste heat recovery systems, due to their economic advantage over other methods of producing electricity from low-grade heat. Their thermal efficiency is unfortunately typically low, in the 10-28% range.

Of special interest is the choice of working fluids for ORCs, as a great many refrigerants are available, often with subtly different properties. In finding the most suitable working fluids many competing aspects must be considered such as thermodynamic properties, stability, cost, availability, safety, health, and environmental factors. In contrast to water, a "wet" fluid (*i.e.* with a negative slope of the saturated vapor curve on a Temperature-entropy diagram), "dry" fluids are usually preferable in order to expand the saturated vapor outside the vapor dome and avoid forming liquid droplets. Liquid content must be avoided to limit turbine blade erosion and the need for expensive turbine blade coatings and reinforcing materials.

A novel Organic Flash Cycle (OFC) [81] has been proposed for low-medium temperature heat sources (300-400ºC) to better match temperatures in the heat exchanger streams reducing cycle irreversibilities during the heat addition step. The OFC showed similar utilization efficiencies (total work output over supply work availability) to an optimized ORC, but better results than basic water-ammonia Rankine cycles or transcritical Rankine cycles.

Supercritical ORC (SORC) are being proposed for geothermal and solar thermal applications to further increase the performance efficiency of ORC in the low temperature source range. A SORC can provide a better thermal match between the heat source and the working fluid temperature profiles in the supercritical heat exchanger. Also, a SORC can be used as a bottoming cycle for a combined cycle plant where s-CO_2 is used as the topping cycle.

There are even new projects like a recent one funded by the Australian Renewable Energy Agency (ARENA) [82] that circulate a supercritical organic fluid through a parabolic trough field test-bed coupled to a SORC in what they call Direct Supercritical Fluid Generation (DSFG).

The critical point of some common refrigerants is shown in Table **7** as potential working fluids in SORC. In addition to considering a low critical pressure to reduce pumping requirements, the environmental impact of such refrigerants should be taken into account, since many of them have high ozone depletion potential.

Table 7. Potential working fluids for SORC.

Fluid	**Tcritical (ºC)**	**Pcritical (Bar)**
R125	66	36.2
R134a	101.1	40.6
R143a	72.7	37.6

(Table 7) cont.....

Fluid	Tcritical (°C)	Pcritical (Bar)
R170	32.2	48.7
R218	71.9	26.4
R245fa	153.8	36.5
R32	78.1	57.8

Supercritical Rankine Cycle

Standard subcritical steam Rankine cycles operate up to about 600°C and are limited to gross thermal efficiencies around 42%. Higher operating temperatures are not possible in many cases due to material limitations of standard steels or to the HTFs thermal degradation limits. Using steam at supercritical conditions will allow higher temperatures (up to 700°C), yielding thermal efficiencies above 45%. They also benefit from not requiring inefficient and complex multiphase heat exchangers.

The main advantage of using s-H_2O Rankine cycles is that commercial technology already exists from conventional fossil fuel plants and it is well-known (low risk). Current supercritical steam Rankine cycles are designed for larger power outputs than typical concentrated solar power plants (+800 MWe). The s-steam turbines will need to be scaled down and redesigned for lower volumetric flow rates, leading to smaller blades and overall more compact turbines.

Although s-H_2O are the closest to commercialization of the supercritical fluid power block technologies, there are presently two major hurdles for their implementation in solar facilities: 1) finding a HTF to operate in the 600-700°C range, 2) evaluating the use of standard ferritic steels (or other low cost materials) at temperatures above 630°C to avoid needing the use of expensive higher temperature nickel alloys. A quick glance to Fig. (**13**) and Table **5** shows that current molten salts cannot be used above 600°C and, with the exception of a novel multicomponent salt based mainly on chlorides (saltstream 700°C®), there is no other molten salt covering the 200-700°C range. Liquid sodium is accompanied by too high a risk increase and particles as a HTM are not easily moved throughout the plant with current designs. In other words, the immediate solutions to these two material limitations (HTF and inexpensive container materials) are not entirely clear and they will increase cost and risk, delaying the path to commercialization.

Supercritical CO_2 Cycle

Closed supercritical CO_2 Brayton cycles have been considered for nuclear and

CSP plants [83], concentrating most of the recent power cycle R&D activities. It seems as the most promising cycle option in the 400-800°C temperature range (and probably the only option in this temperature range as shown in Fig. **15**) capable of achieving efficiencies over 50%, besides a combined cycle. Additionally, due to the fluid's compressibility above the critical point, low flow rates are required which lead to smaller turbine blades. Having compact turbomachinery is a very interesting feature for CSP plants that might allow efficient new designs with a small power block inside the central receiver tower.

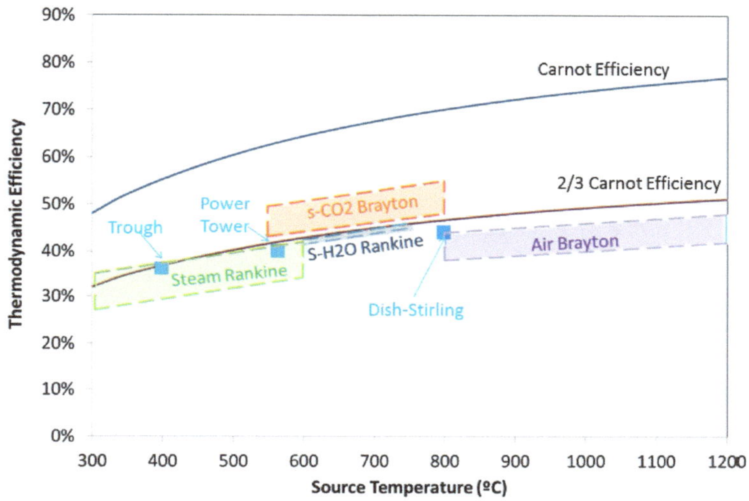

Fig. (15). Typical thermal efficiency and source temperature for common CSP power cycles.

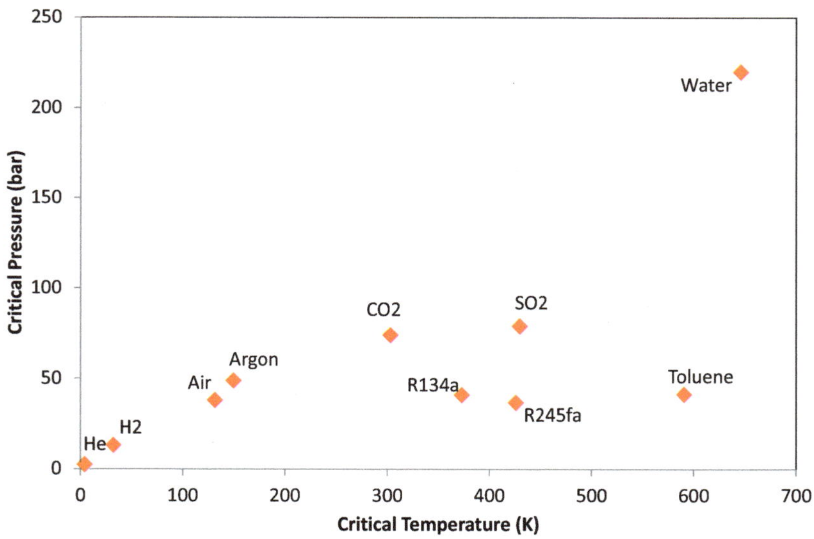

Fig. (16). Critical properties of some common working fluids.

The R&D community seems to agree that the way to achieve higher efficiency is with an s-CO_2 Brayton cycle. Comparing Fig. (**13**) and Fig. (**15**), one can see that there are not many fluids operating at the s-CO_2 source temperature (400-800ºC): molten glass is too viscous, liquid sodium is a fire hazard, current molten salts do not operate above 600ºC, and new molten salt mixtures have yet to be tested for stability, compatibility and corrosion with containing materials. A possibility is to use s-CO_2 as both the HTF and power block working fluid simplifying the system configuration [32], but in order to keep piping costs to a minimum a modular multi-tower approach is proposed.

However, at the moment, there is currently no commercial s-CO_2 turbine available. Small-scale generators and subcomponents for s-CO_2 turbines are still being designed and tested. New developments in materials to withstand the high corrosive nature of CO_2 are also needed.

Auxiliary Hardware for Advanced Power Cycles

The development of metal or ceramic high temperature heat exchangers to work with advanced power cycles is an active area of research. It is a great concern for s-CO_2 cycles, where the TES must use a separate medium from the turbine working fluid and the technology is yet to be developed.

Additionally, high temperature and corrosive resistant auxiliary hardware such as piping, valves, flanges, blowers, pressure and flow measurement devices, *etc.*, is needed with supercritical fluids under practical working conditions (high temperature, high pressure, and thermal cycling).

Power Plant Cooling Trends: Dry Cooling

Concentrated solar power plants are typically built in areas that tend to be desert locations with high solar resource but also where water is scarce. In simple terms, since the thermal conductivity water is approximately twenty-fold that of air, cooling with water is much more efficient. About 99% of the base-load thermoelectric conversion power plants are water cooled [84], impacting the local aquatic environment. A standard power block in a central receiver tower using wet cooling utilizes approximately 2500 liters of water to produce one megawatt-hour of solar electricity [23]. Consequently, in the past few years, many R&D groups are exploring dry cooling systems that do not significantly impact the efficiency of the power block to reduce the concentrated solar power plant´s environmental footprint.

Current dry cooling systems in thermoelectric power plants have two main disadvantages: a penalty in performance when ambient (air) temperatures are high and the need for massive and costly heat exchangers due to the poor heat transfer characteristics of air. Dry cooling with ambient air penalizes the power block´s net power production and thermodynamic efficiency by raising the condensation (sink) temperature allowing less power to be extracted from the turbine. For example, in a conventional steam-Rankine cycle, a 3ºC rise in the steam condensation temperature can result in approximately 1% power reduction from the turbine [84].

When dry cooling is necessary, passive cooling techniques can be added to reduce the sink temperature. Some of these techniques include exchanging heat with the ground ("ground coupling") or nocturnal cooling.

As part of the USA´s Department of Energy´s Arpa-E agency a new funding program solely based on Advanced Research in Dry Cooling (ARID) was launched in 2015 with the objective of boosting R&D in ultra-high-performance air-cooled heat exchangers, supplemental cooling systems and/or cool-storage systems that can cost-effectively and efficiently reject waste heat. Among the most innovative ideas in this program is the development of meta-materials and spectrally tuned polymers for radiative cooling during both day and nighttime operation. Other research includes more efficient air-cooled heat exchangers, supplemental cool storage with advanced heat pipes coupled with a large-capacity heat sink, novel cool storage media and low-cost sorption/desorption cooling systems driven by waste heat.

Other Improvement Opportunities

In addition to the development of advanced power cycles and the improvement of dry cooling technology, there are other areas that should be explored to further reduce the power plant cost and increase its overall efficiency. The following ideas have been suggested as potential cost reduction areas for concentrated solar power plants [23]:

- Reduction of parasitic consumption: Commercial scale concentrated solar power plants can consume 10% or more of the gross annual electricity. Receiver pumping power is one of the major sources for this parasitic consumption that should be re-evaluated, but it will also depend on the choice of HTF.
- Hybridization with fossil fuels: Combining existing natural gas combined cycle and coal-fired plants with central receiver tower technology has been recently studied as a cost reduction strategy [23], but at current oil prices and considering the capital investment needed for a concentrated solar power-power tower plant, it no longer seems profitable. Nevertheless, for the same reasons (*i.e.* cheap oil prices), fossil fuel back-up to ensure continuous operation in pure solar central receiver towers can be an interesting strategy if greenhouse gas emission is not

considered.

- Designs for rapid temperature changes and quick turbine restarts: coupling a thermal energy storage system or even batteries can improve the transient response of current commercial concentrated solar power plants.

ALTERNATIVE USES OF SOLAR HEAT BEYOND TRADITIONAL ELECTRICITY GENERATION

In concentrated solar power systems solar radiation is concentrated and converted to heat; this heat is used to raise the temperature of a heat transfer fluid which then transfers the heat to a conventional heat engine, which converts heat to electricity through a thermodynamic cycle. The thermodynamic efficiency of a heat engine is limited in theory by Carnot, but in the case of current commercial concentrated solar power plants using a standard Rankine cycle, this efficiency is around 40%. In other words, the heat engine is penalizing the overall solar-t--electric conversion efficiency of a concentrated solar power plant by 40%.

Two very different approaches to solve this problem by avoiding the traditional heat engine that converts heat-to-mechanical-to-electrical energy have been proposed: 1) an engineering approach, where the incident solar radiation is used in a thermochemical cycle to charge a battery and 2) through solid-state devices that use various physical phenomena to directly convert solar radiation into electricity, such as thermoelectric, thermoionic, and thermophotovoltaic devices. Both of these approaches try to increase the overall solar-to-electric conversion by eliminating the thermodynamic cycle of the heat engine.

Moreover, the fierce competition of other types of renewable resources such as photovoltaics has lead the solar thermal community to diversify and look for alternative markets for CSP such as desalination, process heat, enhanced oil recovery, solar fuel production, industrial process heat or hybridization with fossil fuels, among other processes. Some of these alternatives are briefly explored below.

Thermochemical Cycles Based on High Temperature Batteries

A novel thermochemical cycle has been proposed to use the thermal energy from concentrated solar power to charge an electrochemical battery [85]. More specifically, the suggested cycle is based on liquid Na-S batteries, but it could potentially be used with metal/air batteries such as Zn/air, Mg/air, for example.

The sodium-sulfur battery discharge cycle works at temperatures ranging between 300 and 350°C, at which the sodium, sulfur and the reaction product (sodium polysulfide) coexist in liquid state. Concentrated solar radiation at very high temperatures (>1500ºC) is used to charge a Na-S battery, decomposing sodium polysulfide into Na and S (with a decomposition temperature around 1500- 1700º C). Recharging of the battery is therefore achieved through solar radiation. This concept is yet to be demonstrated in a prototype.

Thermoelectric, Thermoionic, and Thermophotovoltaic Generators

Thermoeletric, thermoionic, and thermo-photovoltaic devices are not entirely new and they have traditionally had very low conversion efficiencies. However, they are currently an active area of research driven by new developments in materials science that will allow higher solar-to-electric conversions. They can be used as a stand-alone concept or as a "topping" cycle on solar thermal, *i.e.* receiving the solar radiation on the top of a central receiver type tower and the waste heat is then dumped to the concentrated solar power central receiver. These are all solid state devices; therefore, they have no moving parts, no need for high temperature moving fluids. They are modular and scalable. Their main advantage is that they are potentially more reliable, robust, and with less maintenance requirements than a traditional heat engine with mechanical moving parts.

A solar thermoelectric generator (STEG) is a solid-state device that converts sunlight to electricity through the thermoelectric (Seebeck) effect. Sunlight is absorbed by a receiver and converted to heat. The heat is then delivered to the thermoelectric generator (TEG) that produces electricity.

Until recently thermoelectric generators have not received much attention and were relegated to waste heat recovery systems because their conversion efficiency has been below 10%. Advances in materials and devices for TEGs have been able to reach efficiencies above 15% [86] making them now an interesting technology for standalone solar-to-electricity generation.

A STEG has three main elements: 1) the solar absorber; 2) the TEG; and 3) the heat management system comprised of insulation, heat exchanger, vacuum enclosure and similar components. As with solar thermal absorbers, multilayer solar selective coatings optimized at high temperatures (1000ºC) are being developed for such purposes [86].

The main challenge of this technology at the moment is increasing the conversion efficiency. The thermoelectric efficiency (η_{TE})depends on zT, the figure of merit of the thermoelectric materials, $zT = (S2\sigma/k)T$, where S is the Seebeck coefficient, σ is the electrical conductivity, k is the thermal conductivity, and T is the absolute temperature. The thermoelectric efficiency is also a function of the thermal efficiency across the device, *i.e.* the Carnot efficiency, $1\text{-}T_C/T_H$, where T_C is the cold temperature and T_H the hot temperature side of the device.

$$\eta_{TE} = \left(1 - \frac{T_C}{T_H}\right) \frac{\sqrt{1 + zT} - 1}{\sqrt{1 + zT} + \frac{T_C}{T_H}} \quad (1)$$

In a STEG, the overall efficiency is the product of the thermoelectric efficiency (η_{TE})by the absorber efficiency (which depends on the solar energy absorbed and on the thermal losses). Improving the efficiencies of STEG involves improving the temperature difference across the hot and cold sides of the device, improving the materials for higher figures of merit (zT), and improving absorber efficiencies with better coatings and/or reducing thermal losses.

Moreover, the thermoelectric figure of merit zT is temperature dependent, presenting an optimum temperature. Stable operation depends on a stable hot side temperature; consequently, recent advances are proposing adding a phase change material (PCM) to keep the hot surface isothermal [87]. In other words, the system generates electricity while operating almost isothermally: charging when the PCM melts and discharging when the PCM solidifies. This prototype has been operated successfully, although the achieved conversion efficiencies of 11% still leave room for improvement.

Solar thermophotovoltaics (STPV) or photovoltaic thermal (PVT) systems are another type of solid-state devices where solar radiation is directly converted to electricity without the intermediate step of converting heat into mechanical energy traditionally performed by a conventional heat engine. Solar radiation is absorbed in a receiver and directed to a single-junction thermophotovoltaic cell.

The challenge of STPV conversion is that an intermediate absorption/re-emission process is necessarily introduced into the system and this step has proven to be inefficient due to the high operating temperature requirements [8].

The generation of heat from the incident solar radiation is based on the same principle as the solar receivers in concentrated solar power plants or the absorber in a solar thermoelectric device: low cost, effective high temperature solar selective multilayer coatings are very much needed for all these applications.

The efficiencies of current demonstrated STPV prototypes are still very low, around 3% [8]. However, efficiencies of approximately 20% are predicted by scaling to larger absorber/emitter areas, by using higher quality PV cells and sub band-gap filtering [88].

Solar Fuels

The thermal energy from concentrated solar power can be used as the high temperature source for endothermic thermochemical processes, storing the solar energy in an energy carrier such as liquid or gaseous fuels. Such fuels are commonly referred to as "solar fuels" and are mainly hydrogen and synthesis gas (syngas). Thermochemical cycles have efficiencies above 60% in theory; when combined in a solar power plant, efficiencies up to 25% are expected [1]. Additionally, solar thermochemical processes may have an environmental benefit, if water and atmospheric CO_2 is used to produce hydrogen and syngas.

Concentrated solar power directly applied to the splitting of water and CO_2 has been successfully demonstrated [92], however there are still significant obstacles to overcome for cost efficient implementation of solar fuels. In the past few years, research efforts have intensified to advance in the production of solar assisted redox-pair based thermochemical cycles for solar fuels. Challenges exist on both the material aspects and on the components and processes elements. On the material aspects, finding an efficient redox-pair with long term and stable production yields has been a major obstacle. Novel reactor/ heat exchanger concepts are being designed to improve efficiencies and yields. Also, process heat recovery has been found to be of crucial importance [93]. Finally, the investment costs for such thermochemical cycle installations are high, similar to those of concentrated solar power plants.

The upgrading of fossil fuels can also be coupled to concentrated solar power. For example, solar-assisted steam reforming of methane to produce syngas has also been widely explored [94]. Reforming of methane when reacted with steam or carbon dioxide is a highly endothermic catalytic process. Molten salts from concentrated solar power plants are typically heated to almost 600ºC and can provide the heat required for endothermic reforming reactions. Common catalysts present challenges such as long term deactivation and stability issues, requiring further analysis of viable alternative catalysts for this process. Reactor designs that allow good chemistry (*e.g.* adequate residence times for reaction completion) and are easily integrating the heat from concentrated solar power systems should continue to be investigated.

Moreover, traditional production of syngas involves burning part of biomass or carbonaceous waste material to produce process heat. Utilizing solar thermal energy for this high temperature process heat would bring an additional environmental benefit.

Solar Thermal Desalination

Fresh water demands are an increase problem in our current society. The desalination of water can have a severe environmental impact by increasing CO_2 emissions and pollution of aquatic resources if brines are not properly disposed. As a consequence, the integration of renewable energy technologies such as solar thermal energy in desalination processes is receiving growing attention. Also, solar driven desalination is of particular interest in arid and remote regions where other types of energy are not widely available.

Solar thermal desalination processes [95] can be direct (*i.e.* solar still) or indirect (*i.e.* non-membrane technologies such as solar multistage flash desalination or solar multi-effect distillation). The solar still is the simplest, cheapest technology. It consists of a basin where brackish or seawater is kept. It is covered by an inclined glass that helps radiative heating and evaporation. The water is then condensed on the glass (due to the temperature and partial pressure difference on the surface) and collected at the bottom. The water produced is of high quality but only small quantities are produced. It is not suitable for large scale production. Many methods have been tested in order to increase the yield: coupling solar stills to flat plate collectors, to dish type collectors, solar ponds, *etc.* Also, optimizing the glass inclination and water depth can increase yields.

In non-membrane desalination processes, the feed water is heated and then evaporated in distillation units to produce steam that is condensed in a condenser. The water produced is of high quality. The feed water temperature, the condensing surface and the process pressure play an important role in determining the yield. By incorporating vacuum during the distillation process, the water production can increase. Of these non-membrane technologies, only two are used at large scale: the multi-stage flash evaporation (MSF) and multi-effect distillation (MED). MSF is the most widespread, but it uses a large volume of water. MSF is based on the evaporation and subsequent condensation of seawater ad different vacuum levels. Multi-effect distillation (MED) has a series of vessels (called "effects") at low pressures where the feed water is sprayed, heated by the solar resource, and subsequently evaporated. MED has several advantages to MSF: it does not require brine recirculation, it can use smaller volumes of water, it has shorter starting times and is more efficient. It is also easier to couple to solar thermal. Both of these technologies exist at a commercial stage for large scale water production, but their coupling to solar thermal is yet to be demonstrated at such magnitudes. There are other indirect non-membrane desalination technologies that are being explored such as vapor compression desalination, freeze desalination, natural vacuum desalination and adsorption desalination [95], but they are generally more costly and less developed.

Some challenges are common among all these solar thermal desalination technologies. For example, the lifetime of a desalination system can be increased dramatically if corrosion free materials (*e.g.* stainless steel, titanium) are used. As always it will be a trade-off with system costs. Integrating the desalination systems in a CSP plant can help decrease the water production cost.

Solar Thermal Energy in Industrial Processes

Using concentrated solar power to provide high temperature process heat or steam is opening a large number of opportunities. This heat and steam could be used for enhanced oil recovery, mining applications (where concentrated solar power is being used for the first time in Chile [10], smelting of aluminum and other metals, steam production in coal plants, industrial processes in the production of food, beverages, textiles, and pharmaceuticals.

Thermochemical cycles are also being explored as new approaches to use the high concentrated radiation obtained through concentrated solar power for other purposes beyond electricity generation such as metal reforming [22].

Concentrated solar power can also be used in cogeneration plants in various forms. Producing steam for cogeneration through solar "boosters" is an active area of R&D and is expected to be the norm in the next generation of coal and cogeneration plants.

HYBRID CSP-PV SOLUTIONS

The hybridization of CSP and PV technologies can be done following two very different approaches: a) developing new receivers that will absorb the full solar spectrum, and b) combining both technologies separately in a single plant to leverage each of their strengths.

Hybrid CSP-PV Plants

PV is very cheap, but it does not presently have a cost-effective storage solution for 24-hour electricity production. On the other hand, concentrated solar power can produce electricity reliably at any time thanks to its storage technology but the cost of electricity is still significantly higher than that of PV. It has been suggested that installing PV in combination with concentrated solar power technology might lead to larger cost reductions than can be achieved by simply up-scaling the concentrated solar power plant field size and TES system alone [89].

There are currently three different projects under development incorporating this hybrid concept in Chile: a 130 MW concentrated solar power+ 150 MW PV by Solar Reserve and two 110 MW concentrated solar power+100 MW PV by

Abengoa [90] (Fig. **17**).

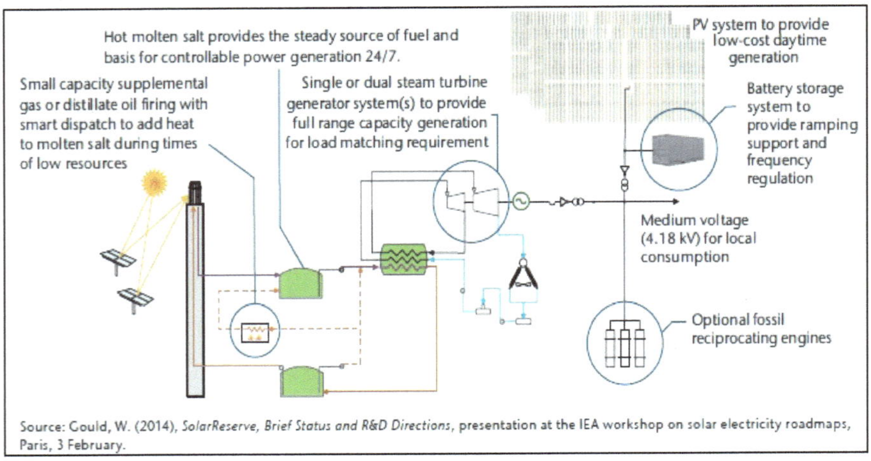

Fig. (17). CSP and PV hybrid solar plant solution example may achieve the lowest cost in high DNI areas [1].

Utilizing the Full Solar Spectrum

Taking advantage of the full solar spectrum by integrating both solar thermal and photovoltaics in a single receiver has been pointed out as the next generation of solar receivers [1]. PV uses only part of the solar spectrum at high efficiency; however, concentrated solar power uses the whole spectrum all at low efficiencies. Combining both technologies, it is possible to generate cheap electricity and produce excess heat to charge a storage system for later use within the same system.

To push R&D in this direction, the US Department of Energy through its Advanced Research Projects Agency–Energy (ARPA-E) has created the Full-Spectrum Optimized Conversion and Utilization of Sunlight (FOCUS) program [91]. Full use of the solar spectrum by combining PV and concentrated solar power technology can be achieved in several ways:

a. Splitting the solar spectrum in the path between collectors and receivers: for example, by replacing a parabolic trough curved mirror by high temperature solar PV cells, which collect a portion of the incident sunlight (only certain wavelengths) and reflect the remaining portion to a traditional absorber for heat conversion like in a standard concentrated solar power parabolic trough plant [91].
b. Collecting losses from high temperature solar cells

Not all PV and concentrated solar power technology can be easily adapted for hybridization. Both trough and FL are the concentrated solar power systems best suited for the combination of PV and concentrated solar power due mainly to their large collector surface area. Consequently, more progress in LFR is to be expected in this area in the following years. On the other hand, not all PV cells are capable of withstanding high temperatures. New materials development will lead the way forward in this area.

CONCLUSIONS AND FUTURE OUTLOOK

This chapter has focused on the generation of electricity from solar thermal energy. The market and current deployment of Solar Thermal Energy (STE) or Concentrated Solar Power (CSP) systems have been discussed, highlighting the space where CSP plants could be profitable and compete with other renewable and non-renewable energy sources. The different CSP technologies have been described, discussing in detail the main elements in a CSP system and where the current technological challenges reside. Recent research and developments in the different areas have been analysed and, finally, the current trends and opportunities towards achieving further efficiency improvements and cost reductions have been discussed.

Solar thermal power plants need advances beyond incremental improvements to continue to survive and to reduce costs to be competitive with other energy sources. For the current technology, it seems clear that new developments in materials and manufacturing technology can improve the reflector optics with low-cost, light weight solutions. New HTF targeting higher temperatures will be able to complement the developments in advanced high temperature cycles such as supercritical steam or supercritical CO2. Hardware for these cycles must still be developed. Storage systems will be an indispensable component of every future solar plant.

Traditional electricity generation through CSP is a relatively mature technology. As discussed above, its main advantage is the possibility of inexpensive and efficient long-term, large-scale thermal energy storage. Batteries are still not able to produce storage inexpensively for 8-20 hours; TES using molten salts is a commercial viable solution that can store this heat to shift the hours of electricity production. However, this advantage may not be enough in the future as other storage technologies evolve and get more competitive.

However, it is time to find new complementary markets for solar thermal technology. Some novel areas of research with future potential are the following:

- New concepts might emerge from rethinking the way solar energy is converted to electricity. For example, novel methods to convert solar energy directly into electricity without using a traditional heat engine are being explored.

- Progress in high temperature receivers and in thermochemical storage can be leveraged to design systems to produce solar fuels (*i.e.* hydrogen)
- Hybrid solutions such as incorporating solar thermal energy as "solar boosters" in coal plants or other fossil fuel plants might be interesting options in today's market.
- New collector /receiver concepts which use the full solar spectrum by hybridizing solar thermal with photovoltaic technology will be further developed and implemented that decouple the useful wave lengths from the heat. This way photovoltaic electric production can be combined with reliable long term thermal energy storage using CSP plant designs.
- Using the thermal energy for industrial processes, whether they require heat at low (below 150ºC), medium (150-400ºC) or high (above 400ºC) temperatures is another very active area of recent developments.
- Finally, effectively using solar thermal energy for desalinization processes is another current technological challenge.

CONSENT FOR PUBLICATION

Not applicable.

CONFLICT OF INTERESTS

The authors declare no conflict of interests.

ACKNOWLEDGEMENT

Declared none.

REFERENCES

[1] International Energy Agency, *IEA Technology Roadmap: Solar Thermal Electricity - 2014 edition,* 2014. [Online]. Available: https://www.iea.org/publications/freepublications/publication/technology-roadmap-solar-thermal-electricity---2014-edition.html

[2] "Solarpaces," [Online]. Available: http://www.solarpaces.org/Concentrated solar powers-technology/Concentrated solar powers-projects-around-the-world

[3] *The Future of Solar Energy-An Interdisciplinary MIT Study.* Massachusetts Institute of Technology, 2015.

[4] *Sunshot-CONCENTRATED SOLAR POWERS.* [Online]. Available: http://energy.gov/eere/sunshot/concentrating-solar-power

[5] *NREL CONCENTRATED SOLAR POWERS projects.* [Online]. Available: http://www.nrel.gov/Concentrated solar powers/solarpaces/

[6] "Concentrating Solar Power - Technology Roadmap", *International Energy Agency (IEA),* 2010.

[7] M. Liu, N.H.S. Tay, S. Bell, M. Belusko, R. Jacob, G. Will, W. Saman, and F. Bruno, "Review on

concentrating solar power plants and new developments in high temperature thermal energy storage technologies", *Renew. Sustain. Energy Rev.*, vol. 53, pp. 1411-1432, 2016. [http://dx.doi.org/10.1016/j.rser.2015.09.026]

[8] L.A. Weinstein, J. Loomis, B. Bhatia, D.M. Bierman, E.N. Wang, and G. Chen, "Concentrating Solar Power", *Chem. Rev.*, vol. 115, no. 23, pp. 12797-12838, 2015. [http://dx.doi.org/10.1021/acs.chemrev.5b00397] [PMID: 26511904]

[9] *Archimede solar energy italy*. Accessed June 2015 [Online]. Available: http://www.archimedesolarenergy.it/en_home.asp

[10] *Abengoa Solar*.Accessed 2015 [Online]. Available: http://www.abengoasolar.com/web/en/plantas_solares/plantas_para_terceros/espana/#seccion_1

[11] G. Zhu, T. Wendelin, M.J. Wagner, and C. Kutscher, "History, current state, and future of linear Fresnel concentrating solar collectors", *Sol. Energy*, vol. 103, pp. 639-652, 2014. [http://dx.doi.org/10.1016/j.solener.2013.05.021]

[12] "Compact Linear Fresnel Reflector solar thermal powerplants", *Sol. Energy*, vol. 68, no. 3, pp. 263-283, 2000. [http://dx.doi.org/10.1016/S0038-092X(99)00068-7]

[13] *Areva*. Accessed 2014 [Online]. Available: http://www.areva.com/EN/news-10195/solar-areva-commissions-molten-salt-energy-storage-demonstration.html

[14] G. Morin, M. Karl, M. Mertins, and M. Selig, "Molten Salt as a Heat Transfer Fluid in a Linear Fresnel Collector – Commercial Application Backed by Demonstration", *Energy Procedia*, vol. 69, pp. 689-698, 2015. [http://dx.doi.org/10.1016/j.egypro.2015.03.079]

[15] *Ripasso Energy AB*. Accessed 2015 [Online]. Available: http://www.ripassoenergy.com/

[16] C. Andraka, "Dish Stirling advanced latent storage feasibility", *Energy Procedia*, vol. 49, pp. 684-693, 2014. [http://dx.doi.org/10.1016/j.egypro.2014.03.074]

[17] G. Ragnolo, L. Aichmayer, W. Wang, T. Strand, and B. Laumert, "Technoeconomic design of a micro gas-turbine for a solar dish system", *Energy Procedia*, vol. 69, pp. 1133-1142, 2015. [http://dx.doi.org/10.1016/j.egypro.2015.03.182]

[18] Accessed 2015 [Online]. Available: http://www.torresolenergy.com/TORRESOL/gemasolar-plant/en

[19] *Solar Reserve*.Accessed 2015 [Online]. Available: http://www.solarreserve.com/en

[20] A.H. Slocum, D.S. Codd, J. Buongiorno, C. Forsberg, T. McKrell, J-C. Nave, C.N. Papanicolas, A. Ghobeity, C.J. Noone, S. Passerini, F. Rojas, and A. Mitsos, "Concentrated solar power on demand", *Sol. Energy*, vol. 85, pp. 1519-1529, 2011. [http://dx.doi.org/10.1016/j.solener.2011.04.010]

[21] A. Rabl, "Tower reflector for solar power plant", *Sol. Energy*, vol. 18, pp. 269-271, 1976. [http://dx.doi.org/10.1016/0038-092X(76)90027-X]

[22] A. Segal, and M. Epstein, "Solar ground reformer", *Sol. Energy*, vol. 75, no. 6, pp. 479-490, 2003. [http://dx.doi.org/10.1016/j.solener.2003.09.005]

[23] G.J. Kolb, C.K. Ho, T.R. Mancini, and J.A. Gary, *Power Tower Technology Roadmap and Cost Reduction Plan. SAND2011-2419.* Sandia Laboratories: Albuquerque, NM, 2011.

[24] C.K. Ho, and B.D. Iverson, "Review of high-temperature central receiver designs for concentrated solar power", *Renew. Sustain. Energy Rev.*, vol. 29, pp. 835-846, 2014. [http://dx.doi.org/10.1016/j.rser.2013.08.099]

[25] J. Liu, D. Lei, and Q. Li, "Vacuum lifetime and residual gas analysis of parabolic trough receiver", *Renew. Energy*, vol. 86, pp. 949-954, 2016.

[http://dx.doi.org/10.1016/j.renene.2015.08.065]

[26] R. Bradshaw, J. Cordaro, S. Jayaraman, A. Sun, N. Siegel, D. Gill, J. Hachman, M. Zimmerman, and G. Sartor, *Advanced heat transfer fluid development, Sandia National Labs.* Presentation DOE CONCENTRATED SOLAR POWERS Annual Program Review, 2010.

[27] C. Jung, J. Dersch, A. Nietsch, and M. Senholdt, "Technological perspectives of silicone heat transfer fluids for concentrated solar power", *Energy Procedia,* vol. 61, pp. 663-671, 2015. [http://dx.doi.org/10.1016/j.egypro.2015.03.076]

[28] J.E. Pacheco, *Final test and evaluation results from the solar two project, SAND2002-0120.* Sandia National Laboratories: Albuquerque, NM, 2002.

[29] D.C. Smith, and J.M. Chavez, *A final report on the phase I testing of a molten-salt cavity receiver, SAND87-2290.* Sandia National Laboratories: Albuquerque, NM, 1987.

[30] L. Amsbeck, T. Denk, M. Ebert, C. Gertig, P. Heller, P. Herrmann, J. Jedamski, J. John, R. Pittz-Paal, T. Prosinecki, J. Rehn, W. Reinalter, and R. Uhlig, *Test of a Solar-Hybrid Microturbine System and Evaluation of Storage Deployment* France: SolarPACES, Perpignan, 2010.

[31] M. Quero, R. Korzynietzb, M. Ebert, A.A. Jiménez, A. del Río, and J.A. Brioso, "Solugas – Operation experience of the first solar hybrid gas turbine system at MW scale", *Energy Procedia,* vol. 49, pp. 1820-1830, 2014. [http://dx.doi.org/10.1016/j.egypro.2014.03.193]

[32] Z. Ma, and C.S. Turchi, "Advanced Supercritical Carbon Dioxide Power Cycle Configurations for Use in Concentrating Solar Power Systems", *Supercritical CO_2 Power Cycle Symposium, Boulder, CO,* 2011.

[33] F.J. Miller, and R.W. Koenigsdorff, "Thermal modeling of a small-particle solar central receiver", *J. Sol. Energy Eng.,* vol. 122, pp. 23-29, 2000. [http://dx.doi.org/10.1115/1.556277]

[34] H. Benoit, L. Spreafico, D. Gauthier, and G. Flamant, "Review of heat transfer fluids in tube-receivers used in concentrating solar thermal systems:Properties and heat transfer coefficients", *Renew. Sustain. Energy Rev.,* vol. 55, pp. 298-315, 2016. [http://dx.doi.org/10.1016/j.rser.2015.10.059]

[35] N.P. Siegel, C.K. Ho, S.S. Khalsa, and G.J. Kolb, "Development and Evaluation of a Prototype Solid Particle Receiver: On-Sun Testing and Model Validation", *Journal of Solar Energy Engineering-Transactions of the ASME,* vol. 132, no. 2, 2010. [http://dx.doi.org/10.1115/1.4001146]

[36] K.M. Armijo, C. Ho, R. Anderson, J. Christian, S. Babiniec, and J. Ortega, *Magnetic Field Flow Phenomena in a Falling Particle Receiver* SolarPACES, Cape Town: South Africa, 2015.

[37] A.S. Oles, and G.S. Jackson, "Modeling of a concentrated-solar, falling-particle receiver for ceria reduction", *Sol. Energy,* vol. 122, pp. 126-147, 2015. [http://dx.doi.org/10.1016/j.solener.2015.08.009]

[38] P. Bermel, J. Lee, J.D. Joannopoulos, I. Celanovic, and M. Soljacie, "Selective Solar Absorbers", *Annual Reviews of Heat Transfer,* vol. 15, pp. 231-254, 2012. [http://dx.doi.org/10.1615/AnnualRevHeatTransfer.2012004119]

[39] X. Xu, K. Vignarooban, B. Xu, K. Hsu, and A.M. Kannan, "Prospects and problems of concentrating solar power technologies for power generation in the desert regions", *Renew. Sustain. Energy Rev.,* vol. 53, pp. 1106-1131, 2016. [http://dx.doi.org/10.1016/j.rser.2015.09.015]

[40] C. Atkinson, C.L. Sansom, H.J. Almond, and C.P. Shaw, "Coatings for concentrating solar systems – A review", *Renew. Sustain. Energy Rev.,* vol. 45, pp. 113-122, 2015. [http://dx.doi.org/10.1016/j.rser.2015.01.015]

[41] J.W. Raade, and D. Padowitz, "Development of Molten Salt Heat Transfer Fluid With Low Melting Point and High Thermal Stability", *J. Sol. Energy Eng.,* vol. 133, no. 3, pp. 031013-031016, 2011. [http://dx.doi.org/10.1115/1.4004243]

[42] E.V. Timofeeva, M.R. Moravek, and D. Singh, "Improving the heat transfer efficiency of synthetic oil with silica nanoparticles", *J. Colloid Interface Sci.,* vol. 364, no. 1, pp. 71-79, 2011. [http://dx.doi.org/10.1016/j.jcis.2011.08.004] [PMID: 21889163]

[43] L. Moens, and D.M. Blake, "Mechanism of Hydrogen Formation in Solar Parabolic Trough Receivers", *J. Sol. Energy Eng.,* vol. 132, no. 3, pp. 031006-031009, 2010. [http://dx.doi.org/10.1115/1.4001402]

[44] T. Bauer, N. Pfleger, N. Breidenbach, M. Eck, D. Laing, and S. Kaesche, "Material aspects of Solar Salt for sensible heat storage", *Appl. Energy,* vol. 111, pp. 1114-1119, 2013. [http://dx.doi.org/10.1016/j.apenergy.2013.04.072]

[45] T. Bauer, N. Pfleger, D. Laing, W.D. Steinmann, M. Eck, and S. Kaesche, *"Chapter20- High-Temperature Molten Salts for Solar Power Application," in Molten Salts Chemistry: From Lab to Applications.* Elsevier, 2013.

[46] G. Janz, *Molten Salt Handbook.* Elsevier, 2013.

[47] *Halotechnics,* 2015. [Online]. Available: http://www.halotechnics.com/products/

[48] R.W. Bradshaw, and N. Siegel, "Molten nitrate salt development for thermal energy storage in parabolic trough solar power systems", In: *Proceedings of ES2008, Energy Sustainability 2008*, Jacksonville, FL, 2008. [http://dx.doi.org/10.1115/ES2008-54174]

[49] R.I. Olivares, "The thermal stability of molten nitrite/nitrates salt for solar thermal energy storage in different atmospheres", *Sol. Energy,* vol. 86, pp. 2576-2583, 2012. [http://dx.doi.org/10.1016/j.solener.2012.05.025]

[50] E. Zarza, L. Valenzuela, J. León, K. Hennecke, M. Eck, H.D. Weyers, and M. Eickhoff, "Direct Steam Generation in Parabolic Troughs: Final Results and Conclusions of the DISS Project", *Energy,* vol. 29, pp. 635-644, 2004. [http://dx.doi.org/10.1016/S0360-5442(03)00172-5]

[51] D. Laing, T. Bauer, D. Lehmann, and C. Bahl, "Development of a Thermal Energy Storage System for Parabolic Trough Power Plants With DIrect STeam Generation", *J. Sol. Energy Eng.,* vol. 132, pp. 1-8, 2010.

[52] R. Guédez, M. Arnaudo, M. Topel, R. Zanino, Z. Hassar, and B. Laumert, *Techno-economic performance evaluation of direct steam generation solar tower plants with thermal energy storage systems based on high-temperature concrete and encapsulated phase change materials* SolarPACES: Cape Town, South Africa, 2015.

[53] J. Pacioa, A. Fritsch, C. Singer, and R. Uhlig, "Liquid metals as efficient coolants for high-intensity point-focus receivers: implications to the design and performance of next-generation CONCENTRATED SOLAR POWERS systems", *Energy Procedia,* vol. 49, pp. 647-655, 2013. [http://dx.doi.org/10.1016/j.egypro.2014.03.070]

[54] J.P. Kotzé, T.W. Backström, and P.J. Erens, *NaK as a primary heat transfer fluid in thermal solar power installations.* SolarPACES, 2012.

[55] C. Yuan, C. Jarrett, W. Chueh, Y. Kawajiri, and A. Henry, "A new solar fuels reactor concept based on a liquid metal heat transfer fluid: Reactor design and efficiency estimation", *Sol. Energy,* vol. 122, pp. 547-561, 2015. [http://dx.doi.org/10.1016/j.solener.2015.08.019]

[56] N.J. Bridges, A.E. Visser, and E.B. Fox, "Potential of Nanoparticle-Enhanced Ionic Liquids (NEILs) as Advanced Heat Transfer Fluids", *Energy Fuels,* vol. 25, pp. 4862-4864, 2011.

[http://dx.doi.org/10.1021/ef2012084]

[57] *Solvionic*. Accessed 2015 [Online]. Available: http://en.solvionic.com/.

[58] E.B. Fox, L.T. Smith, T.K. Williamson, and S.E. Kendrick, "Aging Effects on the Properties of Imidazolium, Quaternary Ammonium, Pyridinium, and Pyrrolidinium-Based Ionic Liquids Used in Fuel and Energy Production", *Energy Fuels,* vol. 27, pp. 6355-6361, 2013. [http://dx.doi.org/10.1021/ef401148j]

[59] B. Elkin, L. Finkelstein, T. Dyer, and J. Raade, "Molten oxide glass materials for thermal energy storage", *Energy Procedia,* vol. 49, pp. 772-779, 2014. [http://dx.doi.org/10.1016/j.egypro.2014.03.083]

[60] L. Heller, B. Mesnil, and J. Hoffmann, "Initial investigation of liquid glass as a thermal medium in CONCENTRATED SOLAR POWERS plants", *Third Southern African Solar Energy Conference,* 2015 Kruger National Park, South Africa.

[61] Z. Knez, E. Markocic, M. Leitgeb, M. Primozic, M. Knez Hrncic, and M. Skerget, "Industrial applications of supercritical fluids: A review", *Energy,* vol. 77, pp. 235-243, 2014. [http://dx.doi.org/10.1016/j.energy.2014.07.044]

[62] J. Spelling, A. Gallo, M. Romero, and J. González-Aguilar, "A High-efficiency Solar Thermal Power Plant using a Dense Particle Suspension as the Heat Transfer Fluid", *Energy Procedia,* vol. 69, pp. 1160-1170, 2015. [http://dx.doi.org/10.1016/j.egypro.2015.03.191]

[63] P.K. Nagarajan, J. Subramani, S. Suyambazhahan, and R. Sathyamurthy, "Nanofluids for solar collector applications: A Review", *Energy Procedia,* vol. 61, pp. 2416-2434, 2014. [http://dx.doi.org/10.1016/j.egypro.2014.12.017]

[64] R.A. Taylor, P.E. Phelan, T.P. Otanicar, C.A. Walker, M. Nguyen, S. Trimble, and R. Prasher, *Applicability of nanofluids in high flux solar collectors.* vol. Vol. 3. Journal Renewable and Sustainable Energy, 2011.

[65] D. Shin, H. Tiznobaik, and D. Banerjee, *Specific heat mechanism of molten salt nanofluids.* vol. Vol. 104. Applied Physics Letters, 2014.

[66] D. Brosseau, J.W. Kelton, D. Ray, M. Edgar, K. Chisman, and B. Emms, "Testing of thermocline filler materials and molten-salt heat transfer fluids for thermal energy storage systems in parabolic trough power plants", *Solar Energy Engineering- Transactions of the ASME,* vol. 127, 2005no. 1, pp. 19-116. [http://dx.doi.org/10.1115/1.1824107]

[67] Z. Yang, and S.V. Garimella, "Thermal analysis of solar thermal energy storage in a molten-salt thermocline", *Sol. Energy,* vol. 84, pp. 974-985, 2010. [http://dx.doi.org/10.1016/j.solener.2010.03.007]

[68] Y. Tian, and C.Y. Zhao, "A review of solar collectors and thermal energy storage in solar thermal applications", *Appl. Energy,* vol. 104, pp. 538-553, 2013. [http://dx.doi.org/10.1016/j.apenergy.2012.11.051]

[69] S. Kuravi, J. Trahan, D.Y. Goswami, M.M. Rahman, and E.K. Stefanakos, "Thermal energy storage technologies and systems for concentrating solar power plants", *Pror. Energy Combust. Sci.,* vol. 39, pp. 285-319, 2013. [http://dx.doi.org/10.1016/j.pecs.2013.02.001]

[70] B. Zalba, J.M. Marin, L.F. Cabeza, and H. Mehling, "Review on thermal energy storage with phase change: materials, heat transfer analysis and applications", *Appl. Therm. Eng.,* vol. 23, pp. 251-283, 2003. [http://dx.doi.org/10.1016/S1359-4311(02)00192-8]

[71] M. Kenisarin, "High-temperature phase change materials for thermal energy storage", *Renew. Sustain. Energy Rev.,* vol. 14, pp. 955-970, 2010. [http://dx.doi.org/10.1016/j.rser.2009.11.011]

[72] A. Sharma, V. Tyagi, C.R. Chen, and D. Buddhi, "Review on thermal energy storage with phase change materials and applications", *Renew. Sustain. Energy Rev.*, vol. 13, pp. 318-345, 2009. [http://dx.doi.org/10.1016/j.rser.2007.10.005]

[73] M. Liu, W. Saman, and F. Bruno, "Review on storage materials and thermal performance enhancement techniques for high temperature phase change thermal storage systems", *Renew. Sustain. Energy Rev.*, vol. 16, pp. 2118-2132, 2012. [http://dx.doi.org/10.1016/j.rser.2012.01.020]

[74] P. Pardo, A. Deydier, Z. Anxionnaz-Minvielle, S. Rougé, M. Cabassud, and P. Cognet, "A review on high temperature thermochemical heat energy storage", *Renew. Sustain. Energy Rev.*, vol. 32, pp. 591-610, 2014. [http://dx.doi.org/10.1016/j.rser.2013.12.014]

[75] D. Aydin, S.P. Casey, and S. Riffat, "The latest advancements on thermochemical heat storage systems", *Renew. Sustain. Energy Rev.*, vol. 41, pp. 356-367, 2015. [http://dx.doi.org/10.1016/j.rser.2014.08.054]

[76] S. Pendyala, "Macroencapsulation of Phase Change Materials for Thermal Energy Storage", In: *Graduate Theses and Dissertations* University of South Florida, 2012.

[77] *Terrafore Technologies LLC.* Accessed 2015 [Online]. Available: http://www.terraforetechnologies.com/

[78] M. T. Dunham, and B. Iverson, *High-Efficiency Thermodynamic Power Cycles for Concentrated Solar Power Systems* All Faculty Publications, 2014. [http://dx.doi.org/10.1016/j.rser.2013.11.010]

[79] G. Barigozzi, G. Bonetti, G. Franchini, A. Perdichizzi, and S. Ravelli, "Thermal performance prediction of a solar hybrid gas turbine", *Sol. Energy,* vol. 86, pp. 2116-2127, 2012. [http://dx.doi.org/10.1016/j.solener.2012.04.014]

[80] H.U. Frutschi, *Closed-Cycle Gas Turbines.* ASME PRESS: New York, 2005. [http://dx.doi.org/10.1115/1.802264]

[81] T. Ho, S.S. Mao, and R. Greif, "Comparison of the Organic Flash Cycle (OFC) to other advanced vapor cycles for intermediate and high temperature waste heat reclamation and solar thermal energy", *Energy,* vol. 42, pp. 213-223, 2012. [http://dx.doi.org/10.1016/j.energy.2012.03.067]

[82] *Australian Renewable Energy Agency (ARENA).* Accessed 2015 [Online]. Available: http://arena.gov.au/project/solar-supercritical-organic-rankine-cycle-for-power-and-industrial-heat/

[83] V. Dostal, P. Hejzlar, and M.J. Driscoll, "The Supercritical Carbon Dioxide Power Cycle: Comparison to Other Advanced Power Cycles", *Nucl. Technol.,* vol. 154, no. 3, pp. 283-301, 2006. [http://dx.doi.org/10.13182/NT06-A3734]

[84] Accessed 2016 [Online]. Available: http://arpa-e.energy.gov/?q=arpa-e-programs/arid

[85] *Weizmann Institute of Technology.*Accessed 2015 [Online]. Available: http://www.yedarnd.com/technologies/thermal-generation-electricity-without-turbine-0

[86] M.L. Olsen, E.L. Warren, P.A. Parilla, E.S. Toberer, C.E. Kennedy, G.J. Snyder, S.A. Firdosy, B. Nesmith, A. Zakutayev, A. Goodrich, C.S. Turchi, J. Netter, M.H. Gray, P.F. Ndione, R. Tirawat, L.L. Baranowski, A. Gray, and D.S. Ginley, "A high-temperature, high-efficiency solar thermoelectric generator prototype", *Energy Procedia,* vol. 49, pp. 1460-1469, 2014. [http://dx.doi.org/10.1016/j.egypro.2014.03.155]

[87] M.L. Olsen, J. Rea, G.C. Glatzmaier, C. Hardin, C. Oshman, J. Vaughn, T. Roark, J.W. Raade, R.W. Bradshaw, J. Sharp, A.D. Avery, D. Bobela, R. Bonner, R. Weigand, D. Campo, P.A. Parilla, N.P. Siegel, E.S. Toberer, and D.S. Ginley, *Solar Thermoelectricity via Advanced Latent Heat Storage* SolarPACES: Cape Town, South Africa, 2015.

[88] A. Lenert, D.M. Bierman, Y. Nam, W.R. Chan, I. Celanović, M. Soljačić, and E.N. Wang, "A nanophotonic solar thermophotovoltaic device", *Nat. Nanotechnol.*, vol. 9, no. 2, pp. 126-130, 2014. [http://dx.doi.org/10.1038/nnano.2013.286] [PMID: 24441985]

[89] W.J. Platzer, *Combined Solar Thermal and Photovoltaic Power Plants– an Approach to 24h Solar Electricity?* SolarPACES: Cape Town, South Africa, 2015.

[90] *CONCENTRATED SOLAR POWERS Today,* March 17, 2015. Accessed 2015 [Online]. Available: powerstoday.com/technology/integration-Concentrated

[91] *FOCUS - ARPA-e,*. Accessed 2015 [Online]. Available: http://arpa-e.energy.gov/?q=arpa- e-programs/focus

[92] D.A. Marxer, P. Furler, J.R. Scheffe, H. Geerlings, C. Falter, V. Batteiger, A. Sizmann, and A. Steinfeld, "Demonstration of the Entire Production Chain to Renewable Kerosene via Solar Thermochemical Splitting of H2O and CO2", *Energy Fuels,* vol. 29, pp. 3241-3250, 2015. [http://dx.doi.org/10.1021/acs.energyfuels.5b00351]

[93] M. Roeb, A. Steinfeld, G. Borchardt, C. Feldmann, M. Schmücker, C. Sattler, and R. Pitz-Paal, *SolarSyngas: Results from a Virtual Institute developing Materials and Key Components for Solar Thermochemical Fuel Production* SolarPACES: Cape Town, South Africa, 2015.

[94] S.A.M. Said, M. Waseeuddin, and D.S.A. Simakov, "A review on solar reforming systems", *Renew. Sustain. Energy Rev.,* vol. 59, pp. 149-159, 2016. [http://dx.doi.org/10.1016/j.rser.2015.12.072]

[95] H. Sharon, and K.S. Reddy, "A review of solar energy driven desalination technologies", *Renew. Sustain. Energy Rev.,* vol. 41, pp. 1080-1118, 2015. [http://dx.doi.org/10.1016/j.rser.2014.09.002]

CHAPTER 2

Thermal Energy Storage Systems for a Global Sustainable Growth: Current Status and Future Trends

Irene P. Koronaki[*], **Michael T. Nitsas** and **Efstratios G. Papoutsis**

Laboratory of Applied Thermodynamics, National Technical University of Athens, Zografou, 15780, Greece

Abstract: Modern societies are called upon to face the consequences of their persistence on using fossil fuels. Energy resources shortage and environment pollution urge us to find alternative advanced technologies for viable and sustainable growth. Thermal Energy Storage (TES) systems supported by renewable energy sources (mainly solar energy) can be viewed as an effective means of achieving the aforementioned goal. This study reviews the available TES systems. In this context, sensible TES systems which utilize liquid and solid storage media and their applications are presented. Furthermore, the usage of ice and other solid-liquid phase change materials in latent heat storage systems is investigated in terms of materials, applications and future trends. Finally, the utilization of thermochemical reactions in TES systems is presented.

Keywords: Aquifers, Boreholes, Energy Storage, Latent Heat, Microencapsulation, Molten Salts, Phase Change Materials, Sensible Heat, Stratification, Thermochemical Storage.

INTRODUCTION

Nowadays, the humanity is leading a life of increased energy consumption, progressing energy resources depletion and environment pollution due to the large amount of wastes. The consequences of this consumption profile are obvious when the extreme weather conditions that have stricken our planet are considered. The International Energy Agency through studies [1] has already pointed out the worldwide ongoing dependency on fossil fuels and its consequences on climate change. The same studies have shown an extreme increase of 37.8% in CO_2 emissions in just 17 years from 20.9 Gt back in 1990 and through projection the emissions are expected to increase up to 34.5 Gt in 2020 and 40.2 Gt in 2030 [1].

[*] **Corresponding author Irene P. Koronaki:** Laboratory of Applied Thermodynamics, National Technical University of Athens, Zografou, 15780, Greece; Tel: +302107721581; Email: koronaki@central.ntua.gr

Emmanuel D. Rogdakis & Irene P. Koronaki (Eds.)

Therefore, advanced technologies need to be adopted so as to face a potential energy crisis and its environmental issues with a sustainable growth planning. The technology of the thermal energy storage (TES) can be viewed as an effective means towards this direction. TES refers to the interim collecting of thermal energy in hot or cold materials in the form of sensible or latent heat for a later utilization. TES can find implementation in various applications including solar energy storage at high temperatures, heating, ventilation, air conditions and refrigeration.

Many factors should be taken under consideration when selecting a TES system. The thermal and physical properties (density, specific heat, latent heat, *etc.*) of the material used in TES systems, the storage capacity, the heat losses as well as the operating conditions are among the most important.

The TES technologies found in most applications include sensible and latent TES. The sensible TES systems have been used commercially for many years and refer to the storage of thermal energy in the form of temperature difference. The prevailing material in sensible TES systems is water. Its characteristics, that is nontoxicity, availability, utilization in a wide range of temperatures, render it as a suitable storage medium in both domestic and industrial applications.

Latent TES systems utilize phase change materials (PCMs) which at a suitable temperature range can be melted and thus store thermal energy. The stored energy is removed during the reverse process which solidifies the PCM. Due to the superiority of high latent heat compared to sensible heat, PCMs can contribute to the reduction of the storage systems' size. However, phase separation, corrosion, supercooling, low thermal conductivity and manufacturing cost make PCMs uneconomic for small scale domestic usage.

Regarding the sorption TES [2, 3] in specific the adsorption TES, the insufficient heat and mass transfer in the adsorbent beds, the low thermal conductivities and porosities of the adsorbent materials make these TES systems commercially not viable.

The utilization of TES can bring significant energy savings and CO_2 emissions reduction. Indicatively, Arce *et al.* [4] considering a 10 year TES scenario estimated for EU-25 a load reduction of 5854139 MWth, thermal energy savings of 9527227 GWhth and electrical energy savings of 17526 GWhe. Finally, for EU-25 a reduction in CO_2 emissions of 2579088559 t is estimated. From an economic point of view, a TES system is substantially affected by the type of application and operational needs as well as the number and frequency of storage cycles [5]. In general, latent (10-50 €/kWh [6]) and thermochemical (8-100 €/kWh [6]) heat storage systems are more expensive than sensible (0.1-10 €/kWh

[6] heat storage systems and are economically viable when a high number of storage cycles is considered [5].

In this chapter, the existing sensible and latent TES systems are reviewed in terms of operation principles, materials used, efficiency and applications. Moreover, thermochemical heat storage systems are introduced.

SENSIBLE HEAT STORAGE SYSTEMS

Sensible heat storage refers to the thermal energy that can be stored in the form of temperature increase in storage materials. The amount of stored heat depends on the density, the heat capacity, the volume, and the temperature variation of the storage materials. A material can be appropriate for sensible heat storage if it fulfills the following [7]:

- High energy density (stored heat over system volume)
- High thermal conductivity
- Availability and low cost
- Chemical stability
- Environmental friendly

Table **1** shows the main thermal properties of a number of storage materials.

Table 1. Thermal properties of storage material.

Material	Specific Heat [kJ/kgk]	Volumetric Heat Capacity [MJ/m^{3}K]
Water	4.2	4.2
Oil	2.0	1.7
Ice	2.0	1.8
Paraffin	2.9	2.6
Sand	0.8	1.2
Steel	0.5	3.6
Concrete	0.8	2.1
Brick	0.8	1.2
Glass	0.8	2.2
Copper	0.4	3.5
Gold	0.1	2.5
Wood	1.8	0.9

WATER IN SENSIBLE TES SYSTEMS

From the above table, it appears that the material that combines effectively the desired properties of a storage material is water as it has a very high heat storage density both per weight and per volume compared to other potential heat storage materials. Moreover, water is relatively cheaper, readily available and requires no manufacturing process. Furthermore, water gives the opportunity of utilizing it as a heat storage material in its whole temperature interval, from its freezing point to its boiling point.

Hot water is stored in tanks made of steel, stainless steel, concrete or plastic. Equipment such as heat exchanger spirals, auxiliary electric heating elements, stratification devices for enhancement of thermal stratification in the hot water stores, are usually built into the tanks. The hot water tanks, commonly met in solar applications, are normally insulated with an insulation material of low thermal conductivity in order to minimize heat losses to the ambience and thus ensure the stability of the system's performance.

One of the most widely known applications is the hot water tank charged by a solar collector, as shown in Fig. (**1**). There are two configurations widely used, the direct and the indirect for both natural and forced circulation. In the first configuration (open loop), the solar energy is transformed into heat in the collector's absorber and the produced useful thermal energy is transferred directly through the working fluid to the storage tank. In this configuration, no intermediate heat exchanger is utilized as both working fluid and storage medium are pressurized water. In the indirect (closed loop) configuration, a circulator is utilized and the heat transfer fluid does not mix with the storage medium in the tank. In this configuration, the working fluid can be, other than water, ethylene or propylene glycol, *etc.* Usually such systems are designed at a constant temperature following the corresponding demand on hot water needs. However, due to the variation of solar irradiance, the energy demand is not always satisfied by the collector and thus, an auxiliary heater, attached to the tank, is utilized.

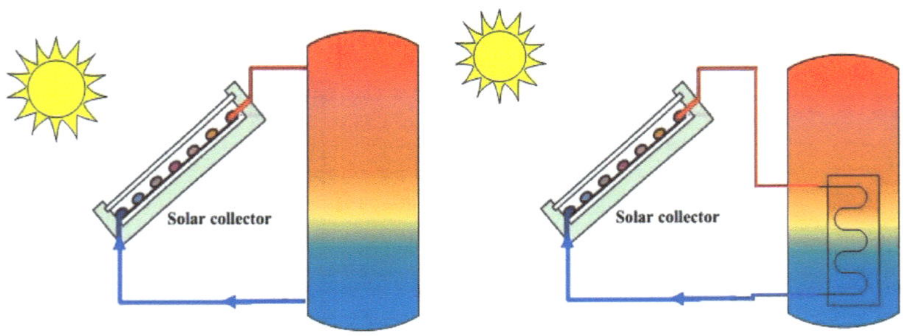

Fig. (1). Configurations for solar thermal energy storage systems.

The efficiency for most energy systems with hot water stores can be increased if a large thermal stratification is built during charge and discharge processes. A highly stratified hot water store for a solar heating system ensures that the solar collector temperature during operation is low and thus the solar collector performance is enhanced. In this case, the need for auxiliary energy supply is minimized due to the high temperatures of water drawn from the top of the water tank.

Water Stratification

Thermal stratification phenomenon in sensible TES systems has a substantial effect on the efficiency of the system and therefore should be taken into consideration when evaluating the performance of such a system. When the water tank is thermally charged, the useful energy from the solar field enters the top of the tank while it is drawn off from the top of the tank during the opposite procedure. The drawn hot water is replaced simultaneously by cold water derived from the bottom section of the tank. The so called thermal stratification phenomenon, occurring in the transient operation, divides the tank content in three layers, the hot water layer which floats to the upper part of the tank, the cold-water layer which sinks to the bottom of the tank due to high density and a layer of an intermediate temperature. The aforementioned layer, called thermocline, acts as a moving barrier between the hot and the cold region. This movement leads to a varying thermocline thickness. In practical applications the thermocline thickness should be as small as possible so as to achieve a maximum hot water volume to the upper part of the tank. Studies show that the stratification plays an important role on the performance of the TES system and a highly stratified water store is 20% more efficient compare to a fully-mixed one [8 - 10]. Parameters such as the tank geometry and the operating condition affect the stratification process substantially. Attention should be also paid to the flow direction as improper directions, may lead to the destruction of thermocline and thus a fully mixed tank of a lower temperature [11 - 13].

Fig. (**2**) depicts the variation of stratification degrees in a storage tank: (a) highly stratified, (b) medium stratification and (c) fully mixed storage. From Fig. (**2**) it can be concluded that the smaller the thickness of the thermocline the higher the gradient of the temperature between the hot and cold regions in the tank.

APPLICATIONS

The utilization of hot water tanks is among the most popular means of thermal energy storage. Hot water tanks are used for energy storage in water systems based on solar energy and in co-generation energy supply systems. Researches [14] have proved that water tank storage is a cost-effective option and that its efficiency can be further enhanced if the tank water is optimally stratified and the tank is sufficiently insulated.

Fig. (2). Stratification degrees in hot water tank (from left to right): highly stratified, medium stratification, non-stratified (fully mixed).

Hot water storage systems are used mostly for covering domestic hot water needs (DHWN) with a total system volume varying from few liters to numerous m^3. The same technology is often met in solar thermal installations combined with building heating systems and thus creating Solar-Combi systems the volume of which can reach several m^3. In the latter case, sensible TES systems are charged thermally by a solar field up to the temperature of 80-90°C. The installation of Solar-Combi systems is becoming more and more frequent within the residential market; however, when used for a daily storage, they cannot utilize more than 50% of the available solar energy [15]. Greater solar fractions are achieved when a seasonal TES is implemented. In such cases, a yearly solar fraction close to 100% can be obtained [16].

A characteristic application of a Solar-Combi system is the "Am Ackermann-bogen" (Munich, Germany) which provides hot water and space heating for a multi-residential block covering a total area of about 30400m^2. The system is driven by 2761 m^2 of flat plate collectors and the useful thermal energy produced by the solar field is either consumed directly or stored in a 6000 m^3 underground storage tank. The system is able to supply sufficient heat at a yearly level covering more than half of the demand (approx. 2000 MWh per annum). During the second year of operation, "Am Ackermann-bogen", Fig. (**3**), achieved a 45% solar energy fraction which is expected to increase above 50% after further optimization [17].

Future trends in water stores should adopt modular concepts which are ideal for building stock, flexible in design and easy to install. The modules tend to be pre-fabricated polymeric containers VIP insulated that achieve optimum stratification temperature. The modular concept of water stores will lead to a minimum cost and an efficient cost-manufacturing relationship.

Fig. (3). Solar Thermal District Heating "Am Ackermann-bogen"-Construction and final state.

OTHER LIQUID SENSIBLE STORAGE MEDIA IN TES

If heat is needed to be stored in temperatures above 100 °C there are several liquid materials that can compete water. For medium grade heat (between 100-300 °C) heat transfer oils can be used as storage medium. There are examples of mineral (*e.g.*, Caloria) and synthetic oils (*e.g.*, Dowtherm, Therminol *etc.*) that can be utilized for such purposes. Heat transfer oils present lower heat capacities compared with water. Their main drawback is that they tend to degrade over time especially when operate above their recommended temperature limit [18]. Safety issues may also arise due to the rather high vapor pressure of oils. There is danger of ignition if they operate above their flash point and for this reason their application in combination with an inert gas cover is recommended [18]. Finally, synthetic and silicon oils are considered as very expensive materials to be used as storage medium.

Molten salts are appropriate for heat storage at higher temperature (above 300 °C). Liquid molten salts offer a huge potential to be used as storage medium in solar power plants because of their exceptional thermal stability under high temperature levels, low vapor pressure, low viscosity, high thermal conductivities, non-flammability and non-toxicity [19]. The main drawback of molten salt mixtures is their relatively high freezing point which is typically above 100 °C [20]. Due to this fact special anti-freezing protection should be considered for smooth operation of the storage system. The first molten salt that was used for storage purposes is a nitrate salt mixture consisting of 60% $NaNO_3$ and 40% KNO_3 by weight. This binary mixture is called Solar Salt and has been utilized in several heat storage applications at both pilot and commercial Concentrating Solar Power (CSP) plants. Solar Salt exhibits the highest melting point and highest thermal stability at the lowest cost [20]. Other commercially available molten salts are the Hitec and the Hitec XL. Hitec is a ternary mixture of 53 wt.% KNO_3, 40 wt.% $NaNO_2$ and 7 wt.% $NaNO_3$ and Hitec XL of 48 wt.% $Ca(NO_3)_2$, 45 wt.% KNO_3 and 7 wt.% $NaNO_3$. Their melting temperature is 142 °C and 133 °C respectively.

The research recently has been focused on finding a suitable molten salt mixture that covers certain criteria, such as [20]:

- Low melting temperature (ideally below 100 °C);
- Low vapor pressure;
- High chemical stability at high operating temperatures;
- Advantageous thermal properties (high specific heat, high thermal conductivity, low viscosity *etc.*).

Peng *et al.*, [21] examined multi-component molten salts composed of KNO_3, $NaNO_2$ and $NaNO_3$ with 5% additive A of chlorides. They observed that the additive A lowers the freezing point (around 138 °C) and improve the high temperature thermal stability of the mixture (up to 550 °C). Zhao and Wu [20] developed and tested ternary nitrate salt mixtures consisting of KNO_3, $LiNO_3$ and $Ca(NO_3)_2$. Their mixtures presented melting temperatures below 100 °C, lower viscosities compared with synthetic oils and common molten salts and high chemical stability at high operating temperatures. Wang *et al.*, [22] developed quaternary eutectic mixtures consisting of $LiNO_3$, $NaNO_3$, KNO_3 and $NaNO_2$ and they achieved a melting point of 99.02 °C.

In Table **2**, the main characteristics of the most popular liquid materials in sensible TES systems are presented.

Table 2. Main characteristics of molten salts and high temperature oils [19, 23].

Storage Material	Working Temperature (°C)	Density (kg/m³)	Thermal Conductivity (W/(m K))	Specific Heat (kJ/(kg °C))	Costs (US$/kg)
Solar Salt	220-600	1899	n.a.	1.5	0.93
HitecXL	120-500	1992	0.52	1.4	1.19
Mineral oil	200-300	770	0.12	2.6	0.30
Synthetic oil	250-350	900	0.11	2.3	3.00
Silicone oil	300-400	900	0.10	2.1	5.00
Nitrite salts	250-450	1825	0.57	1.5	1.00
Liquid sodium	270-530	850	71.0	1.3	2.00
Nitrate salts	265-565	1870	0.52	1.6	0.50

(Table 2) cont.....

Storage Material	Working Temperature (°C)	Density (kg/m^3)	Thermal Conductivity (W/(m K))	Specific Heat (kJ/(kg °C))	Costs (US$/kg)
Carbonate salts	450-850	2100	2.0	1.8	2.40

The above-mentioned oils and molten salts have been utilized in thermal energy storage systems installed in CPS plants. The main heat storage technology consists of two tanks. One of them stores the fluid at a high temperature level and the other at a lower one. These systems can be divided in direct and indirect storage systems. In direct storage systems the working fluid of the primary circuit is also the storage medium. In the indirect storage system, the heat contained in the primary heat transfer fluid must be transferred to another suitable for heat storage medium. In this case, an extra heat exchanger is required in order to transfer the heat from the heat transfer fluid to the storage medium, thus making the system more expensive.

A direct storage system with two tanks was installed in the SEGS I CSP plant in the Mojave Desert in California. The plant was operating commercially between 1985 and 1989. The storage system utilized the mineral oil Caloria as both heat transfer fluid and storage medium. The hot tank stored the hot oil at a temperature of 307 °C after it was heated on the solar field and the cold tank stored the oil at 240 °C after it had delivered its heat to the power generation block. The storage system had the ability to operate the plant for 3 h at full load. Caloria oil exhibited high costs of utilization reaching a 42% of the total thermal energy storage cost. Due to its relatively high vapor pressure, Caloria cannot be used as storage medium at high temperature levels.

Another CSP plant with a two-tank direct storage system is Gemasolar located within the city limits of Fuentes de Andalucía in the province of Seville, Spain. The plant was originally called Solar Tres (built in 2008) and later renamed to Gemasolar. This system makes use of the solar power tower technology and utilizes Solar Salt (7900 tons) as both working and heat storage fluid. It is the first commercially available solar plant with central tower receiver and molten salt TES technology. It presents a maximum output of 19.9 MW and it can operate for 15 h supplied by the storage system which has a thermal capacity of 600 MWh. The hot tank stores Solar Salt at a temperature of 565 °C and the cold one at 290 °C. The diameter of the storage tanks is 23 m and the height is 10.5 m each. Due to the storage system the plant can operate for 6500 h every year and can achieve an annual capacity factor of about 75% [24].

Andasol solar power station consists of Andasol I (2008), Andasol II (2009) and Andasol III (2011) CSP power plants. Andasol I was Europe's first commercial

CSP plant utilizing parabolic trough technology. Each plant has an output capacity of 50 MW and comprises a two-tank indirect storage system. They utilize steam as heat transfer fluid and Solar Salt as storage medium. The storage tanks are 36 m in diameter and 14 m in height. The hot storage tank contains molten salt at a temperature of 384 °C and the cold one at 291 °C. They make use of 28500 tons of molten salt and their thermal capacity is 1010 MWh. The storage system can operate the plant for 7.5 h at its peak of electricity production. The plant can achieve a peak 25% and annual 14.7% solar to electric efficiency [25]. The performance curve of the solar plant on a clear summer day is illustrated in Fig. (**4**).

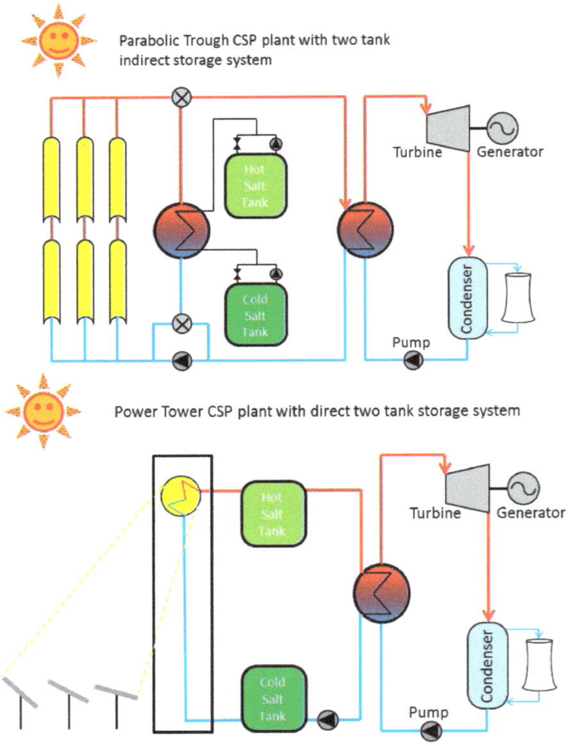

Fig. (4). CSP plants with two tank thermal energy storage systems.

An Italian research center, ENEA, has proven the technical viability of molten salts utilization as heat transfer fluid in a parabolic trough solar field [23]. This allows the operating temperature of the HTF to be increased compared to the heat transfer oils resulting in significant benefits at plant's operation. The first parabolic trough CPS plant that uses molten salt as HTF is Archimede, located in Priolo Gargallo, Sicily, Italy. The plant utilizes a two-tank direct heat storage system and it is in operation since 2010. The hot tank stores molten salt at 550 °C and the colder one at 290 °C. The tanks are 6.5 m high and 13.5 m in diameter.

They store 1580 tons of molten salt (Solar Salt) exhibiting a thermal capacity of 100 MWh and thus enabling the continuous plant operation for 8 h.

SOLID STORAGE MEDIA IN TES

Thermal energy can be stored in various solid materials like rocks, concrete, gravels, metals, sand, bricks *etc.* The main advantages of using solid materials as heat storage media is their relatively low cost, their high availability, the fact that they are chemically inert and their durability in high temperatures. Their main disadvantage is their relatively low specific heat of capacity which leads to lower energy densities and by extension to higher in volume energy storage systems. In Table **3**, the main properties of some solid materials that can be used in TES systems are presented. The main application of these solid storage media is in pebbled beds (thermoclines) or in concrete thermal storages.

Table 3. Properties of solid materials [18, 26].

Material	ρ (kg/m^3)	c_p (kJ/(kg K))	c_v (MJ/(m^3 K))	k (W/(m K))
Brick	1600	0.840	1.344	1.20
Sandstone	2200	0.712	1.566	1.83
Wood	800	2.093	1.674	0.16
Concrete	2240	1.130	2.531	0.9-1.3
Aluminium	2707	0.896	2.425	204
Iron	7897	0.452	3.569	73
Steel Slab	7800	0.502	3.916	503
Limestone	2500	0.741	1.853	2.2
Soil (clay)	1450	0.880	1.276	1.28
Soil (gravelly)	2040	1.840	3.754	0.59

The thermocline of TES systems consists of a tank packed with a low-cost filler material (see Fig. (**5**)). Both the hot and the cold heat transfer fluid are stored in the same tank. The hot fluid due to a lower density is at the top of the storage tank while the cold fluid sinks at the bottom of the tank. The temperature gradient separates the hot from the cold fluid. The filling material provides the majority of the thermal capacitance of the system [27] replacing the relatively expensive heat transfer fluid. The buoyancy forces along with the filler material help in the maintaining of the thermal gradient, preventing the convective mixing.

When the system is charged, cold fluid is pumped from the bottom of the vessel, is heated in the energy source and returns hot at the top of the vessel. When it is time to exploit the stored heat, the opposite procedure is followed. Hot heat transfer fluid is drawn from the top of the tank, is cooled down delivering heat through a heat exchanger and returns back at the bottom of the vessel.

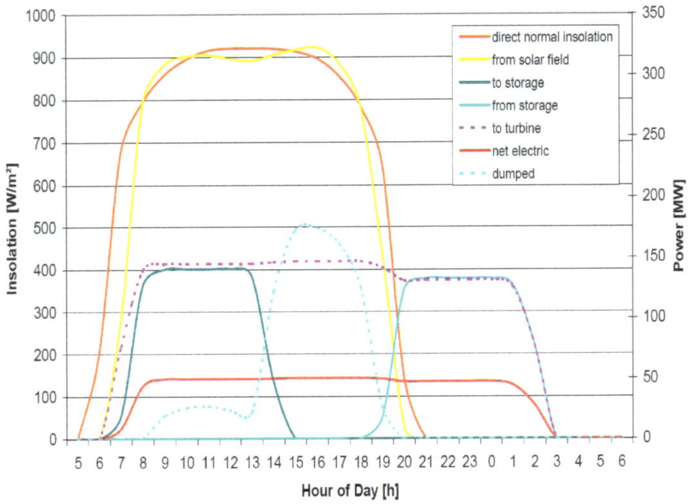

Fig. (5). Performance curve of Andasol I power plant on a clear summer day [25].

A wide range of filler materials can be used in this type of TES system depending on the application and the heat transfer fluid. The ideal filler material must be cheap, widely available, have low void fraction and high heat capacitance. Such materials are gravels, marble, iron, concrete, bricks, ceramics *etc.* The commonest heat carriers are air, water, synthetic oils and molten salts.

This type of thermal energy storage is very promising for Concentrated Solar Power (CSP) plants (Fig. **6**). There are several studies both experimental and computational that deal with the development of thermocline energy storage systems. (Fig. **7**). Molten salts like Solar Salt (60% $NaNO_3$/40% KNO_3) and Hitec XL (42% $Ca(NO_3)_2$/15% $NaNO_3$/ 43% KNO_3) and synthetic oils like Therminol VP-1 are standard heat transfer fluids for such applications. Quartzite rock, marble, silica sand, taconite and limestone are the most promising filler materials.

Solar One pilot solar tower plant was the first CSP station which utilized a thermocline for thermal energy storage. This system used the synthetic oil Caloria® as heat transfer fluid and a mixture of rock and sand as filler material. Due to the limitations of Caloria® (liquid at atmospheric pressure when its temperature is below 315 °C [29]) the system was operating at a range of temperatures between 218 °C and 302 °C. The thermal capacity of the thermocline was 182 MWh_{th} and is used 6170 tonnes of filler material and 906 m^3 of Caloria® [29].

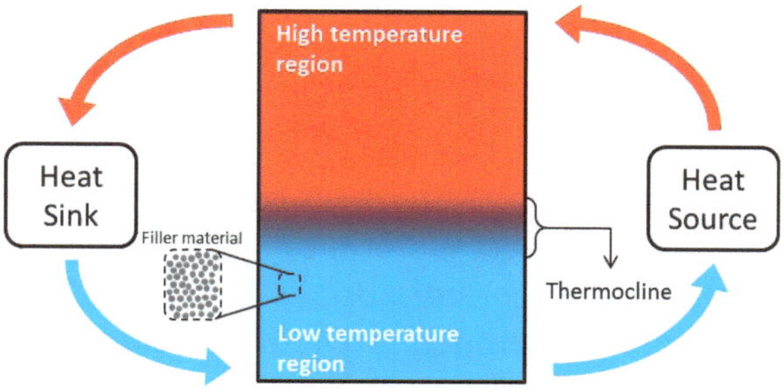

Fig. (6). Sketch of a thermocline energy storage system.

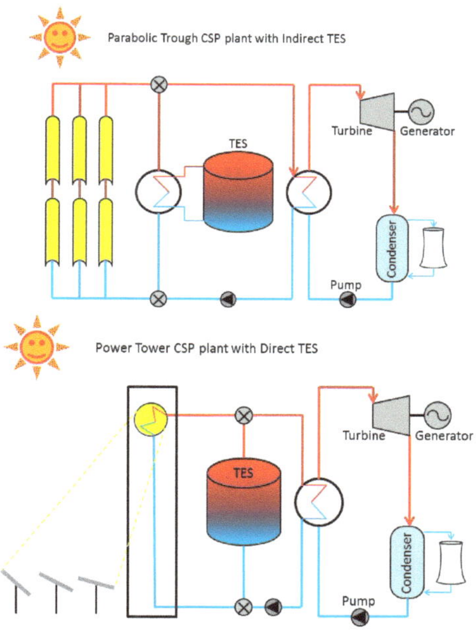

Fig. (7). CSP plants with thermocline energy storage system [28].

The performance improvement of the thermocline energy storage system (higher operating temperatures) required several researches to be carried out testing molten salts as heat carrier fluids in the packed bed. Pacheco *et al.*, [29] tested various candidate filler materials in order to check their compatibility with nitrate salts. They observed that silica sand, quartzite rock and taconite were appropriate for working with common molten salts.

An innovative packed-bed TES system utilizing air as heat transfer fluid and stone gravel as filler material was designed and operates at Ait-Baha CSP pilot plant in Morocco. The low cost and high availability of filler material along with the non-polluting and non-corrosive heat carrier characteristics make the idea very promising. Zavattoni *et al.*, [30] modelled and experimentally tested a similar storage system. The experimental prototype vessel was made by insulated concrete and it was completely buried. It was packed with 25 m^3 of homogenous natural rocks (quartzite, limestone, calcareous sandstone, gabbro and helvetic siliceous limestone) with an average diameter of 0.03 m. Another CSP plant that uses a similar technology for energy storage is the Power Plant Jülich. In this system a packed bed filled with ceramic bricks is used as heat storage. The heat carrier is air and the operating temperatures of the storage system are between 120 °C-680 °C.

Apart from CSP plants, packed bed energy storage systems are used in buildings solar heating applications. A typical solar air heater with heat storage is illustrated in Fig. (**8**). In a system like this, the storage unit receives heat from the collector during the charging period and delivers this heat to the building at the discharging process. Rock [31], brick [32], concrete and sand [33] are appropriate materials for utilization as filler materials in solar air heater applications.

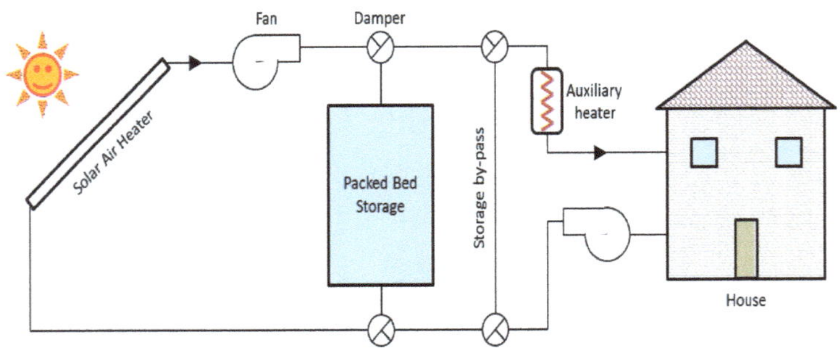

Fig. (8). Schematic of packed bed TES system for solar air heaters.

Concrete is usually selected as storage material due to its low cost, high availability and its uncomplicated processing [34]. The main characteristics of concrete as storage medium are [35]:

- high specific heat;
- desirable mechanical properties
- a thermal expansion coefficient close to that of steel (pipe material);
- high mechanical resistance to cyclic thermal loading.

In Fig. (**9**), a typical concrete heat storage system is depicted. The system comprises of tubes going through a concrete block. During the charging process, the high temperature working fluid passes through the tubes delivering heat to the concrete by conduction. At the discharging process the cold heat transfer fluid flows at the opposite direction receiving heat by the concrete. This type of TES system is very promising for applications in solar power plants and industrial heat recovery systems.

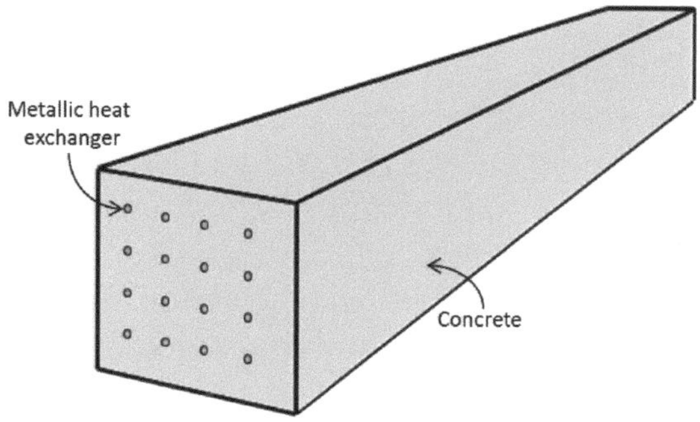

Fig. (9). Concrete thermal energy storage module.

A concrete thermal storage unit was constructed at the Plataforma Solar de Almeria in Spain and it was compared with a similar storage unit built by castable ceramic. The properties of the two examined materials are furnished in Table **4**. The storage capacity of the two units is 350 kWh for the castable ceramic and about 280 kWh for the concrete [36]. The maximum operating temperature of the system is 390 °C. High temperature concrete is considered more appropriate solution for TES applications due to its low cost, higher material strength and easier handling [36].

Table 4. Material properties of storage materials developed at DLR [36].

Material	Castable Ceramic	High Temperature Concrete
Density [kg/m^3]	3500	2750
Specific heat capacity at 350 °C [J/kg K]	866	916
Thermal conductivity at 350 °C [W/m K]	1.35	1.0
Coeff. of thermal expansion at 350 °C [10^{-6}/K]	11.8	9.3
Material strength	Low	Medium
Crack initiation	Hardly no cracks	Several cracks

Underground Thermal Energy Storage (UTES) Systems

UTES is widely used when there are large quantities of thermal loads to be stored usually for a long-time frame. In this case the underground is used as storage medium for both heat and cold storage. UTES technologies can be divided in three main groups:

1. Aquifer Thermal Energy Storage (ATES)

2. Borehole Thermal Energy Storage (BTES)
3. Cavern Thermal Energy Storage (CTES)

In ATES systems thermal energy is stored in the underground in sandy aquifers or water-bearing layers (20-500 m below the surface) [37]. A typical ATES system is depicted in Fig. (**10**) for both summer and winter operation. This technology makes use of two wells: a cold well (*e.g.* 7 °C) and a hot well (*e.g.* 15 °C). During summer, cold groundwater (7 °C) is pumped from the cold well so as to remove heat from the building. The hot water (15 °C) is then injected to the hot well. During winter the flow follows the opposite direction. Hot water is pumped from the warm well, it delivers heat to the building through a heat exchanger and it returns back to the cold well. A typical distance between the cold and the warm well is about 100 m [38] in order to avoid thermal mixing within the aquifer. The main characteristics of a successful application of ATES are the following [39]:

1. An appropriate aquifer of low flow velocities and thus reduced energy losses;
2. High quality, high efficiency groundwater production and injection wells;
3. A low cost thermal energy source with a later demand for stored energy.

ATES systems are mainly used in Sweden, Denmark, Belgium and Netherlands. In the Table below (Table **5**) the basic characteristics of the ATES applications in different countries are presented. In Table **6**. Some indicative ATES applications are presented along with their thermal storage capacity.

Table 5. Major ATES application in Europe [40].

Country	# of ATES Applications	Aquifer Type	Type of Application
Belgium	10	Sand	Hospitals
Denmark	10	Sand/gravel, chalk	Industrial
Netherlands	700	Sand	Large buildings
Sweden	70	Chalk, sand/gravel	Large buildings

Table 6. Selected ATES systems for heating and cooling applications [40, 41].

Building	Location	Initial Operating Date	Capacity (kW)
IBM office	Zoetermeer (NL)	1992	700
Nike office	Hilversum (NL)	1999	2000
Maria hospital	Overpelt (BE)	2005	1500
IKEA store	Duiven (NL)	1999	750
Westway housing project	London (the United Kingdom)	2006	250

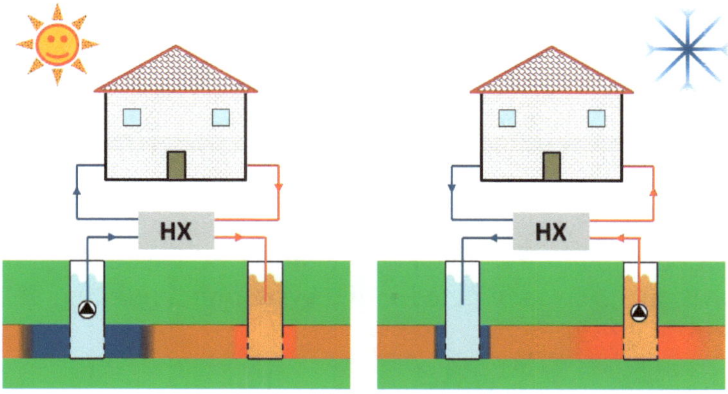

Fig. (10). ATES storage system in summer and winter operation [38].

In BTES, vertical heat exchangers are installed underground in order to transfer heat towards and from the ground layers. A heat transfer fluid is used to deliver thermal energy from the surface to the underground and there is no pumping of water from the underground like the ATES systems. The drilling depth in BTES systems can reach 250 m but in general is lower in comparison with ATES systems.

A schematic of a BTES is illustrated in Fig. (**11**) for summer and winter operation. As it can be seen, during the charging period (summer) heat is delivered from the building to the subsurface. In discharging period (winter) thermal energy is removed from the underground and conveyed to the building. BTES systems can be used for direct cooling (without heat pumps), even though heat pumps sometimes are required. There is also BTES systems that store high temperature thermal energy during summer to deliver low temperature heating during winter [42].

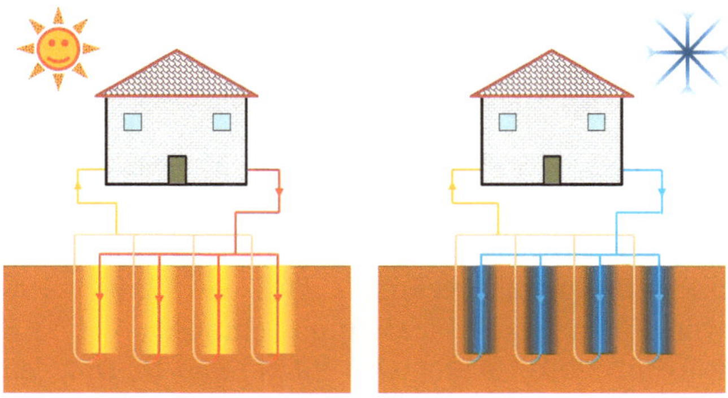

Fig. (11). BTES storage system in summer and winter operation [38].

The first large scale high temperature BTES system was built in Luleå University, Sweden, in 1982 for experimental and demonstration purposes. This system consists of 120 (10×12) vertical boreholes of 65 m deep in granitic and gneissic bedrock [43]. The total storage land area was 1584 m^2 and the total rock storage volume was 120000 m^3. The storage system was charged at a temperature of 70-82 °C during summer period exploiting the waste heat from a gas fired cogeneration plant. In winter the BTES system was discharged delivering its stored heat to a building of the Luleå University. It was calculated that about 2.3 GWh of thermal energy was injected in the system per year and about 1.0 GWh was extracted from it [43].

There are also applications of BETS concerning the storage of solar heat during summer and the exploitation of this thermal energy for heating purposes during winter. A system like this is operational since 1997 in Neckarslum, Germany. The system utilizes the heat from 5263 m^2 of solar thermal collectors and a BETS of 63360 m^3 of ground volume. The highest temperature that is expected to be achieved in the BETS is about 85 °C. A solar fraction of 50% is expected to be reached by this solar assisted district heating system [44].

The Drake Landing Solar Community, at Okotoks, Alberta, Canada is another example of district heating system that stores summer's solar energy to a BTES in order to provide heat in local buildings during winter. The heat is produced by 2295 m^2 of flat plate collectors installed on the roofs of the houses' detached garages. The excess thermal energy that produced during summer is stored in a BTES which consist of 144 boreholes of 35 m deep. The system can achieve a solar fraction of over 90% of space heating with solar energy to 52 detached single-family houses [45].

In CTES, large underground caverns and cavities are used in order to store thermal energy. This type of storage systems is generally expensive so there are a limited number of applications. In Fig. (**12**), a CTES system is illustrated. In such a system the hot water is injected at the top of the cavern and the cold water is extracted from the bottom of it in order to maintain thermal stratification.

The first two CTES systems were built in Sweden in early 80ies. Avesta CTES with a storage volume of 15000 m^3 and Lyckebo CTES with a storage volume of 115000 m^3. Avesta CTES was constructed in 1981 for short-term storage of heat produced at an incineration plant. Lyckebo CTES was built in 1983 in Upssala, Sweden and it is world's first large scale high temperature CTES system. The maximum temperature of stored water was 90 °C and thermal energy of 5500 MWh was stored between seasons [42]. The storage was partly heated by solar collectors.

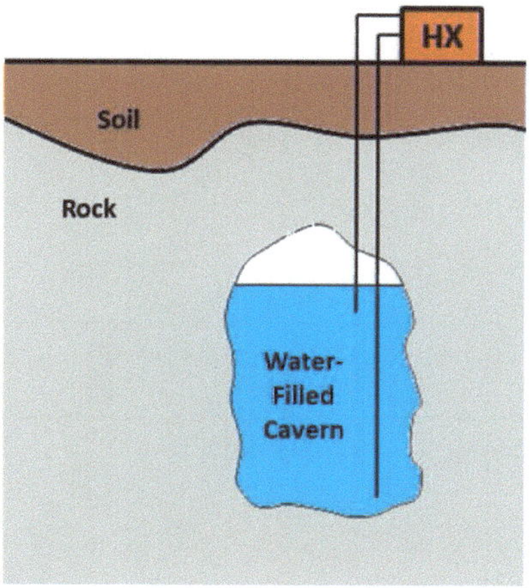

Fig. (12). CTES hot water storage system.

LATENT HEAT STORAGE SYSTEMS

As seen earlier in this chapter, sensible heat storage regards systems that store heat in the form of temperature rise of medium utilized for energy storage. Such materials include water, rocks, concrete and sand. Although widely used, the aforementioned materials exhibit low energy density and thus tanks of large volumes are required for sufficient energy storage. Large volumes lead to a substantial increase of the unit size and overall cost making, therefore, sensible TES systems preferable for small-scale applications where economics is a key parameter.

On the contrary, in latent heat storage, heat is stored (or removed during the opposite process) when a material changes phase from one to another. The materials used in latent TES systems are called phase change materials (PCM) and during their phase change they can absorb or release thermal energy in rather controlled temperature variations [46, 47].

A material can be described as an efficient PCM if it fulfills two basic criteria, a large latent heat of melting at a relatively low temperature and a high thermal conductivity for better heat transfer. A phase change material should also be chemically stable, nontoxic, noncorrosive, and able to recycle. The following Table **7**, summarizes the required properties that a material needs to exhibit so as to be used as a PCM.

Phase change materials have been widely accepted among the most progressive materials, the utilization of which can augment energy efficiency in buildings applications [2, 49], as they can store large amounts of thermal energy in both sensible and latent heat regions. This asset of PCMs can be seen in Fig. (**13**). It must be pointed out that latent heat regards only the melting and solidification processes since other phase changes (condensation/evaporation) cannot be exploited in TES systems due to large volume variations during phase change.

Table 7. Preferable properties of PCM [48].

Thermal	Kinetic	Chemical	Physical	Other
Desired Phase change temperature	No super-cooling	Stable over a no. of cycles	Small vapor pressure	Low cost
High latent heat	No sub-cooling	Non-corrosive	low volume expansion coefficient during phase change	Easily available
High thermal conductivity	No phase segregation	Non-toxic	High density	
High specific heat capacity	Good nucleating properties	Non-flammable		
		Compatible with tank material		

The latent heat method of storage has become widely popular and it can be implemented in numerous applications, as it exhibits higher energy storage density compared to that of the sensible TES systems. Nevertheless, serious shortcomings become evident when implementing the latent heat method due to the materials low thermal conductivity, volume expansion, stability over a number of cycles and sometimes phase segregation.

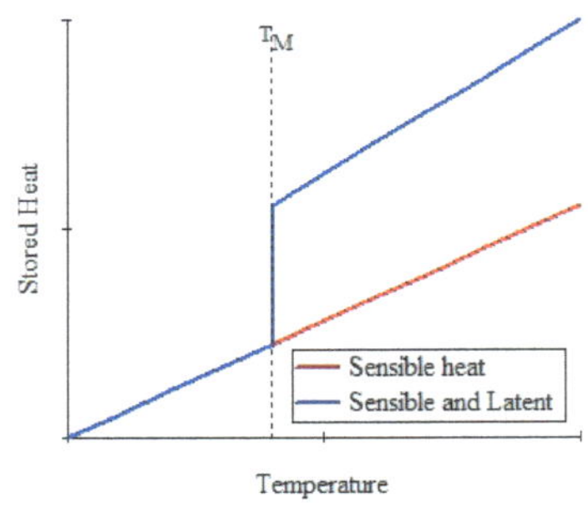

Fig. (13). Heat stored in sensible and latent TES systems.

Categories and Materials

There are four types of phase change materials, the solid to liquid, the solid to solid, the solid to gas and the liquid to gas, (Fig. **14**). The category of materials that attracts most applications and it has been used since 1800s [50] is the solid to liquid phase change. Cardenas and Leon [51] have given a sorting list of phase change materials used for latent heat storage.

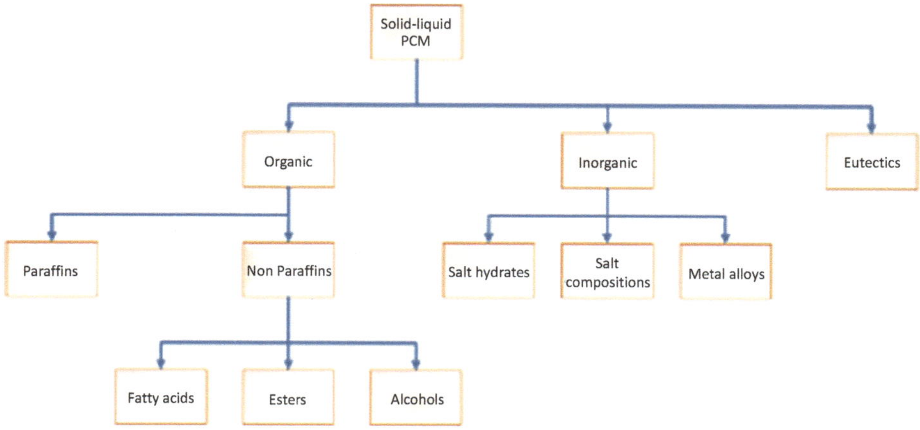

Fig. (14). Categories of solid to liquid PCM.

Organic PCM

Organic PCMs cover an extensive range including pure n-alkanes, fatty acids and esters. They have drawn attention due to their advantages which refer to their high latent heat, a phase change temperature appropriate for many applications, and stable physical and chemical properties. However, organic PCMs exhibit relatively low thermal conductivity (approximately 0.2 W/mK [52]) and high-volume variation [53], characteristics which limit their utilization in practice. Organic PCMs are divided into two groups, paraffin and non-paraffin materials.

Parrafins are among the most common PCMs since, apart from all advantages mentioned above, they need no sub-cooling. Sub-cooling concerns a condition in which a PCM requires temperatures adequately below the melting point before its solidification process begins. Moreover, paraffin materials are also regarded to be safe, reliable, inexpensive [54], making them ideal for domestic applications. On the other hand, nevertheless, the low thermal conductivity of paraffin is the key parameter that hinders its extensive application (0.21- 0.24 W/mK) [55, 56].

Unlike paraffin PCMs, which share similar features, non-paraffin PCMs cover a wide variety with different properties. Due to their vast variety, they are regarded as the most appropriate category for TES applications [57]. Alcohols, esters, glycols and fatty acids are the most popular non-paraffin PCMs [58]. In this category of organic PCMs, fatty acids find direct implementation in building cooling applications under different climates and conditions [59] since they demonstrate appropriate characteristics, such as high latent heat, no phase segregation and various melting temperatures. Compared to paraffin PCMs, however, they are more expensive, corrosive and highly flammable.

The heat of melting of some organic PCMs are illustrated in Fig. (**15**).

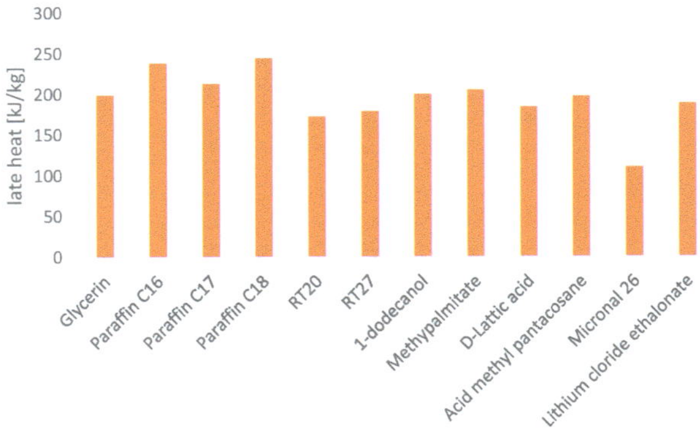

Fig. (15). Heat of fusion of organic PCM [57, 60 - 63].

Inorganic PCM

Inorganic PCMs have higher latent heat and lower cost per used kilogram than organic ones. However, they are deficient in thermal stability and they demonstrate phase segregation, corrosion and decomposition, properties which distract from their advantages [64 - 66]. This category of PCM includes salt hydrates, salt solutions and metals [62] with salt hydrates being the most used PCM in thermal storage applications due to their higher storage density, higher thermal conductivity and lower utilization cost when compared with organic PCM [62]. Fig. (**16**) depicts the latent heat of commonly used inorganic PCMs.

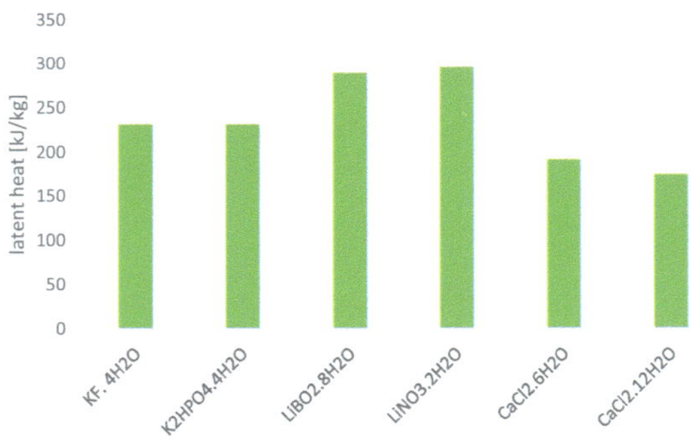

Fig. (16). Heat of fusion of inorganic PCM [57, 61, 67].

Eutectics PCM

A mixture of two or more organic or inorganic PCMs forms an Eutectic PCM [65]. After the formation of a blend crystal during the freezing procedure, this kind of PCM is unlikely to be separated into its components as it changes phase without demonstrating ant segregation and moreover all components turn into liquid simultaneously [68]. Some of the eutectic PCMs that can be applied in passive cooling in buildings are shown in Fig. (**17**).

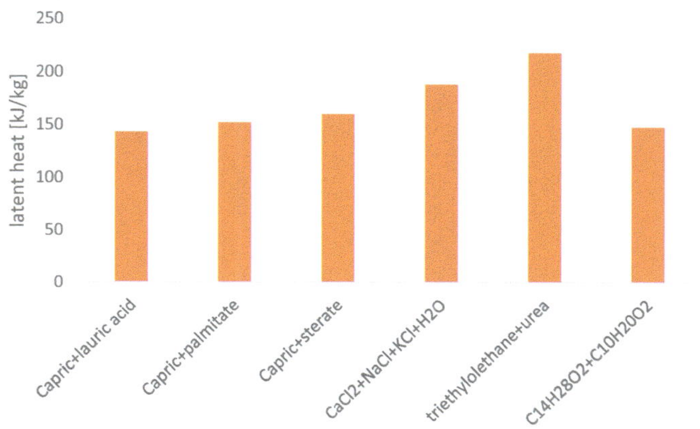

Fig. (17). Heat of fusion for eutectics PCM [57, 61, 26, 67].

As already mentioned above, the main disadvantage of PCMs is their low thermal conductivity. Therefore, the need to enhance thermal conductivity of the PCMs is apparent. According to current researches, additives are seemed as a potential solution to the low PCMs conductivity. Conventional additives are divided into carbon type (exfoliated graphite, graphite powder, carbon nanotubes, graphene) and metal type (metal foam, metal nanoparticles, metal salts). The following tables, Table **8** and Table **9**, present recent researches on PCM thermal conductivity enhancement.

Table 8. Researches on PCM thermal conductivity enhancement with carbon type additives.

Carbon Type Additives	Research	Main Findings
	Jeong *et al.* [69]	The thermal conductivity of Bio-based PCMs with exfoliated graphite nano platelets exhibited a 375% enhancement
	Shi *et al.* [70]	A paraffin PCM with a 10 wt% loading of exfoliated graphite nano platelets led to a 10-times augmentation in thermal conductivity

(Table 8) cont.....

Carbon Type Additives	Research	Main Findings
	Huang *et al.* [71]	A LiNO3/KCL-exfoliated graphite (EG) composite PCM for solar TES application was prepared. The experiments showed that the thermal conductivity of the PCM can be improved 185% by using 10% EG while a 30% EG led to a 6.65-fold increase of the thermal conductivity.
	Wang *et al.* [72]	The thermal conductivity of microencapsulated PCM increased from 0.43 W/mK to 4.59 W/mK after the addition of 20 wt% exfoliated graphite.
	Li [73]	The carbon powder (10 wt%)/paraffin composite exhibits a maximum thermal conductivity of 0.9362 W/mK.
	Johansen *et al.* [74]	Sodium acetate trihydrate and graphite powder mixtures were prepared for the conduction of thermal conductivity tests. The mixture with 5 wt% graphite powder exhibits a thermal conductivity of 0.746 W/mK.
	Li *et al.* [75, 76]	A stearic acid/multi-wall CNTs (MWCNTs) composite was prepared and it was proved that the thermal conductivity can be improved by 5.7% by using 3.0 wt% of MWCNTs. A composite PCM with grafted CNTs and paraffin has a thermal conductivity of 2.42 times the conductivity of pure paraffin.
	Wang *et al.* [77]	The thermal conductivity of palmitic acid/treated CNTs was determined in both liquid and solid phase. They proved that the thermal conductivity is nearly 30% higher than that of the pure palmitic acid.
	Zhong *et al.* [78]	A mixture of graphene aerogel (GA) and octadecanoic acid (OA) was prepared. The composite exhibited a thermal conductivity of about 2.635 W/mK for 20% vol GA, a 14-fold increase compared to that of the OA (0.184 W/mK).
	Fu *et al.* [79]	Graphene sheets were used as additives to create an epoxy/graphene composite. The results of the tests showed an astonishing 22-fold enhancement of the thermal conductivity of the epoxy resins at a 10 wt% loading.

Table 9. Researches on PCM thermal conductivity enhancement with metal type additives.

Metal Type Additives	Research	Main findings
	Xiao *et al.* [80]	The addition of copper foam of various porosities to a paraffin PCM led to an enhancement of the thermal conductivity varying from 13 to 44 times that of the pure PCM. They also tested a nickel foam enhanced PCM composite which exhibited a conductivity 3-5 times higher than that of the paraffin.
	Thapa *et al.* [81]	The thermal conductivity of the icosane wax PCM reached the value of 3.7 W/mK through the addition of copper foam.
	Sahan *et al.* [82]	The mixtures of paraffin and nano magnetite (Fe_3O_4) show 48% and 60% enhancement of thermal conductivity for 10 wt% and 20 wt% of magnetite, respectively.

(Table 9) cont.....

Metal Type Additives	Research	Main findings
	Yu *et al.* [83]	Fe3O4 incorporated in kerosene (1.0 vol%) yield a thermal conductivity improvement of 134%
	Fauzi *et al.* [66]	The incorporation of 10% sodium laurate into a Myristic acid (MA)/palmitic acid (PA) eutectic mixture led to a 4.4 increase of the thermal conductivity.
	Wang *et al.* [84]	A composite PCM was created by blending polyethylene glycol, silica gel and β-Aluminum nitride (β-AIN) powder. The thermal conductivity augmented from 0.2985 W/mK (pure polyethylene glycol) to 0.7661 W/mK when 30wt% β-AIN were utilized.
	Sharma *et al.* [85]	Ethylene glycol was used as a solvent PCM and silver nitrate nanoparticles were utilized for thermal conductivity enhancement. Thermal conductivity increased 18% with 1 wt% silver particles.

PCM in TES Systems: Applications

Phase change materials can be utilized for both diurnal and seasonal energy storage by applying a variety of methods.

Solar Water-heating Systems

Prakesh *et al.* [86] analysed a solar water storage system which contained PCM at the bottom of the storage tank. During daylight, the water was heated up energy was transferred to the PCM in the form of latent heat. When no irradiance was available, the hot water was drawn and replaced by cold water, which received energy released from the PCM during its solidification process.

Chaurasia [87] compared the energy storage of two systems, one based on latent heat and another on sensible heat. Two identical storage tanks were utilized. One storage unit contained 17.5 kg paraffin wax as the storage medium packed in a heat exchanger made of the aluminium tubes and the other was filled only with water. Both units were independently charged during daylight by identical flat plate solar collectors. This study showed that the latent TES system produces more hot water on the next day morning compared to that of the sensible TES system.

Kamiz Kayguz [88] examined both experimentally and theoretically the performance of PCM used in solar water-heating systems. $CaCl_2 \cdot 6H_2O$ was utilized as a phase change material. Solar energy was harvested, transformed into heat and transferred to a storage tank that was filled with 1500 kg of PCM. The test rig comprised of a horizontal vessel with cylindrical tubes which contained the PCM while the working fluid flowed outside the tubes transferring heat to the storage material.

Mettawee *et al.* [89] investigated the thermal performance of a PCM-integrated solar collector. In examined collector, the absorber plate (PCM container) absorbs the available solar irradiance and stores it directly to the PCM. The heat stored in the paraffin wax was then transferred to the working fluid. The test results showed that during the charging process, the average heat transfer coefficient increased as the melting front moved further. In the opposite procedure, the heat gain was enhanced with the increase of the water mass flow rate.

Cabeza *et al.* [90] built a pilot solar plant at the University of Lleida so as to evaluate the PCM performance in conditions of continuous operation either with the solar system or an auxiliary electrical heater. A PCM-graphite composite of 90 vol.% of sodium acetate trihydrate and 10 vol.% graphite was utilized as the PCM. The results led to the conclusion that the integration of PCM in water tanks for covering DHW needs is a propitious TES method as it provides availability in hot water for longer periods of time even without solar irradiance or grid energy.

PCMs in Greenhouses

A quite interesting application of PCM regards their utilization in green houses for harvesting the solar energy in the form of heat so as to be used for curing and drying processes and plant production [91].

Kern and Aldrich [92] used 1650 kg of $CaCl_2{\cdot}6H_2O$ PCM so as to examine the potential of energy storage inside and outside of a 36 m^2 greenhouse covered with fiberglass, Fig. (**18**). PCM cans were placed in rows inside and outside the greenhouse. The TES unit inside the greenhouse stored energy through the warm air during day and released it at night through the reversion of the air flow direction.

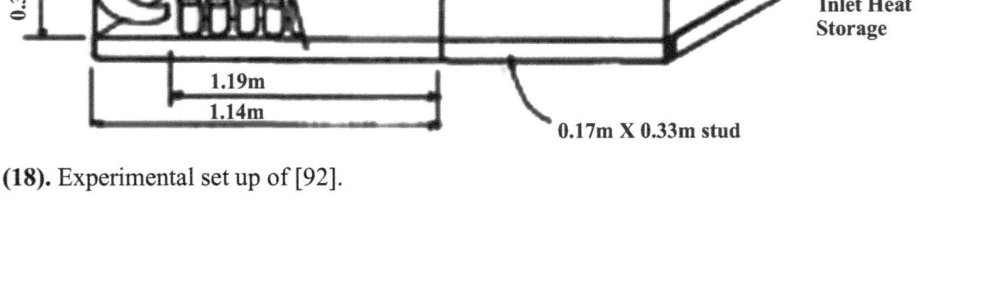

Fig. (18). Experimental set up of [92].

Experiments with Na2SO4·10H_2O as PCM were also conducted in a 500 m^2 single glazed roses greenhouse located in France. The available solar irradiance was harvested and stored inside the greenhouse by reprocessing the air through an underground flat heat exchanger filled with the PCM. The performance of the PCM-integrated greenhouse was compared with that of a conventional greenhouse of the same geometry and plants. The PCM greenhouse exhibited an 80% increase in propane gas savings when compared with the conventional one. The stored heat was used when no solar irradiance was available so as to preserve the required temperature inside the greenhouse [93].

Takakura and Nishina [94] tested polyethylene glycol and $CaCl_2·6H_2O$ as PCMs utilized for heating a 7.2 m^2 greenhouse in Japan. The integrated system consisted of the aforementioned PCM thermally charged by a solar collector. Through this, the greenhouse exhibited an overall efficiency of 59%. Moreover, due to the presence of the PCM, the temperature of the greenhouse was maintained in the required range even at night when the outside temperature was down to -0.6°C.

PCMs in Buildings

The most promising utilization of PCMs is their integration in buildings. The envelope of a building affects the quality and controls the indoor conditions regardless the transient outdoor conditions. PCM can be installed into literally all components of the building envelope. However, the most common parts for PCM integration in the envelope are the walls, floors, ceilings, roofs and windows due to easy installation and more effective heat transfer. In the following pages, researches on PCM in buildings are presented along with the main findings of each work.

Researcher	Investigated Structure	Findings
Neeper [95]	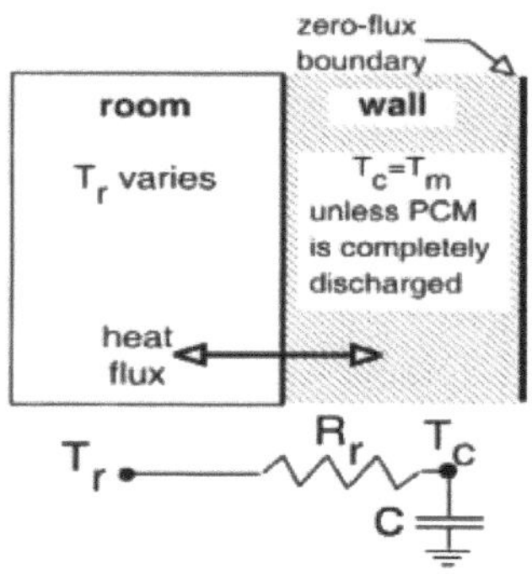	The maximum daily energy storage occurs at a phase change temperature that is close to the average room temperature. The diurnal storage achieved in practice may be limited to the range of 300-400 kJ/m^3, regardless the greater latent capacity of the wallboard.

Cont.....

Researcher	Investigated Structure	Findings
Ahmad *et al.* [96]	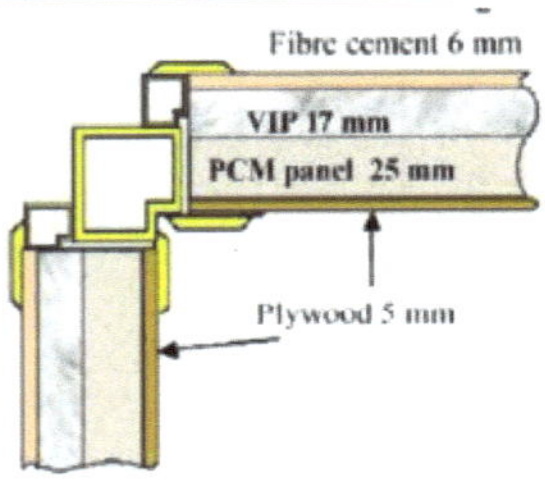	During summer the amplitude of the temperature inside the cell with PCM is reduced by 20 °C on a diurnal cycle. In winter, the presence of PCM avoids indoor ambience reaching negative temperatures.
Chen *et al.* [97]	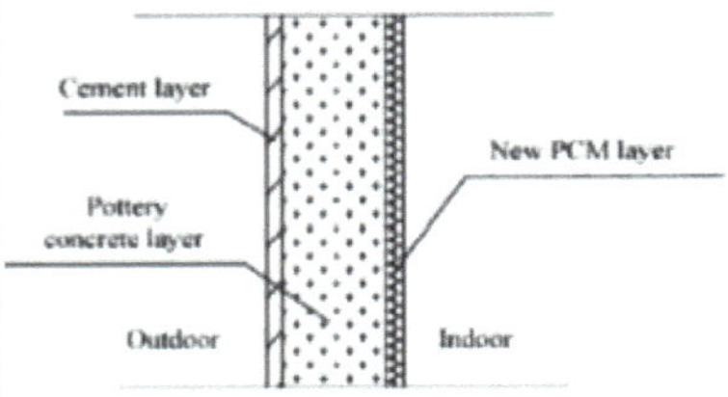	Integrating suitable PCM in the inner surface of the north wall in a usual room not only augments the indoor thermal comfort radically, but also improves the utilization rate of the solar irradiance. The decrease in energy consumption for heating leads to energy savings which for a seasonal operation can reach up to 17% or higher.
Kuznik *et al.* [98]	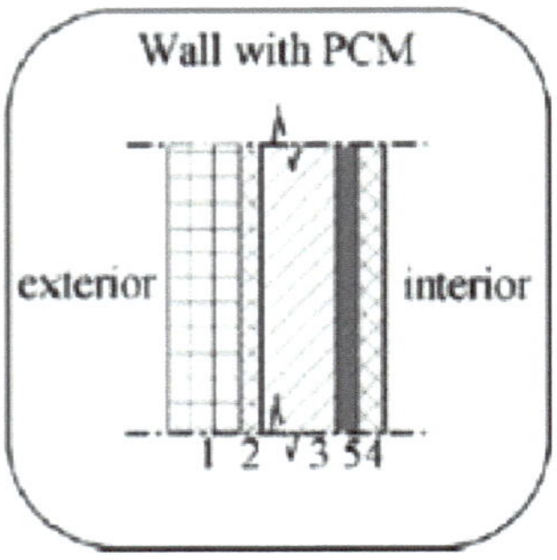	The energy stored in the PCM prevents high room temperatures during hot days, and during the night it keeps the temperature within the comfort range. Generally, the presence of PCM reduces large temperature variations inside rooms.
Xiao *et al.* [99]	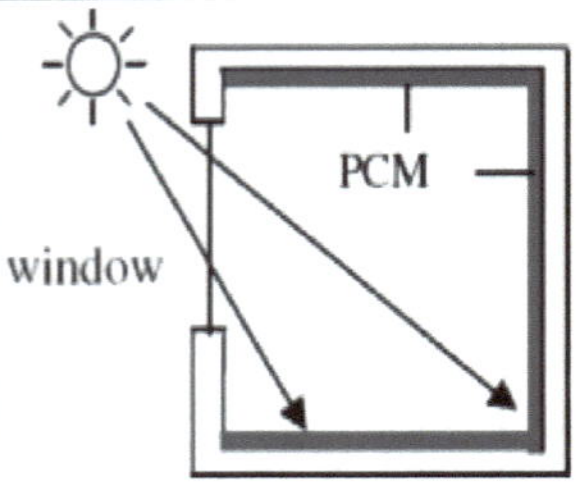	The simulation results show that the optimum phase change temperature depends strongly on the average indoor temperature and the solar irradiance absorbed by the phase change material.
Lee *et al.* [100]	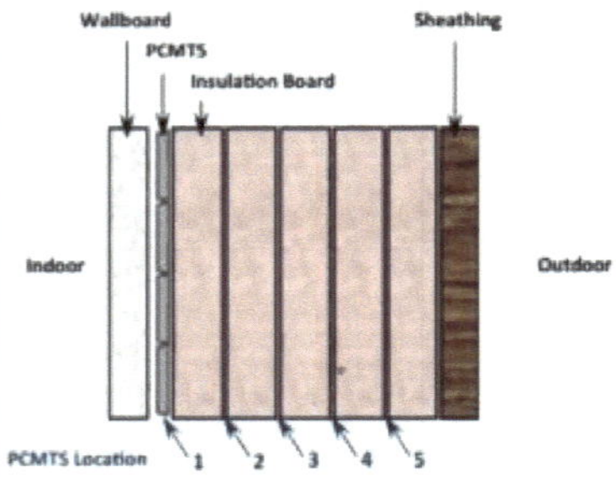	PCM panels, when installed at optimum locations, can contribute to an astonishing reduction of heat fluxes for the tested wall directions. The reduced heat fluxes occur with a multi-hours delay due to the presence of PCM.

Cont.....

Researcher	Investigated Structure	Findings
[101]	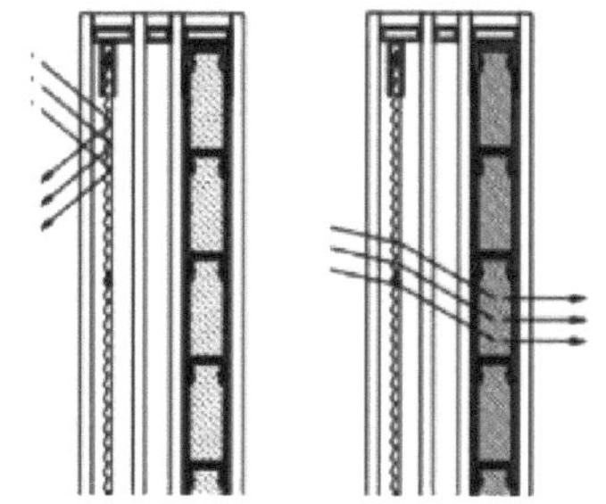	Up to 1186 W/m^2 of latent thermal storage. PCM in windows reduces the room temperatures by 4 °C-12 °C. In the winter, it leads to an average reduction of heating load by 150-200 kWh/ m^2 annually.
Xu *et al.* [102]	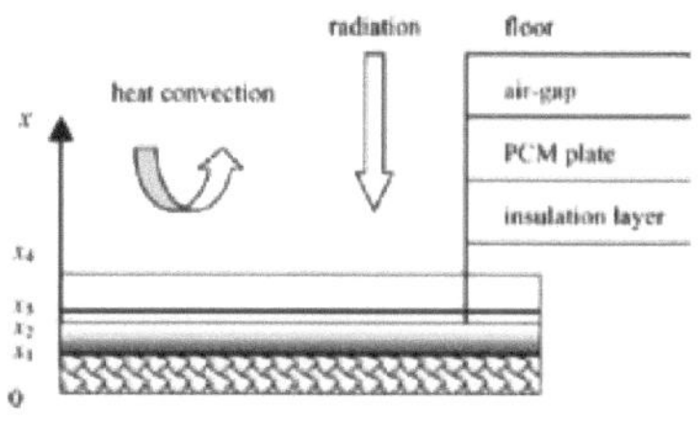	Lower limits of PCM latent heat and thermal conductivity were determined for achieving effective TES; the air-gap between PCM and the floor should be minimum since it acts as further thermal resistance.

PCM MICROENCAPSULATION

The utilization of PCMs in traditional manner necessitates the existence of latent heat exchanger which increases the overall cost and thermal resistance between the PCM and its ambience [103]. A solution to this is the development of form-stable or shape-stabilized PCMs which are produced by mixing and further treating a PCM with a supporting material, commonly polymers [59, 103, 104]. Although these compounds can retain their shape during the melting process, the PCM tends to diffuse to the surface and leak [105].

The prevention of the melted PCM leakage during the phase change process, the enhancement of the heat transfer rate, the reduction of PCM tendency to react with the outside environment as well as the controlling of the volume expansion as phase change occurs, can be achieved through microencapsulation [106]. Microencapsulation is a procedure of covering individual particles or droplets with a continuous film in order to produce capsules of micro to millimeter in size, known as a microcapsule [107]. Microencapsulated PCMs consist of two main parts: a PCM as core and a polymer or inorganic shell as PCM vessel. Microcapsules may vary in shape presenting either regular geometry shapes or irregular. The typical shape of the microcapsules depends mainly on the PCM type and the shell formation procedure.

The microencapsulation method depends strongly on the physical and chemical properties of the materials to be utilized. There are various physical and chemical approaches used for the production of microcapsules [107, 108]. The most often used microencapsulation methods are illustrated in the following figure, Fig. (**19**).

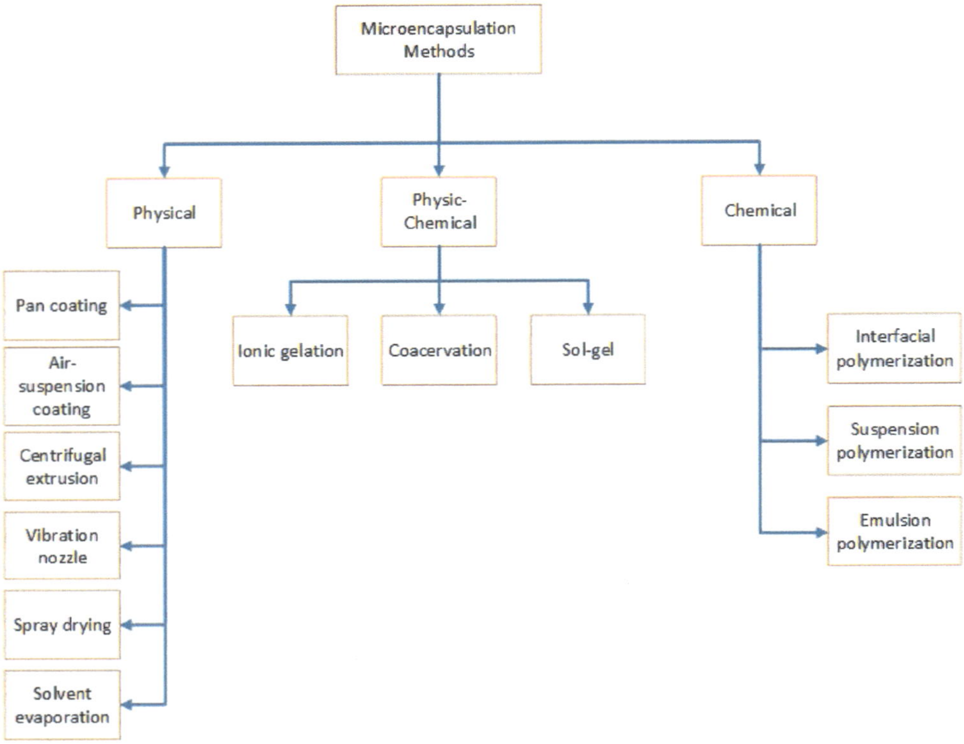

Fig. (19). Microencapsulation methods.

A wide variety of microencapsulated PCMs of various core and shell materials and variable thermal and mechanical properties is already manufactured, some of them are readily available commercially. However, thorough testing of the produced microencapsulated PCMs is essential in order to assure their quality. Researches for future works may take into consideration the following:

- Long-term instability; especially when it is used to enhance heat transfer coefficient of a fluid in a pumping cycle or in buildings.
- Encapsulation in nanometer-sized shells
- A further testing for microcapsules is needed, as most of the produced microencapsulated PCMs are not checked for leakage.
- Supercooling; it is a major parameter that hinders the industrial application of microencapsulated PCMs, as most PCMs tend to supercool when encapsulated.

THERMOCHEMICAL HEAT STORAGE SYSTEMS

Sensible and latent heat storage systems present two major drawbacks: 1) they lose thermal energy during the storage time making them impractical for long-period energy storage; 2) they offer low energy density leading to storage systems of high volume. Thermochemical energy storage systems can be a promising alternative to conventional (sensible, latent) thermal energy storage systems as they can overcome the two above mentioned disadvantages. First of all, they can provide high energy densities (Fig. **20**). As a consequence, these systems can store large amounts of energy in smaller volumes. Furthermore, the heat losses to the environment over time are negligible making these systems ideal for seasonal heat storage. Another benefit of thermochemical energy storage systems is that they can cover a wide range of working temperatures (from -50 to over 1000 °C) [109]. Despite the favourable characteristics of thermochemical energy storage the state of development is in a relatively low level in comparison with the other two storage technologies (Fig. **21**).

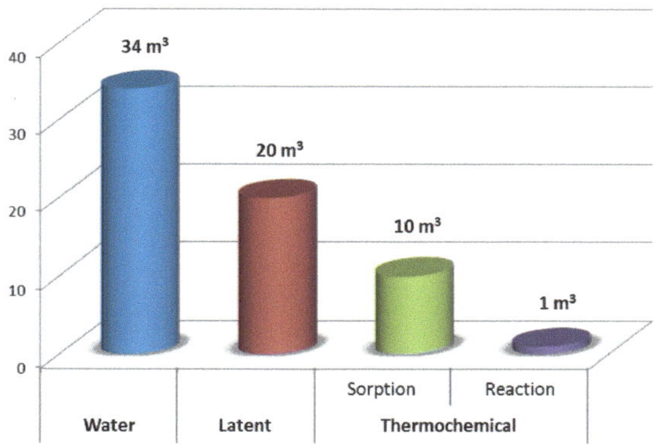

Fig. (20). Volume required to store 6.7 GJ (ΔT=70°C for water) [110].

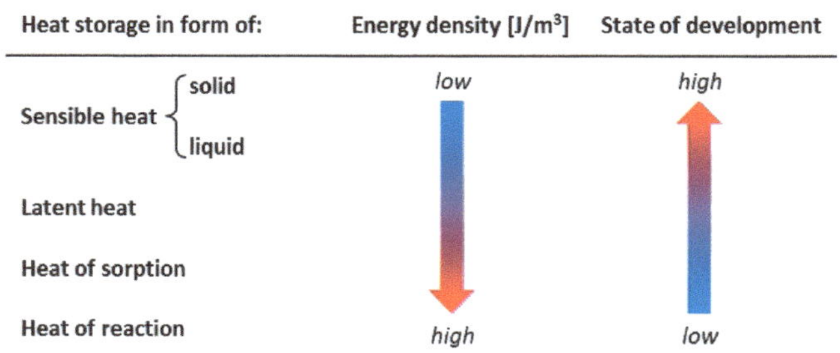

Fig. (21). Energy density and state of development of heat storage systems [111].

In general, thermochemical energy storage systems use a reversible chemical process or reaction to store heat. In these reactions, heat is stored in chemical or physical bonds of specific materials and the charging/discharging process of the storage is actually an endothermic/exothermic reaction (Eq. **1**) [112].

$$AB + Heat \rightleftharpoons A + B \quad \mathbf{(1)}$$

AB is the thermochemical material. When heat is provided the chemical compound, AB is separated into components A and B (endothermic process) which can be separately stored. When it is time to exploit the stored heat the two components A and B are come in contact releasing heat to the environment (exothermic process). By this way, thermal energy can be stored with a nearly lossless way due to the fact that heat is in neither sensible nor latent form but it is expressed as a chemical potential [111] (Fig. **22**).

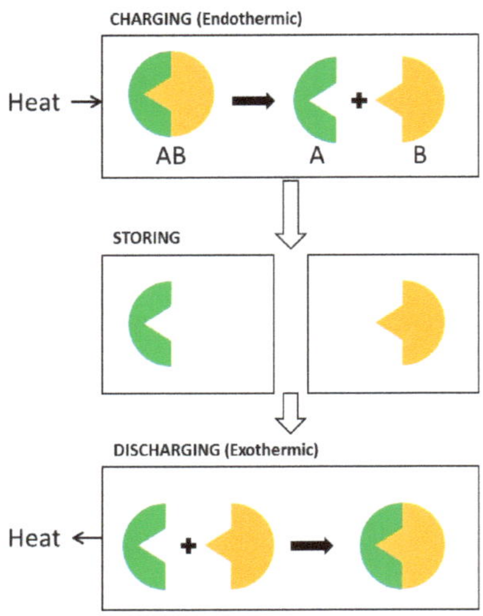

Fig. (22). Processes involved in a thermochemical energy storage cycle [113].

Thermochemical energy storage systems can be divided into two main groups (Fig. **23**): sorption processes and chemical reactions [109].

Sorption storage processes include both absorption and adsorption. Absorption is the process that occurs when a substance is taken up by a liquid or solid forming a solution (Fig. **23**). Concerning storage applications, absorption mainly refers to absorption of gases (absorbate) by liquids (absorbent). Adsorption on the other hand is a surface phenomenon that occurs when a substance accumulates at the surface of a liquid or solid phase. In most cases the phenomenon is related with the coupling of a gas (adsorbate) to the surface of a solid porous material (adsorbent) (Fig. **23**) [111]. Adsorption processes can be categorized into two groups: 1) physical adsorption (physisorption) and 2) chemical adsorption (chemisorption) (Fig. **24**). In physisorption the forces between the adsorbate and the adsorbent are weak intermolecular forces (van der Waals) while in chemisorption the forces that prevail are valence forces. Chemisorption processes can provide larger amounts of heat (heat of sorption) compared to physisorption but may exhibit irreversibilities [111].

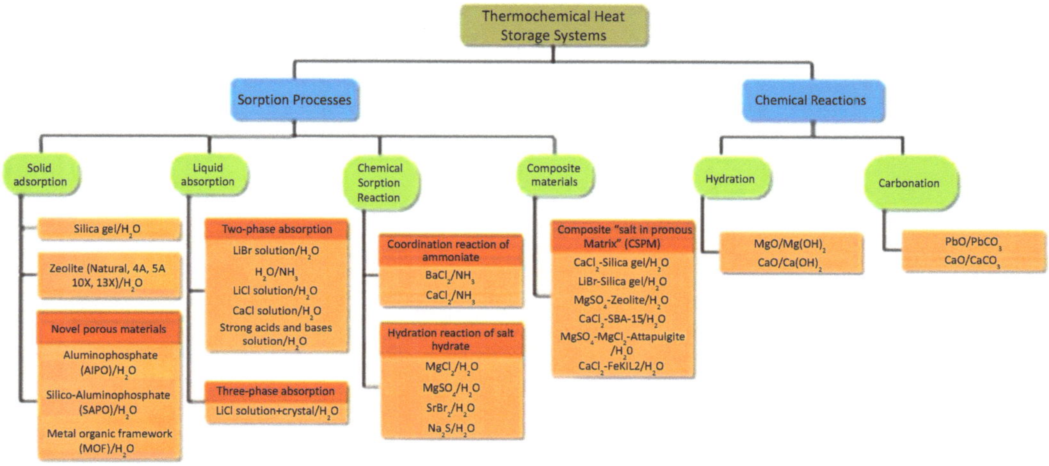

Fig. (23). Thermochemical heat storage classification [99, 104].

Chemical sorption reactions comprise of two kinds of reactions: coordination reactions of ammoniate with ammonia and hydration of salt hydrate with water [114].

Composite sorbents are developed and studied so as to achieve two main goals: 1) improve heat and mass transfer mechanisms in chemical sorbents and 2) increase the adsorption quantity of physical adsorbents [115]. This new family of composite sorbents is called Composite "Salt in Porous Matrix" (CSPM) [116]. If the sorbate is water, these materials are known as Selective Water Sorbents [114, 116]. The composite sorbents are made from porous media used as host matrix and chemical sorbents which are impregnated into the matrix pores. Typical porous materials that used as host matrices include silica gel, zeolite, alumina, aerogel *etc.* As chemical sorbents are mainly used inorganic salts like LiCl, $CaCl_2$, $MgCl_2$, $MgSO_4$, $Ca(NO_3)_2$, $LiNO_3$, *etc* [114]. Water sorption process which occurs in selective water sorbents involves two key mechanisms: 1) chemical reaction between salt and water and 2) liquid absorption [114].

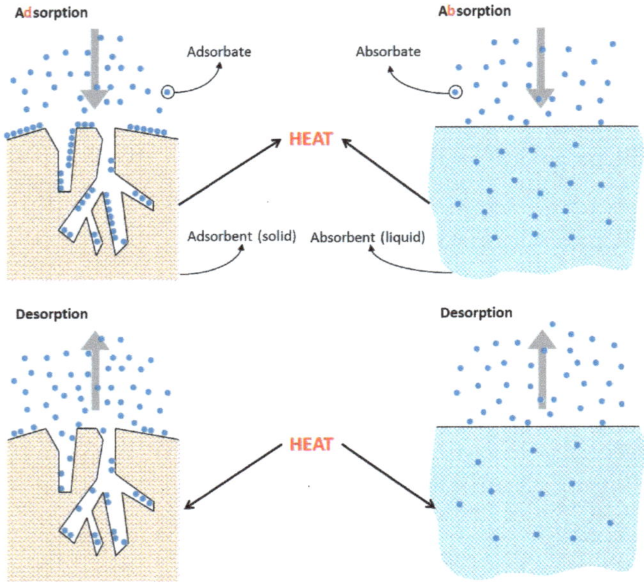

Fig. (24). Mechanism of gas-solid adsorption and gas-liquid absorption.

Finally, chemical reactions mainly consist of hydration and carbonation of metal oxides [117]. Thermochemical energy storage systems based in chemical reactions are suitable for medium and high-grade heat storage (above 400 °C).

Sorption Heat Storage Systems

In a sorption thermal energy storage system, the heat of sorption is provided to the system during the charging process. This heat can be provided by harvesting solar energy or waste heat in order not to burden the environment with CO_2 emissions and prevent the consumption of the limited amount of fossil fuels. During the discharging process the sorbent material come in touch with the sorbate and the heat of sorption is released. This heat can be exploited in various applications depending on its temperature (*e.g.* space heating, CSP plants *etc.*). These storage systems can be split into two types: 1) open loop and 2) closed loop storage systems (Fig. **25**). In closed loop systems there is no matter exchange between the environment and the system. In this case the sorbate must be condensed during the charging process and then evaporate in the opposite process [16]. On the other hand, in open loop thermochemical systems there is exchange of mass between the ambient and the system. In a system like this, the sorbent material is typically located in a packed bed or impregnated in a honeycomb structure in which the ventilation air of a building is distributed [16]. During the discharging phase the humid ambient air passes through the system and as it interacts with the sorbent its humidity is extracted due to the sorption process. As a result, the heat of sorption is released increasing the temperature of the ambient air. When the sorbent material is saturated it is unable to remove water vapour from the air anymore. In order to recharge it the heat of sorption is provided to the system and the sorbent releases the water vapour to the environment and dried. The characteristics of an ideal sorption material include [111, 114, 118]:

- high energy density;
- high affinity between the sorbent and the sorbate;
- high heat transfer rate from the sorbent to the heat exchanger;
- low regeneration (charging temperature);
- Non-toxicity, low GWP and ODP indexes;
- Non-corrosive behaviour
- Good thermal and chemical stability under operating conditions;
- Low material cost

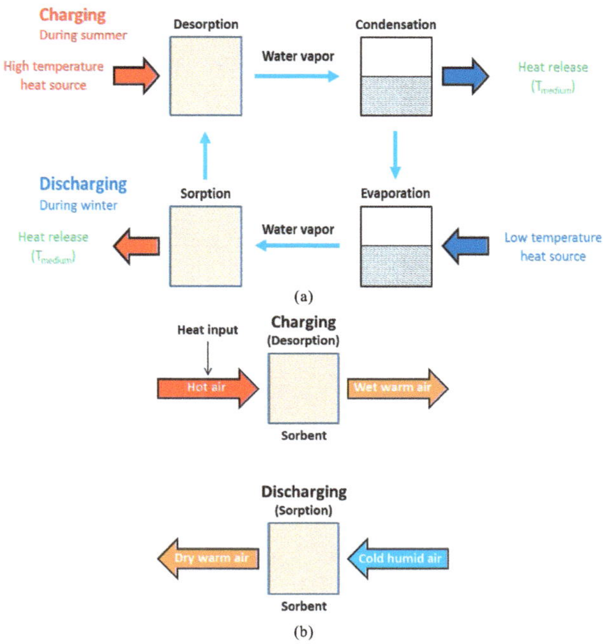

Fig. (25). Operating principle of (a) closed loop (b) open loop sorption storage [114, 118].

Sorption thermal energy systems can reach high energy densities especially when chemical sorption reactions are used. Specifically, considering solid adsorption (physisorption) systems the lower energy densities are reported for the zeolite 13X/water pair (86 kWh/m^3) and for the silica gel/water pair (94 kWh/m^3) [118]. When composite sorbent materials are used the systems can provide energy densities of 226-309 kWh/m^3 [118]. Concerning absorption storage systems, the higher energy densities are observed for the LiCl/H_2O (4387 kJ/kg) and KOH/H_2O (2618 kJ/kg) pairs [119]. Chemical sorption reactions present the highest energy densities with Sodium Sulphide (Na_2S) /water (H_2O) pair reaches almost 1980 kWh/m^3 for heat [111].

Solid adsorption storage systems are mainly about heating applications exploiting solar energy. An open loop adsorption energy storage system using zeolite 4A was developed at ITW (Institute of Thermodynamics and Thermal Engineering) in the University of Stuttgart (Germany) through the Monosorp project (2005-2007). The system (Fig. **26**) utilizes the excess solar heat during summer to heat the ambient air in an air to water heat exchanger. The heated air, flow through the storage, delivers the heat of adsorption to the zeolite which desorbs the collected water vapours and dries (charging process). During winter, the storage is discharged providing the stored heat to rise the temperature of the incoming ambient air. Wet indoor air passes through the storage system. The dry zeolite adsorbs humidity from the indoor air stream and releases the heat of adsorption increasing its temperature. The indoor air (dry and hot) passes through an air to air heat exchanger and warms the incoming ambient air. The resulting temperature rise relies on the amount of water vapour in the exhaust air and it usually varies between 15 K and 25 K [112]. In order zeolite to desorb the collected humidity high regeneration temperatures are required. These high temperatures can be achieved by high efficiency solar thermal collectors (*e.g.* evacuated tube collectors) while a pressurized collector loop is required when water acts as working fluid [112]. An energy density of 160 kWh/m^3 is achieved by this system [118] while its main drawback is the relatively high charging temperature (around 180 °C).

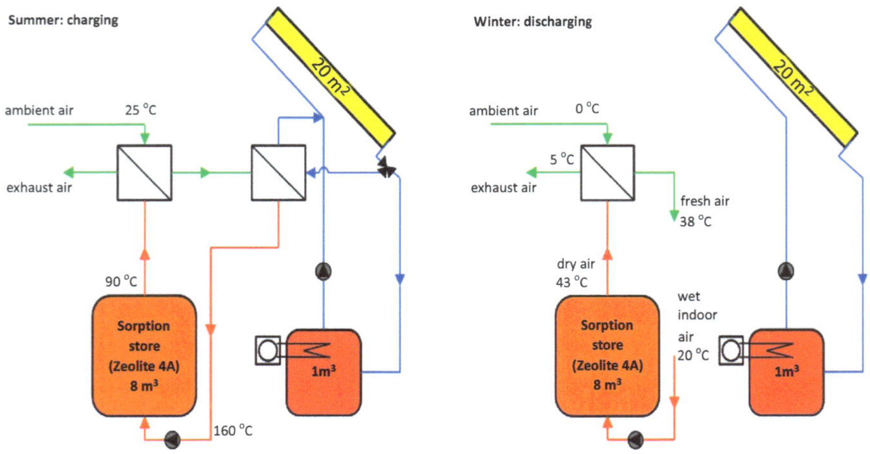

Fig. (26). Operation principle of the Monosorp concept [111].

Another prototype adsorption thermal storage system was built within the framework of HYDES project and installed in the laboratory of AEE-INTEC in Gleisdorf/Austria [120]. It was a closed loop sorption heat storage system utilizing the silica-gel water pair coupled with 20.4 m^2 of solar collectors. The produced heat was used for residential heating and hot water generation. The operating principle of the system is illustrated in Fig. (**24a**) and described briefly below: During the charging process, heat from the solar collectors or/and from an auxiliary heater is used to dry the silica-gel. The desorbed water vapours are condensed in the condenser releasing heat to the environment at a lower temperature. The charging period is followed by the storage period where the dry silica-gel is separated from the water. In this period, heat storage can be achieved with marginal heat losses. Finally, during the opposite process the water evaporates in the evaporator absorbing heat from a low temperature level. The produced water vapour is adsorbed by the silica-gel and the useful heat of adsorption is released. The experimental data obtained by this prototype were 20% lower than simulation results regarding energy density. An energy output up to 123 kWh/m^3 was achieved experimentally [120]. A "second generation" prototype was developed through MODESTORE where all the key components (evaporator/condenser and the reactor) were integrated into one single container [120]. The new system exhibited an energy density of approximately 50 kWh/m^3 which was by far below the expectations. This value corresponds to only 25% of the theoretically expected energy storage of the material [118] and is about 30% less efficient than water storage for a temperature difference of 25-85 °C.

Thermal energy storage systems based on chemical sorption reactions are mainly in laboratory scale. The most studied pair used in these prototypes is Sodium Sulphide (Na_2S)/water (H_2O) which exhibits a high sorption capacity and a high heat of sorption (300 kJ/mol of Na_2S salt=1.1 kWh/kg) [111]. The Modular Chemical Energy Storage (MCES) prototype that was designed and constructed at Chiang Mai University (Thailand) utilized the aforementioned pair in order to store both heat and cold [121]. Another prototype that utilized the Na_2S/H_2O pair was developed under the SWEAT (Sal-Water Energy Accumulation and Transformation) project at ECN by de Boer *et al.* [122]. They focused on designing a modular storage system prototype intended for residential and industrial applications.

Absorption heat storage systems have been built in both experimental and commercial level using mainly the pairs $NaOH/H_2O$, $LiCl/H_2O$ and $LiBr/H_2O$. In Table **11**, the thermodynamic characteristics of the most promising absorption pairs are presented. A prototype for long-term energy storage based on the principles of absorption was built by Weber and Dorer from EMPA using the $NaOH/H_2O$ pair [123]. In Fig. (**27**), the charging and discharging process of the

NaOH storage is illustrated. During the charging process the weak NaOH solution is regenerated by solar heat desorbing water vapour in the condenser where it condenses releasing the heat of condensation to the environment. At the discharging process, the condensed water is evaporated using a low temperature heat source. The water vapour is then absorbed by the concentrated NaOH solution in the absorber releasing the heat of absorption. Another example of absorption thermal storage is the Thermo-Chemical Accumulator (TCA). This is an absorption machine utilizing the LiCl/H_2O pair with the capacity of energy storage. It was developed by a Swedish company named ClimateWell and it is considered as one of the most successful prototypes since it combined both short term absorption thermal storage and solar cooling systems. This innovative absorption heat storage system uses lithium chloride crystals to increase energy density and it is known as three-phase absorption [114].

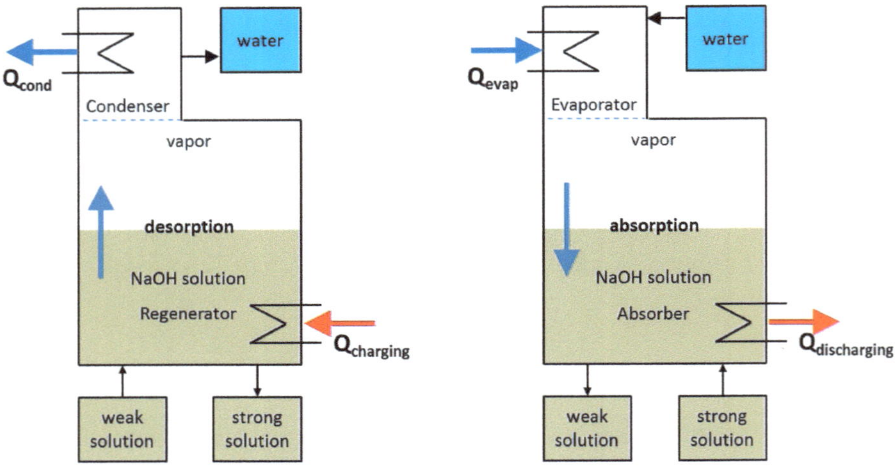

Fig. (27). Operation principle of the NaOH storage [123].

Several composite materials have been examined as potential mediums in TES systems (Table **10**). In Table **11**, some promising composite materials are presented. The highest energy density is observed for the hybrid zeolite 13X and activated alumina impregnated with LiCl (309 kWh/m^3). This performance represents an approximately 50% increase over the unmodified adsorbent [118].

Table 10. Evaluation points for the choice of the absorption couples [119].

	$CaCl_2$/ H_2O	**Glycerin/ H_2O**	**KOH/ H_2O**	**LiBr/ H_2O**	**LiCl/ H_2O**	**NaOH/ H_2O**	**NH_3/ H_2O**
Mass fraction of absorbent after desorption (%)	39.8	90.0	50.8	58.8	44.3	33.5	0

(Table 10) cont.....

	$CaCl_2$/ H_2O	Glycerin/ H_2O	KOH/ H_2O	LiBr/ H_2O	LiCl/ H_2O	NaOH/ H_2O	NH_3/ H_2O
Storage capacity (kJ/kg material)	914	193	2618	2019	4387	1558	1390
Temperature required for desorption [°C]	44.8	53.0	63	72	65.6	50	186.6
Volume of solution [m^3/1000 kWh]	8.4	20.0	3.16	3.2	2.5	6.5	9.6
Pressure absolute [kPa]	1.2-4.2	1.2-4.2	1.2-4.2	1.2-4.2	1.2-4.2	1.2-4.2	615-1167
Efficiency	0.909	0.545	0.83	0.85	0.95	0.75	0.658
Price [€/ton] (purity ≥ 99%)	350	500	1200	5000	6000	3000	400

Table 11. Energy densities of various composite materials [114, 118].

Adsorbent	Reported Energy Density (kWh/m^3)
Hybrid of zeolite 13X and AA impregnated with LiCl	309 [124]
Silica gel impregnated with $CaCl_2$	228 [125]
Activated alumina with high alkaline addition	226 [126]
ZM10 (Zeolite 4A impregnated with 10 wt.% $MgSO_4$)	178 [127]
AS/$CaCl_2$ (Impregnated aluminosilicate)	172 [128]
Zeolite 13X + $MgSO_4$ (Impregnation)	166 [129]
BT-$CaCl_2$ (Bentonite impregnated with 40 wt.% $CaCl_2$)	135 [127]
MgNaX (impregnated zeolite)	128 [128]

Chemical Heat Storage Systems

The hydration and carbonation of metal oxides are the commonest chemical reactions that utilized in this category of thermochemical energy storage systems. The reaction enthalpy for these reactions is typically in the range of 80 to 180 kJ/mol [109]. In Table **12**, potential materials that can be used in chemical reaction storage systems are presented. Due the relatively high charging temperature these reactions can be used in CSP plants for solar heat storage [117].

Table 12. Promising materials for heat storage systems based on chemical reactions [113].

Thermochemical Material (AB)	Solid Reactant (A)	Working Fluid (B)	Energy Storage Density of Thermochemical Material (GJ/m^3)	Charging Reaction Temperature (°C)
$FeCO_3$	FeO	CO_2	2.6	180

(Table 11) cont.....

Thermochemical Material (AB)	Solid Reactant (A)	Working Fluid (B)	Energy Storage Density of Thermochemical Material (GJ/m^3)	Charging Reaction Temperature (°C)
$CaCO_3$	CaO	CO_2	3.3	837
$Fe(OH)_2$	FeO	H_2O	2.2	150
$Ca(OH)_2$	CaO	H_2O	1.9	479

CONCLUSIONS

Thermal Energy Storage (TES) systems can help to balance the mismatch between supply and demand of energy. This Chapter investigated the current situation in TES technology and systems. Three different technologies for TES (sensible, latent and thermochemical) are reviewed.

The utilization of TES can bring significant energy savings and CO_2 emissions reduction. From an economic point of view, a TES system is substantially affected by the type of application and operational needs as well as the number and frequency of storage cycles. In general, latent and thermochemical heat storage systems are more expensive than sensible heat storage systems and are economically viable when a high number of storage cycles is considered.

Taking into account the sensible thermal storage solutions, the technological research as well as the applications with liquids and solids could be claimed as being in relatively mature stage. For buildings applications, the liquids thermal storage are normally utilized as water storage; while the solids could be implemented not only as a separate storage unit, such as packed bed storage, but also as envelope integrated energy storage.

A material can be appropriate for sensible heat storage if it fulfills, high energy density (stored heat over system volume), high thermal conductivity, availability and low cost, chemical stability and environmental friendly.

Thermochemical energy storage systems based in chemical reactions are suitable for medium and high-grade heat storage (above 400 °C). Solid adsorption storage systems are mainly for heating applications exploiting solar energy.

Related applications and projects are also pointed out. TES systems based on sensible heat are rather inexpensive and are applicable to both domestic and industrial systems. The main disadvantage of this technology is the relatively low energy density of the storage medium which leads to large volumes. Latent and thermochemical heat storage systems have great potential due to their higher energy densities but present difficulties (*e.g.* thermal stability, high cost of materials *etc.*) and there is need to be further researched.

CONSENT FOR PUBLICATION

Not applicable.

CONFLICT OF INTERESTS

The authors declare no conflict of interests.

ACKNOWLEDGEMENT

Declared none.

REFERENCES

[1] "World energy outlook", *Agency International Energy,* 2009.

[2] G. Li, *Review of thermal energy storage technologies and experimental investigation of adsorption thermal energy storage for residential application* University of Maryland at College Park.

[3] G. Li, S. Qian, H. Lee, Y. Hwang, and R. Radermacher, "Experimental investigation of energy and exergy performance of short term adsorption heat storage for residential application", *Energy,* vol. 65, pp. 675-691, 2014.
[http://dx.doi.org/10.1016/j.energy.2013.12.017]

[4] P. Arce, M. Medrano, A. Gil, E. Oró, and L. Cabeza, "Overview of thermal energy storage (TES) potential energy savings and climate change mitigation in Spain and Europe", *Appl. Energy,* vol. 88, no. 8, pp. 2764-2774, 2011.
[http://dx.doi.org/10.1016/j.apenergy.2011.01.067]

[5] "Thermal Energy Storage Technology", In: *IEA-ETSAP and IRENA© Technology Brief E17*Abu Dhabi, 2013.

[6] S. Kalaiselvam, and R. Parameshwaran, *Thermal Energy Storage Technologies for Sustainability: Systems Design, Assessment and Applications.* Elsevier: Amsterdam, 2014.

[7] J. Duffie, and W. Beckman, *Solar engineering of thermal processes.* John Wiley & Sons: New York, 2013.
[http://dx.doi.org/10.1002/9781118671603]

[8] C. Van Koppen, J. Thomas, and W. Veltkamp, "The actual benefits of thermally stratified storage in a small and medium size solar system", *Proceedings ISES Biennal Meeting,* 1979 Atlanta

[9] N. Ghaddar, "Stratified storage tank influence on the performance of solar water heating systems tested in Beirut", *Renew. Energy,* vol. 4, no. 8, pp. 911-925, 1994.
[http://dx.doi.org/10.1016/0960-1481(94)90225-9]

[10] S. Furbo, *Educational note U-067.* Department of Civil Engineering, Technical University of Denmark, 2004.

[11] S. Furbo, and L. Shah, "How mixing during hot water draw-offs influence the thermal performance of small solar domestic hot water systems", *Proceedings ISES Solar World Congress,* 2005 Orlando

[12] U. Jordan, and S. Furbo, "Thermal stratification in small solar domestic storage tanks caused by draw-offs", *Sol. Energy,* vol. 78, no. 2, pp. 291-300, 2005.
[http://dx.doi.org/10.1016/j.solener.2004.09.011]

[13] L. Shah, and S. Furbo, "Entrance effect in solar hot water stores", *Sol. Energy,* vol. 75, no. 4, pp. 337-

348, 2003.
[http://dx.doi.org/10.1016/j.solener.2003.04.002]

[14] "Energy Conservation through Energy Storage (ECES) Programme", *International Energy Agency*. [Online]. Available: http://www.iea-eces.org/

[15] S. Edwards, Sensitivity analysis of two solar combisystems using newly developed hot water draw profles-M.A.Sc. thesis, Carleton University: Ottawa, 2014.

[16] P. Pinel, C. Cruickshank, I. Beausoleil-Morrison, and A. Wills, "A review of available methods for seasonal storage of solar thermal energy in residential applications", *Renew. Sustain. Energy Rev.*, vol. 15, pp. 3341-3359, 2011.
[http://dx.doi.org/10.1016/j.rser.2011.04.013]

[17] M. Reuss, "Solar District Heating in Germany - Findings and Prospects", *Proceedings of the ISES Solar World Congress,* 2011 Kassel, Germany
[http://dx.doi.org/10.18086/swc.2011.21.05]

[18] A. Ercan, STORAGE OF THERMAL ENERGY.*Energy Storage Systems," in Encyclopedia of Life Support Systems (EOLLS)*. Eolss Publishers: Oxford, 2006.

[19] Y. Tian, and C. Zhao, "A review of solar collectors and thermal energy storage in solar thermal applications", *Appl. Energy,* vol. 104, pp. 538-553, 2013.
[http://dx.doi.org/10.1016/j.apenergy.2012.11.051]

[20] C. Zhao, and Z. Wu, "Thermal property characterization of a low melting-temperature ternary nitrate salt mixture for thermal energy storage systems", *Sol. Energy Mater. Sol. Cells,* vol. 95, no. 12, pp. 3341-3346, 2011.
[http://dx.doi.org/10.1016/j.solmat.2011.07.029]

[21] Q. Peng, J. Ding, X. Wei, J. Yang, and X. Yang, "The preparation and properties of multi-component molten salts", *Appl. Energy,* vol. 87, no. 9, pp. 2812-2817, 2010.
[http://dx.doi.org/10.1016/j.apenergy.2009.06.022]

[22] T. Wang, D. Mantha, and R. Reddy, "Novel low melting point quaternary eutectic system for solar thermal energy storage", *Appl. Energy,* vol. 102, pp. 1422-1429, 2013.
[http://dx.doi.org/10.1016/j.apenergy.2012.09.001]

[23] A. Gil, M. Medrano, I. Martorell, A. Lazaro, P. Dolado, B. Zalba, and L. Cabeza, "State of the art on high temperature thermal energy storage for power generation. Part 1—Concepts, materials and modellization", *Renew. Sustain. Energy Rev.*, vol. 14, no. 1, pp. 31-55, 2010.
[http://dx.doi.org/10.1016/j.rser.2009.07.035]

[24] *NATIONAL RENEWABLE ENERGY LABORATORY (NREL)*. Accessed 29 10 2016 [Online]. Available: http://www.nrel.gov/csp/solarpaces/project_detail.cfm/projectID=40

[25] U. Herrmann, and R. Kistner, "The AndaSol Project", *Workshop on Thermal Storage for Trough Power Systems,* 2002

[26] P. Tatsidjodoung, N. Pierres, and L. Luo, "A review of potential materials for thermal energy storage in building applications", *Renew. Sustain. Energy Rev.,* vol. 18, pp. 327-349, 2013.
[http://dx.doi.org/10.1016/j.rser.2012.10.025]

[27] J. Sattler, B. Hoffschmidt, M. Günther, and M. Joemann, "Chapter 9 Thermal Energy Storage", *Advanced CSP Teaching Materials..*

[28] Z. Ma, G. Glatzmaier, and M. Mehos, "Fluidized Bed Technology for Concentrating Solar Power With Thermal Energy Storage", *J. Sol. Energy Eng.,* vol. 136, no. 3, 2014.
[http://dx.doi.org/10.1115/1.4027262]

[29] J. Pacheco, S. Showalter, and W. Kolb, "Development of a molten-salt thermocline thermal storage system for parabolic trough plants", *in Proceedings of Solar Forum 2001 Solar Energy: The Power to Choose,* 2001 Washington, DC

[30] S. Zavattoni, M. Barbato, A. Pedretti, G. Zanganeh, and A. Steinfeld, "Effective thermal conductivity and axial porosity distribution of a rock-bed TES system: CFD modeling and experimental validation", *SolarPACES Conference 2012,* 2012 Marrakech

[31] J. Jurinak, and S. Abdel-Khalik, "Sizing phase-change energy storage units for air-based solar heating systems", *Sol. Energy,* vol. 22, no. 4, pp. 355-359, 1979. [http://dx.doi.org/10.1016/0038-092X(79)90189-0]

[32] G. Rizzi, and V. Sharma, "An inexpensive solar collector-storage system for space heating-I. Design Methodologies", *Solar Wind Technol.,* vol. 7, no. 4, pp. 447-455, 1990. [http://dx.doi.org/10.1016/0741-983X(90)90029-2]

[33] A. Saxena, and V. Goel, "Solar Air Heaters with Thermal Heat Storages", *Chinese Journal of Engineering,* 2013. [http://dx.doi.org/10.1155/2013/190279]

[34] R. Boonsu, S. Sukchai, S. Hemavibool, and S. Somkun, "Performance Analysis of Thermal Energy Storage Prototype in Thailand", *Journal of Clean Energy Technologies,* vol. 4, no. 2, pp. 101-106, 2016. [http://dx.doi.org/10.7763/JOCET.2016.V4.261]

[35] I. Dincer, and M. Rosen, *THERMAL ENERGY STORAGE, SYSTEMS AND APPLICATIONS.* John Wiley & Sons: Chichester, 2011.

[36] D. Laing, W-D. Steinmann, R. Tamme, and C. Richter, "Solid media thermal storage for parabolic trough power plants", *Sol. Energy,* vol. 80, no. 10, pp. 1283-1289, 2006. [http://dx.doi.org/10.1016/j.solener.2006.06.003]

[37] A. Willemsen, and M. Godschalk, *PA EUROPE.*Accessed 30 10 2016 [Online]. Available: http://pa-europecom.webhosting.be/www.pa-international.org/htdocs//documents/MicrosoftWord-managementofhotandcoldwater-GuusWillemsen.pdf

[38] R. Caljé, Future use of Aquifer Thermal Energy Storage below the historic centre of Amsterdam-M.Sc Thesis, Delft University of Technology: Amsterdam, 2010.

[39] *ClimateTechWiki A CLEAN TECHNOLOGY PLATFORM.*Accessed 30 10 2016 [Online]. Available: http://www.climatetechwiki.org/technology/jiqweb-utes

[40] A. Snijders, *Toronto and Region: Conservation for The Living City.*Accessed 30 10 2016 [Online]. Available: http://www.trca.on.ca/dotAsset/16551.pdf

[41] I. Dincer, and M. Rosen, *Exergy Analysis of Heating, Refrigerating, and Air Conditioning Methods and Applications.* Elsevier: Amsterdam, 2015.

[42] B. Nordell, *"Large-scale Thermal Energy Storage," in WinterCities'2000.* Energy and Environment: Lulea, 2000.

[43] B. Nordell, *Borehole heat store design optimization-Doctoral Tsesis.* Lulea University of Technology: Lulea, 1994.

[44] J. Nussbicker, W. Heidemann, and H. Mueller-Steinhagen, "Monitoring results and operational experiences for a central solar district heating system with borehole thermal energy store in Neckarsulm (Germany)", In: *ECOSTOCK*New Jersey, 2006.

[45] B. Sibbit, D. McClenahan, R. Djebbar, and K. Paget, "Groundbreaking Solar-Drake Landing Solar Community: Okotoks, Alberta, Canada", *High Performing Buildings Magazine,* pp. 36-47, Summer 2015.

[46] T. Mahlia, T. Saktisahdan, A. Jannifar, M. Hasan, and H. Matseelar, "A review of available methods and development on energy storage; technology update", *Renew. Sustain. Energy Rev.,* vol. 32, pp. 532-545, 2014. [http://dx.doi.org/10.1016/j.rser.2014.01.068]

[47] P. Salunkhe, and P. Shembekar, "A review on effect of phase change material encapsulation on the thermal performance of a system", *Renew. Sustain. Energy Rev.,* vol. 16, pp. 5603-5616, 2012. [http://dx.doi.org/10.1016/j.rser.2012.05.037]

[48] S. Chandel, and T. Agarwal, "Review of current state of research on energy storage, toxicity, health hazards and commercialization of phase changing materials", *Renew. Sustain. Energy Rev.,* vol. 67, pp. 581-596, 2017. [http://dx.doi.org/10.1016/j.rser.2016.09.070]

[49] Z. Zhou, Z. Zhang, J. Zuo, K. Huang, and L. Zhang, "Phase change materials for solar thermal energy storage in residential buildings in cold climate", *Renew. Sustain. Energy Rev.,* vol. 48, pp. 692-703, 2015. [http://dx.doi.org/10.1016/j.rser.2015.04.048]

[50] B. He, and F. Setterwall, "Technical grade paraffin waxes as phase change materials for cool thermal storage and cool storage systems capital cost estimation", *Energy Convers. Manage.,* vol. 43, no. 13, pp. 1709-1723, 2002. [http://dx.doi.org/10.1016/S0196-8904(01)00005-X]

[51] B. Cárdenas, and N. León, "High temperature latent heat thermal energy storage: phase change materials, design considerations and performance enhancement techniques", *Renew. Sustain. Energy Rev.,* vol. 27, pp. 724-737, 2013. [http://dx.doi.org/10.1016/j.rser.2013.07.028]

[52] X. Fang, L. Fan, Q. Ding, X. Yao, Y. Wu, and J. Hou, "Thermal energy storage performance of paraffin-based composite phase change materials filled with hexagonal boron nitride nanosheets", *Energy Convers. Manage.,* vol. 80, pp. 103-109, 2014. [http://dx.doi.org/10.1016/j.enconman.2014.01.016]

[53] M. Mehrali, S. Tahan Latibari, T. Mahlia, E. Sadeghinezhad, and H. Metselaar, "Preparation of nitrogen-doped graphene/palmitic acid shape stabilized composite phase change material with remarkable thermal properties for thermal energy storage", *Appl. Energy,* vol. 135, pp. 339-349, 2014. [http://dx.doi.org/10.1016/j.apenergy.2014.08.100]

[54] A. Sarı, C. Alkan, and C. Bilgin, "Micro/nano encapsulation of some paraffin eutectic mixtures with poly(methyl methacrylate) shell: preparation, characterization and latent heat thermal energy storage properties", *Appl. Energy,* vol. 136, pp. 217-227, 2014. [http://dx.doi.org/10.1016/j.apenergy.2014.09.047]

[55] S. Kamali, "Review of free cooling system using phase change material for building", *Energy Build.,* vol. 80, pp. 131-136, 2014. [http://dx.doi.org/10.1016/j.enbuild.2014.05.021]

[56] M. Lachheb, M. Karkri, F. Albouchi, F. Mzali, and S. Nasrallah, "Thermophysical properties estimation of paraffin/graphite composite phase change material using an inverse method", *Energy Convers. Manage.,* vol. 82, pp. 131-136, 2014. [http://dx.doi.org/10.1016/j.enconman.2014.03.021]

[57] A. Sharma, V. Tyagi, C. Chen, and D. Buddhi, "Review on thermal energy storage with phase change materials and applications", *Renew. Sustain. Energy Rev.,* vol. 13, pp. 318-345, 2009. [http://dx.doi.org/10.1016/j.rser.2007.10.005]

[58] M. Kenisarin, "Thermophysical properties of some organic phase change materials for latent heat storage. A review", *Sol. Energy,* vol. 107, pp. 553-575, 2014. [http://dx.doi.org/10.1016/j.solener.2014.05.001]

[59] A. Sarı, "Form-stable paraffin/high density polyethylene composites as solid-liquid phase change material for thermal energy storage: preparation and thermal properties", *Energy Convers. Manage.,* vol. 45, pp. 2033-2042, 2004. [http://dx.doi.org/10.1016/j.enconman.2003.10.022]

[60] Y. Yuan, N. Zhang, W. Tao, X. Cao, and Y. He, "Fatty acids as phase change materials: a review", *Renew. Sustain. Energy Rev.,* vol. 29, pp. 482-498, 2014. [http://dx.doi.org/10.1016/j.rser.2013.08.107]

[61] A. Waqas, and Z. Ud Din, "Phase change material (PCM) storage for free cooling of buildings—a review", *Renew. Sustain. Energy Rev.,* vol. 18, pp. 607-625, 2013. [http://dx.doi.org/10.1016/j.rser.2012.10.034]

[62] P. Tatsidjodoung, N. Le Pierrès, and L. Luo, "A review of potential materials for thermal energy storage in building applications", *Renew. Sustain. Energy Rev.,* vol. 18, pp. 327-349, 2013. [http://dx.doi.org/10.1016/j.rser.2012.10.025]

[63] D. David, K. Johannes, J. Roux, and F. Kuznik, "A review on phase change materials integrated in building walls", *Renew. Sustain. Energy Rev.,* vol. 15, pp. 379-391, 2011. [http://dx.doi.org/10.1016/j.rser.2010.08.019]

[64] R. Baetens, B. Jelle, and A. Gustavsen, "Phase change materials for building applications: a state-o--the-art review", *Energy Build.,* vol. 42, pp. 1361-1368, 2010. [http://dx.doi.org/10.1016/j.enbuild.2010.03.026]

[65] V. Tyagi, and D. Buddhi, "PCM thermal storage in buildings: a state of art", *Renew. Sustain. Energy Rev.,* vol. 11, pp. 1146-1166, 2007. [http://dx.doi.org/10.1016/j.rser.2005.10.002]

[66] H. Fauzi, H. Metselaar, T. Mahlia, and M. Silakhori, "Sodium laurate enhancements the thermal properties and thermal conductivity of eutectic fatty acid as phase change material (PCM)", *Sol. Energy,* vol. 102, pp. 333-337, 2014. [http://dx.doi.org/10.1016/j.solener.2013.07.001]

[67] G. Ferrer, A. Solé, C. Barreneche, I. Martorell, and L. Cabeza, "Review on the methodology used in thermal stability characterization of phase change materials", *Renew. Sustain. Energy Rev.,* vol. 50, pp. 665-685, 2015. [http://dx.doi.org/10.1016/j.rser.2015.04.187]

[68] H. Akeiber, P. Nejat, and M.Z. Abd, "C. Majid, A. Mazlan and e. al, "A review on phase change material (PCM) for sustainable passive cooling in building envelopes", *Renew. Sustain. Energy Rev.,* vol. 60, pp. 1470-1497, 2016. [http://dx.doi.org/10.1016/j.rser.2016.03.036]

[69] S. Jeong, O. Chung, S. Yu, and S. Kim, "Improvement of the thermal properties of Bio-based PCM using exfoliated graphite nanoplatelets", *Sol. Energy Mater. Sol. Cells,* vol. 117, pp. 87-92, 2013. [http://dx.doi.org/10.1016/j.solmat.2013.05.038]

[70] J. Shi, M. Ger, Y. Liu, Y. Fan, N. Wen, C. Lin, and N. Pu, "Improving the thermal conductivity and shape-stabilization of phase change materials using nanographite additives", *Carbon,* vol. 51, pp. 365-372, 2013. [http://dx.doi.org/10.1016/j.carbon.2012.08.068]

[71] Z. Huang, X. Gao, T. Xu, Y. Fang, and Z. Zhang, "Thermal property measurement and heat storage analysis of LiNO3/KCl-EG composite phase change material", *Appl. Energy,* vol. 115, pp. 265-271, 2014. [http://dx.doi.org/10.1016/j.apenergy.2013.11.019]

[72] X. Wang, Q. Guo, J. Wang, Y. Zhong, L. Wang, X. Wei, and L. Liu, "Thermal conductivity enhancement of form-stable phase-change composites by milling of expanded graphite, micro-capsules and polyethylene", *Renew. Energy,* vol. 60, pp. 506-509, 2013. [http://dx.doi.org/10.1016/j.renene.2013.05.038]

[73] M. Li, "A nano-graphite/paraffin phase change material with high thermal conductivity", *Appl. Energy,* vol. 106, pp. 25-30, 2013. [http://dx.doi.org/10.1016/j.apenergy.2013.01.031]

[74] J. Johansen, M. Dannemand, W. Kong, J. Fan, J. Dragsted, and S. Furbo, "Thermal conductivity enhancement of sodium acetate trihydrate by adding graphite powder and the effect on stability of supercooling", *Energy Procedia,* vol. 70, pp. 249-256, 2015. [http://dx.doi.org/10.1016/j.egypro.2015.02.121]

[75] T. Li, J. Lee, R. Wang, and Y. Kang, "Enhancement of heat transfer for thermal energy storage application using stearic acid nanocomposite with multi-walled carbon nanotubeσ", *Energy,* vol. 55, pp. 752-761, 2013. [http://dx.doi.org/10.1016/j.energy.2013.04.010]

[76] M. Li, M. Chen, Z. Wu, and J. Liu, "Carbon nanotube grafted with polyalcohol and its influence on the thermal conductivity of phase change material. Energy", *Energy Convers. Manage.,* vol. 83, pp. 325-329, 2014. [http://dx.doi.org/10.1016/j.enconman.2014.04.002]

[77] J. Wang, H. Xie, Z. Xin, and Y. Li, "Increasing the thermal conductivity of palmitic palmitic acid based phase change materials with carbon nanotubes as fillers", *Sol. Energy,* vol. 84, pp. 339-34, 2010. [http://dx.doi.org/10.1016/j.solener.2009.12.004]

[78] Y. Zhong, M. Zhou, F. Huang, T. Lin, and D. Wan, "Effect of graphene aerogel on thermal behavior of phase change materials for thermal management", *Sol. Energy Mater. Sol. Cells,* vol. 113, pp. 195-200, 2013. [http://dx.doi.org/10.1016/j.solmat.2013.01.046]

[79] Y. Fu, Z. He, D. Mo, and S. Lu, "Thermal conductivity enhancement of epoxy adhesive using graphene sheets as additives", *Int. J. Therm. Sci.,* vol. 86, pp. 276-283, 2014. [http://dx.doi.org/10.1016/j.ijthermalsci.2014.07.011]

[80] X. Xiao, P. Zhang, and M. Li, "Effective thermal conductivity of open-cell metal foams impregnated with pure paraffin for latent heat storage. Int J Therm Sci", *Int. J. Therm. Sci.,* vol. 81, pp. 94-105, 2014. [http://dx.doi.org/10.1016/j.ijthermalsci.2014.03.006]

[81] S. Thapa, S. Chukwu, A. Khaliq, and L. Weiss, "Fabrication and analysis of small-scale thermal energy storage with conductivity enhancement", *Energy Convers. Manage.,* vol. 79, pp. 161-170, 2014. [http://dx.doi.org/10.1016/j.enconman.2013.12.019]

[82] N. Şahan, M. Fois, and H. Paksoy, "Improving thermal conductivity phase change materials-a study of paraffin nanomagnetite composites", *Sol. Energy Mater. Sol. Cells,* vol. 137, pp. 61-67, 2015. [http://dx.doi.org/10.1016/j.solmat.2015.01.027]

[83] W. Yu, H. Xie, L. Chen, and Y. Li, "Enhancement of thermal conductivity of kerosene-based Fe3O4 nanofluids prepared via phase-transfer method", *Colloids Surf. A Physicochem. Eng. Asp.,* vol. 355, pp. 109-113, 2010. [http://dx.doi.org/10.1016/j.colsurfa.2009.11.044]

[84] W. Wang, X. Yang, Y. Fang, J. Ding, and J. Yan, "Enhanced thermal conductivity and thermal performance of form-stable composite phase change materials by using b-aluminum nitride", *Appl. Energy,* vol. 86, pp. 1196-1200, 2009. [http://dx.doi.org/10.1016/j.apenergy.2008.10.020]

[85] P. Sharma, I. Baek, T. Cho, S. Park, and K. Lee, "Enhancement of thermal conductivity of ethylene glycol based silver nanofluids", *Powder Technol.,* vol. 208, pp. 7-19, 2011. [http://dx.doi.org/10.1016/j.powtec.2010.11.016]

[86] J. Prakash, H. Garg, and G. Datta, "A solar water heater with a built-in latent heat storage", *Energy Convers. Manage.,* vol. 25, pp. 51-56, 1985. [http://dx.doi.org/10.1016/0196-8904(85)90069-X]

[87] P. Chaurasia, "Phase change material in solar water heater storage system", *Proceedings of the 8th International Conference on Thermal Energy Storage,* 2000 Stuttgart, Germany

[88] K. Kayugz, "Experimental and theoretical investigation of latent heat storage for water based solar heating systems", *Energy Convers. Manage.*, vol. 36, no. 5, pp. 315-323, 1995. [http://dx.doi.org/10.1016/0196-8904(95)98896-U]

[89] E-B. Mettawee, and G. Assassa, "Experimental study of a compact PCM solar collector", *Energy,* vol. 31, pp. 2958-2968, 2006. [http://dx.doi.org/10.1016/j.energy.2005.11.019]

[90] L. Cabeza, M. Ibanez, C. Sole, J. Roca, and M. Nogues, "Experimentation with a water tank including a PCM module", *Sol. Energy Mater. Sol. Cells,* vol. 90, pp. 1273-1282, 2006. [http://dx.doi.org/10.1016/j.solmat.2005.08.002]

[91] "Development of greenhouse bulk drying systems for solar energy utilization and planted mechanization", In: *ASAE paper no. 75-1018, St. Joseph, MI*, 1975.

[92] M. Kern, and R. Aldrich, Phase change energy storage in a greenhouse solar heating system.," in *ASAE paper no. 79-4028*, St. Joseph, MI, 1979.

[93] K. Hhyum, "Utilization of latent heat storage materials for the high concentrated thermal energy storage," Report no. 1988.5–19990.5;, 1988.

[94] T. Takakura, and H. Nishina, "A solar greenhouse with phase change energy storage and a microcomputer control system", *Acta Hortic.,* no. 115, pp. 583-590, 1981. [Energy in Protected Cultivation]. [http://dx.doi.org/10.17660/ActaHortic.1981.115.65]

[95] D. Neeper, "Thermal dynamics of wallboard with latent heat storage", *Sol. Energy,* vol. 68, pp. 393-403, 2000. [http://dx.doi.org/10.1016/S0038-092X(00)00012-8]

[96] M. Ahmad, A. Bontemps, H. Sallee, and D. Quenard, "Thermal testing and numerical simulation of a prototype cell using light wallboards coupling vacuum isolation panels and phase change material", *Energy Build.,* vol. 38, pp. 673-681, 2006. [http://dx.doi.org/10.1016/j.enbuild.2005.11.002]

[97] C. Chen, H. Guo, Y. Liu, H. Yue, and C. Wang, "A new kind of phase change material (PCM) for energy-storing wallboard", *Energy Build.,* vol. 40, pp. 882-890, 2008. [http://dx.doi.org/10.1016/j.enbuild.2007.07.002]

[98] F. Kuznik, J. Virgone, and J. Noel, "Optimization of a phase change material wallboard for building use", *Appl. Therm. Eng.,* vol. 28, pp. 1291-1298, 2008. [http://dx.doi.org/10.1016/j.applthermaleng.2007.10.012]

[99] W. Xiao, X. Wang, and Y. Zhang, "Analytical optimization of interior PCM for energy storage in lightweight passive solar room", *Appl. Energy,* vol. 86, pp. 2013-2018, 2009. [http://dx.doi.org/10.1016/j.apenergy.2008.12.011]

[100] K. Lee, M. Medina, E. Raith, and X. Sun, "Assessing the integration of a thin phase change material (PCM) layer in a residential building wall for heat transfer reduction and management", *Appl. Energy,* vol. 137, pp. 669-706, 2015. [http://dx.doi.org/10.1016/j.apenergy.2014.09.003]

[101] http://www.glassxpcm.com/footer_posts/how-the-pcm-works Online

[102] X. Xu, Y. Zhang, K. Lin, H. Di, and R. Yang, "Modelling and simulation on the thermal performance of shape-stabilized phase change material floor used in passive solar buildings", *Energy Build.,* vol. 37, pp. 1084-1091, 2005. [http://dx.doi.org/10.1016/j.enbuild.2004.12.016]

[103] Y. Hong, and G. Xin-shi, "Preparation of polyethylene-paraffin compound as a form-stable solid-liquid phase change material", *Sol. Energy Mater. Sol. Cells,* vol. 64, pp. 37-44, 2000. [http://dx.doi.org/10.1016/S0927-0248(00)00041-6]

[104] G. Song, S. Ma, G. Tang, Z. Yin, and X. Wang, "Preparation and characterization of flame retardant form-stable phase change materials composed by EPDM paraffin and nano magnesium hydroxide", *Energy,* vol. 35, pp. 2179-2183, 2010. [http://dx.doi.org/10.1016/j.energy.2010.02.002]

[105] Y. Zhang, K. Lin, R. Yang, H. Di, and Y. Jiang, "Preparation, thermal performance and application of shape-stabilized PCM in energy efficient buildings", *Energy Build.,* vol. 38, pp. 1262-1269, 2006. [http://dx.doi.org/10.1016/j.enbuild.2006.02.009]

[106] G. Fang, Z. Chen, and H. Li, "Synthesis and properties of microencapsulated paraffin composites with SiO2 shell as thermal energy storage materials", *Chem. Eng. J.,* vol. 163, pp. 154-159, 2010. [http://dx.doi.org/10.1016/j.cej.2010.07.054]

[107] V. Tyagi, S. Kaushik, S. Tyagi, and T. Akiyama, "Development of phase change materials based microencapsulated technology for buildings: a review", *Renew. Sustain. Energy Rev.,* vol. 15, pp. 1373-1379, 2011. [http://dx.doi.org/10.1016/j.rser.2010.10.006]

[108] C. Zhao, and G. Zhang, "Review on microencapsulated phase change materials (MEPCMs): fabrication, characterization and applications", *Renew. Sustain. Energy Rev.,* vol. 15, pp. 3813-3832, 2011. [http://dx.doi.org/10.1016/j.rser.2011.07.019]

[109] J. Cot-Gores, A. Castell, and L. Cabeza, "Thermochemical energy storage and conversion: A-state--f-the-art review of the experimental research under practical conditions", *Renew. Sustain. Energy Rev.,* vol. 16, no. 7, pp. 5207-5224, 2012. [http://dx.doi.org/10.1016/j.rser.2012.04.007]

[110] J. Hadorn, "Advanced storage concepts for active solar energy-IEA-SHC Task 32 2003-2007", *1st International Congress on Heating, Cooling and Buildings,* 2008 Lisboa.

[111] K. N'Tsoukpoe, H. Liu, N. Le Pierres, and L. Luo, "A review on long-term sorption solar energy storage", *Renew. Sustain. Energy Rev.,* vol. 13, no. 9, pp. 2385-2396, 2009. [http://dx.doi.org/10.1016/j.rser.2009.05.008]

[112] B. Mette, H. Kerskes, and H. Drück, "Concepts of long-term thermochemical energy storage for solar thermal applications - Selected examples", *Energy Procedia,* vol. 30, pp. 321-330, 2012. [http://dx.doi.org/10.1016/j.egypro.2012.11.038]

[113] A. Abedin, and M. Rosen, "A Critical Review of Thermochemical Energy Storage Systems", *Open Renewable Energy Journal,* vol. 8, pp. 42-46, 2011. [http://dx.doi.org/10.2174/1876387101004010042]

[114] N. Yu, R. Wang, and L. Wang, "Sorption thermal storage for solar energy", *Pror. Energy Combust. Sci.,* vol. 39, no. 5, pp. 489-514, 2013. [http://dx.doi.org/10.1016/j.pecs.2013.05.004]

[115] L. Wang, R. Wang, and R. Oliveira, "A review on adsorption working pairs for refrigeration", *Renew. Sustain. Energy Rev.,* vol. 13, no. 3, pp. 518-534, 2009. [http://dx.doi.org/10.1016/j.rser.2007.12.002]

[116] Y. Aristov, "Challenging offers of material science for adsorption heat transformation: A review", *Appl. Therm. Eng.,* vol. 50, no. 2, pp. 1610-1618, 2013. [http://dx.doi.org/10.1016/j.applthermaleng.2011.09.003]

[117] C. Prieto, P. Cooper, A. Inés Fernández, and L. Cabeza, "Review of technology: Thermochemical energy storage for concentrated solar power plants", *Renew. Sustain. Energy Rev.,* vol. 60, pp. 909--, 2016. [http://dx.doi.org/10.1016/j.rser.2015.12.364]

[118] D. Lefebvre, and F. Handan Tezel, "A review of energy storage technologies with a focus on adsorption thermal energy storage processes for heating applications", *Renew. Sustain. Energy Rev.,*

vol. 67, pp. 116-125, 2017.
[http://dx.doi.org/10.1016/j.rser.2016.08.019]

[119] J. Xu, R. Wang, and Y. Li, *A review of available technologies for seasonal thermal energy storage.* [http://dx.doi.org/10.1016/j.solener.2013.06.006]

[120] G. Gartler, D. Jähnig, G. Purkarthofer, and W. Wagner, *Development of a High Energy Density Sorption Storage System*, 2004.

[121] K. Iammak, W. Wongsuwan, and T. Kiatsiriroj, "Investigation of Modular Chemical Energy Storage Performance", *The Joint International Conference on "Sustainable Energy and Environment (SEE)", Hua Hin,* 2004

[122] R. de Boer, W. Haije, J. Veldhuis, and S. Smeding, *Solid-sorption cooling with integrated thermal storage: The SWEAT prototype*, 2004.

[123] R. Weber, and V. Dorer, "Long-term heat storage with NaOH", *Vacuum,* vol. 82, no. 7, pp. 708-716, 2008.
[http://dx.doi.org/10.1016/j.vacuum.2007.10.018]

[124] D. Dicaire, *"Long Term Thermal Energy Storage in Adsorbent Beds for Solar Heating Applications-Master's thesis, Ottawa"*, University of Ottawa, 2010.

[125] D. Zhu, H. Wu, and S. Wang, "Experimental study on composite silica gel supported CaCl2 sorbent for low grade heat storage", *Int. J. Therm. Sci.,* vol. 45, no. 8, pp. 804-813, 2006.
[http://dx.doi.org/10.1016/j.ijthermalsci.2005.10.009]

[126] B. Ugur, *"Thermal energy storage in adsorbent beds-Master's thesis, Ottawa"*, University of Ottawa, 2013.

[127] H. Kerskes, B. Mette, S. Asenbeck, H. Drück, and H. Müller-Steinhagen, *Experimental and numerical investigation on thermo-chemical heat storage*, 2010.
[http://dx.doi.org/10.18086/eurosun.2010.16.14]

[128] J. Jänchen, D. Ackermann, H. Stach, and W. Brösicke, "Studies of the water adsorption on Zeolites and modified mesoporous materials for seasonal storage of solar heat", *Sol. Energy,* vol. 76, no. 1-3, pp. 339-344, 2004.
[http://dx.doi.org/10.1016/j.solener.2003.07.036]

[129] S. Hongois, F. Kuznik, P. Stevens, and J-J. Roux, "Development and characterisation of a new MgSO4–zeolite composite for long-term thermal energy storage", *Sol. Energy Mater. Sol. Cells,* vol. 95, no. 7, pp. 1831-1837, 2011.
[http://dx.doi.org/10.1016/j.solmat.2011.01.050]

CHAPTER 3

Solar Energy Utilization in Buildings

Christos Tzivanidis* and **Evangelos Bellos**

Solar Energy Laboratory, Thermal department, School of Mechanical Engineering, National Technical University of Athens, Zografou, Heroon Polytechniou 9, 15780 Athens, Greece

Abstract: Solar energy is one of the most promising renewable energy sources. Building sector is one of the most suitable candidates for utilizing this energy source because of its abundance. More specifically, solar energy is able to be used for electricity production, as well as for covering the heating and the cooling loads of the buildings. In this chapter, new and innovative ideas about the adoption of solar energy systems in buildings are presented. Simple and low cost solar collectors which can produce heating in low and medium temperatures levels are analyzed. Emphasis is given in the utilization of Phase Change Materials, as well as in the utilization of solar assisted heat pumps. Moreover, innovative passive heating systems, as Trombe wall are presented with detail.

Keywords: Building Thermal Behaviour, Heat Pumps, PCM, Solar Energy.

INTRODUCTION

The new lifestyle trends lead to increased energy consumption in all the domains of human activity. Energy consumption in the building sector is about the 1/3 [1, 2] of the worldwide energy consumption, a high amount with an important impact on the greenhouse gas emissions. Moreover, the use of fossil fuels and of electricity in the buildings makes more intense the fossil fuel depletion and the increase in electricity price respectively.

The use of sustainable and renewable energy sources in the building sector is vital in order to fulfil the energy reduction goals which have been set by the governments in the last years. A characteristic example is the European Union (EU) agreements for reduction in the greenhouse emissions and in fossil fuels. More specifically, directives as the 2010/31/EU [3] and the 2012/27/EU [4] on the building energy performance and efficiency respectively have been voted among the EU countries. According to these agreements, the primary energy consum-

* **Corresponding Author Christos Tzivanidis:** Solar Energy Laboratory, Thermal department, School of Mechanical Engineering, National Technical University of Athens, Zografou, Heroon Polytechniou 9, 15780 Athens, Greece; Tel: +302107723369; Fax: +302107721260; Email: ctzivan@central.ntua.gr

Emmanuel D. Rogdakis & Irene P. Koronaki (Eds.)

ption in buildings has to be reduced in the next years by using renewable energy sources and utilizing more efficient technologies. Moreover, according to Kyoto's protocol [5] the buildings have to be 20% more efficient up to 2020 by using renewable energy sources mainly.

Buildings energy needs are variable and different during the year period. Electricity, heating and cooling, as well as fresh and heated water are the main needs of the buildings. The proportion of these energy demands are different among the countries, fact that leads to different energy strategies in every location. Countries with high temperature levels demand high amounts of electricity during the summer period, while cold locations need high amounts of fuels in the winter periods. The needs for domestic hot water (DHW) are influenced by the use of the building, with the commercial building to present lower needs than the residential.

Solar energy is a flexible renewable energy source which can be converted to useful heat or to electricity with developed and low cost technologies. Solar thermal collectors are the devices which capture the solar energy and transform it partially to useful heat. Different kinds of solar collectors can be selected in every application and the main decision criterion is the demanded temperature levels. In low temperatures, flat plate collectors or evacuated tube collectors are usually preferred, while concentrating collector can be used in medium and high temperature level applications.

Moreover, the recent designs in buildings aim to utilize and manage the incident solar irradiation properly in order to decrease the building energy needs for heating and cooling. More specifically, passive heating systems, as Trombe walls, are structural components similar to solar collectors which capture the solar irradiation and transfer it in the building with the proper time lag. Other innovative designs include phase change materials or photovoltaics panels in the building envelope.

WATER HEATING AND OTHER PROCESSES

Innovative Solar Collectors

The conventional technologies for hot water production are the flat plate collectors (FPC) and the evacuated tube collectors (ETC). The last years, the use of compound parabolic collectors with low concentration ratio has gained more and more attention in applications for hot water production. Innovative ideas have been applied in order to construct a low cost and simultaneously efficient system.

A compound parabolic collector with glass cover is depicted in Fig. (**1**). This collector is a low cost system which can lead to high thermal performance. The concentration ratio of this collector is usually close to 3, while it can reach the maximum value of 5 in some cases. When the concentration ratio takes greater values there is need of a tracking system in order to keep its optical efficiency in high levels. Generally, this collector is able to operate efficiently with seasonal or monthly tracking in the north-south direction, when its longitudinal axis is located in west-east direction. The absorber operates as storage tank and includes great amounts of water. The storage capacity of the system is determined by the absorber diameter which is fully connected with the concentration ratio; greater diameter leads to lower concentration ratio due to greater absorbing area.

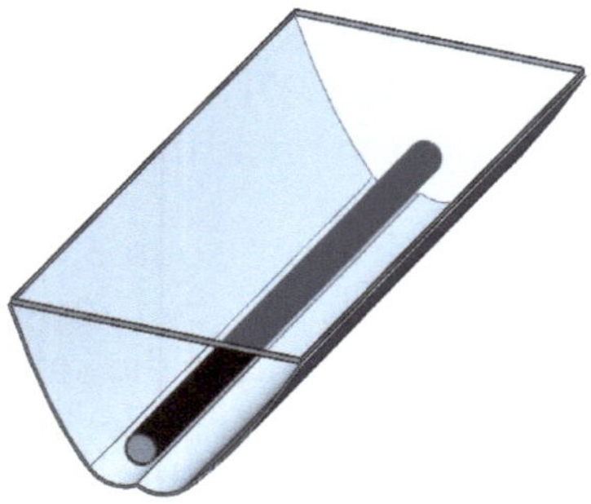

Fig. (1). Glazed compound parabolic collector with tubular absorber.

The use of asymmetric reflectors in a compound parabolic concentrator is an interesting idea which can be applied in stationary collectors. Fig. (**2**) illustrates a simple asymmetric collector. This system has two different reflectors which make the collector to operate for a great range of solar incident angles. These collectors have been studied from Tripanagnostopoulos, Souliotis *et al.* [6, 7] some years before and their design is still under investigation by many researchers.

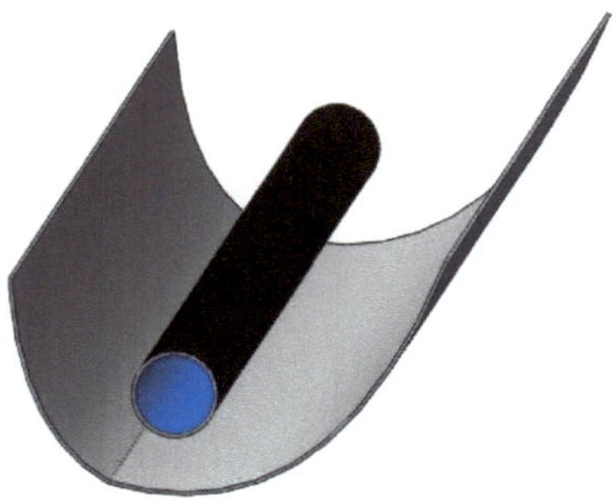

Fig. (2). A simple example of an asymmetric reflector with tubular receiver.

The use of an asymmetric reflector with air gas is way for improving the previous system. Fig. (**3**) shows an innovative system which uses a non-usual reflector with an air trap in order to have greater thermal performance. The air trap is located in the highest place of the collector because the warmest air, due to its low density, is concentrated in this area, something that leads to lower convection thermal losses. Also Tripanagnostopoulos and Souliotis have performed many studies about this idea [6, 7].

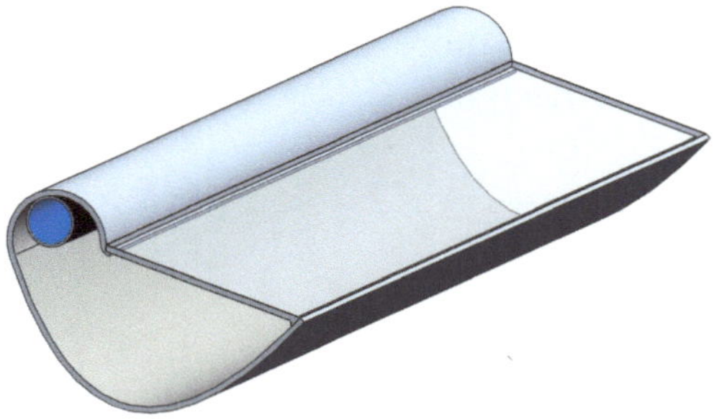

Fig. (3). Asymmetric solar collector with air trap.

The same research team has also proposed the use of two or more storage tank inside the asymmetric collector. Fig. (**4**) exhibits a system with two storage tanks. The advantage of this configuration is the better performance during the year because of the different sun elevation from season to season.

The described solar collectors in Fig. (**1**) to Fig. (**4**) are cheaper technologies than the conventional FPC and their construction is less complex. However, the storage efficiency is lower because of the high thermal losses in the storage during the night.

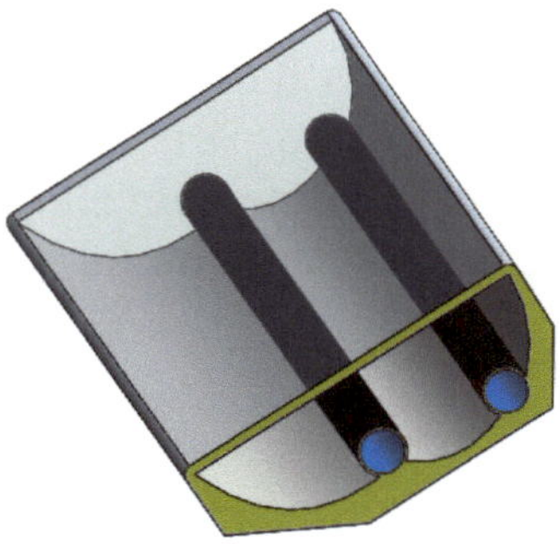

Fig. (4). Asymmetrical design with two storage tanks.

A recent idea for compound parabolic collectors which operate without tracking is suggested by Bellos *et al.* [8]. They proposed that it is better to use two different reflectors than one in order to exploit the solar energy better during the year period. More specifically, the two reflectors have different slopes in order to operate optimally in different seasons. Fig. (**5**) shows the conventional system

(Fig. **5a**) and the new design (Fig. **5b**). In Fig. (**5a**), there is a usual CPC with slope 38°; the optimum value for Athens, while in Fig. (**5b**) there are two reflectors with slopes 20° and 56°. The reflector with 20° slope (low value) performs better the summer period when the sun is high, while the reflector with 56° slope (high value) performs better the winter period. According to Ref [8], the new system is able to create a more uniform profile during the year period, leading to more sustainable design.

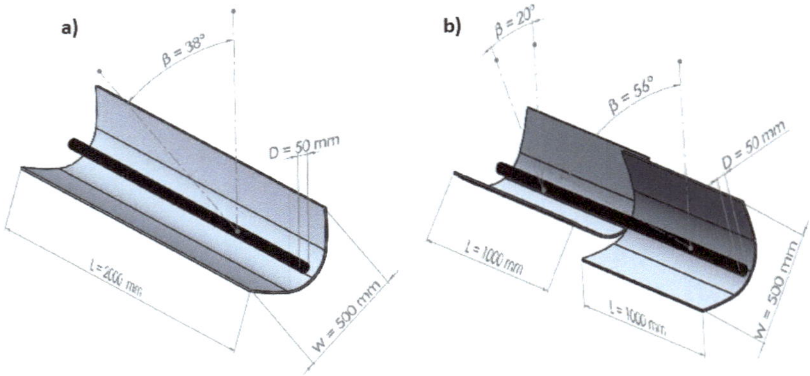

Fig. (5). CPC collectors a) conventional CPC b) innovative CPC with two reflectors.

Two other innovative designs for CPCs are given in Fig. (**6**). Fig. (**6a**) shows a triple reflector which is the improved system of the proposed double previously (Fig. **6b**). Moreover, Fig. (**6b**) shows a parabolic curved reflector which slope changes continuously from the one side to the other. Both these configurations aim to have different optimum parts during the year. The parts with the great slope perform better in the winter, the parts with the small slope are ideal for summer months, while the parts with medium slopes (generally close to latitude of the examined location) are the best solution for spring and autumn.

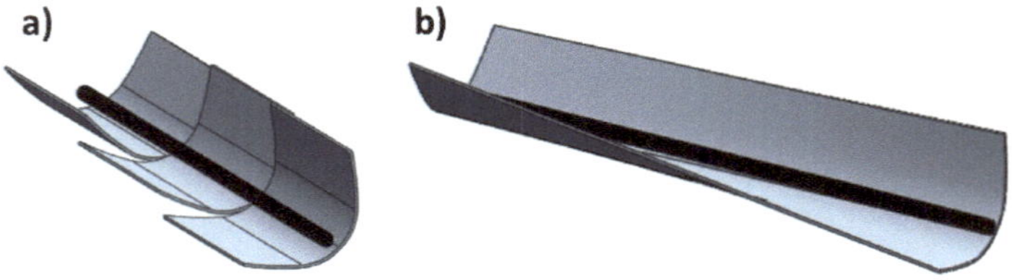

Fig. (6). Innovative reflectors for CPC a) triple reflector b) curved parabolic reflector.

Solar Cookers

In the last years, a lot of research has been focused on the development of efficient solar cookers. These devices utilize solar energy for food production and they can be used in areas without grid or with restriction in electricity consumption. The basic idea is to use reflectors in order to concentrate the solar energy over a great area which is a volumetric absorber. This area is the outer surface of a cooking device and food is included in this object. The total process has a duration which is depended on the available solar irradiation intensity and the energy needs of every case.

Fig. (**7**) shows a solar dish reflector which is used in order to concentrate the solar energy in an object which is located close to its focus. This system needs tracking during the day in order to utilize the solar energy with the optimum way. The weakness of this configuration is the great thermal losses, fact that leads to greater preparation time. This solar cooker has been characterized as innovative, according to a recent study of Regattieri *et al.* [9].

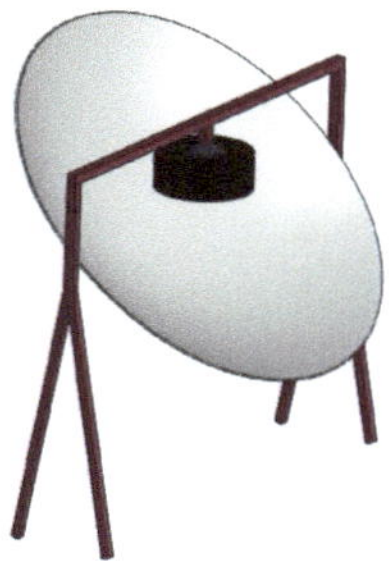

Fig. (7). Solar dish cooker.

Fig. (**8**) illustrates a different solar cooker with a secondary booster reflector. This configuration operates in stationary mode, something with negative impact on the optical performance. On the other hand, this system is simpler than the previous of Fig. (**7**). Moreover, the cooking device is located in an indoor space which is closed; a design with lower convection losses. This solar cooker has been proposed and examined by Harmin *et al.* [10].

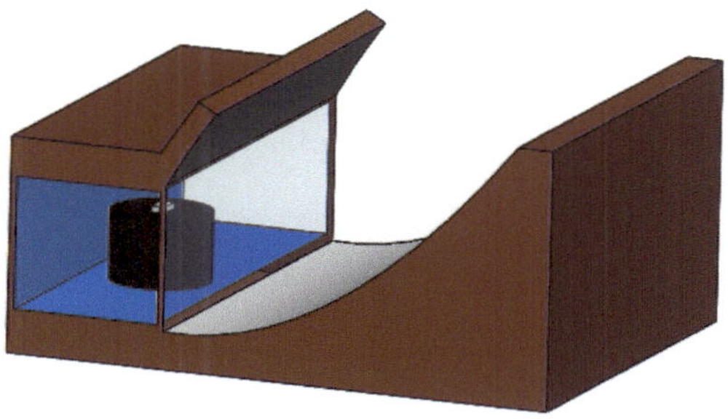

Fig. (8). Solar cooker with boost reflector.

Solar Stills

Solar stills are devices which use solar energy for drinking water production. These devices are more common in arid regions where there is no fresh water supply and the solar energy potential is high. In this technology, brackish water is placed in an enclosed box with a glass cover upper surface. Solar energy passes through the cover and warms up the water. A small quantity of this water is evaporated and it reaches in the cover where it is condensed. The condensed water is fresh water which is collected and it can be used. Fig. (**9**) shows a usual solar still, as described.

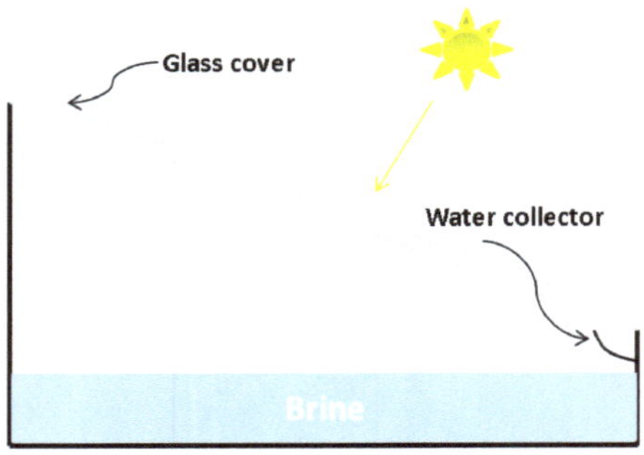

Fig. (9). Usual solar still.

The idea of solar stills has been analyzed by many researchers. New and innovative ideas have been tested experimentally and numerically in order to optimize the solar still efficiency. Dashtban and Tabrizi [11] investigated the use of phase change materials (PCM) in solar stills in order to enhance the performance. The use of PCMs aids the system to operate during the hours without solar irradiation and finally they found 31% improvement. The phase change materials help the system to store great amounts of energy in the desired temperature levels and to utilize this energy when it is needed. Fig. (**10**) illustrates a simple design of a solar still with PCM.

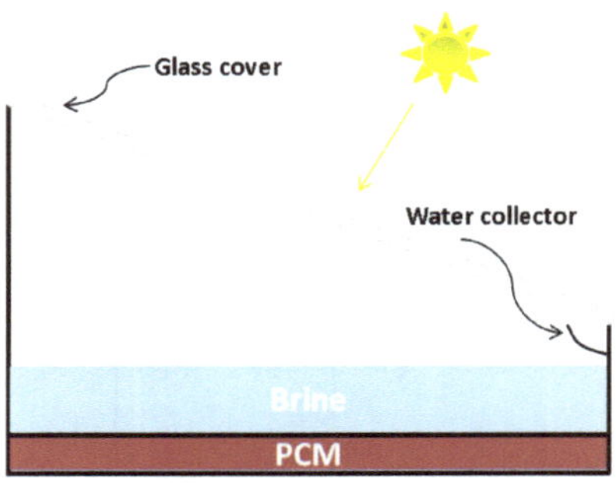

Fig. (10). Solar still with PCM.

Another interesting idea is the use of reflectors in order to utilize greater amounts of solar energy. Various configurations have been proposed in the literature and in any case there are different advantages. Fig. (**11**) depicts a simple solar still with a boosting reflector. The exact position of the reflector and its slope are important parameters that have to be optimized in every design. Tanaka [12] investigated numerically the use of reflectors in solar stills. He finally proved that the reflector has to be inclined backward in the summer and forward in the winter for optimum performance.

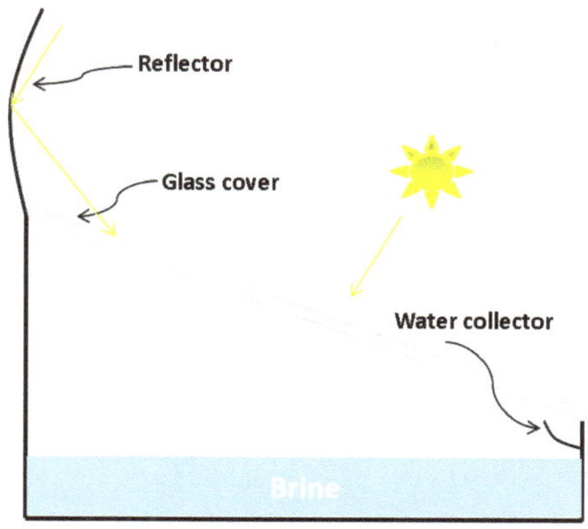

Fig. (11). Solar still with boosting reflector.

SOLAR SPACE HEATING SYSTEMS

Solar Air Heaters

Solar air heaters are a cheap technology for covering the space heating demand partially or totally. Their construction is simple and for this reason their cost is relative low. The thermal efficiency of these collectors varies a lot among the existing technologies from 45% to 90% [13]. The basic idea is to heat the ambient air with solar energy and to send it in the indoor space. Fig. (**12**) depicts various types of solar air heaters. A lot of research has been focused on the increase of the heat transfer from the absorber to the air. For this reason, many strategies have been applied, as the use of inserts in the flow. Fig. (**12**) illustrates a usual air heater with pipes and this collector has been proposed by Alvarez *et al.* [14] in a simpler form. Fig. (**13**) gives an air heater with simple obstacles in the flow in order to increase the turbulence. The use of these obstacles makes the air to make a longer path inside the collector, absorbing greater amounts of heat. Moreover, Fig. (**14**) shows a similar collector with Fig. (**13**), but there are extra obstacles in the flow in order to increase the flow turbulence. Akpinar and Kocyigit [15] investigated and compared similar collectors and proved that the greater number of obstacles has a positive result on the thermal efficiency. Fig. (**15**) is a double flow air heater with a curved absorbing surface in the middle of the collector. Esen [16] examined similar systems energetically and exergetically, taking into account that the existence of obstacles in the flow increases the pressure drop and the exergetic efficiency is getting lower due to this fact. However, Esen proved that the flat absorber has low exergetic performance because of the low thermal performance in this case. As a conclusion from the previous analysis, the obstacles in the flow lead finally to better thermal and exergetic performance in solar air heaters.

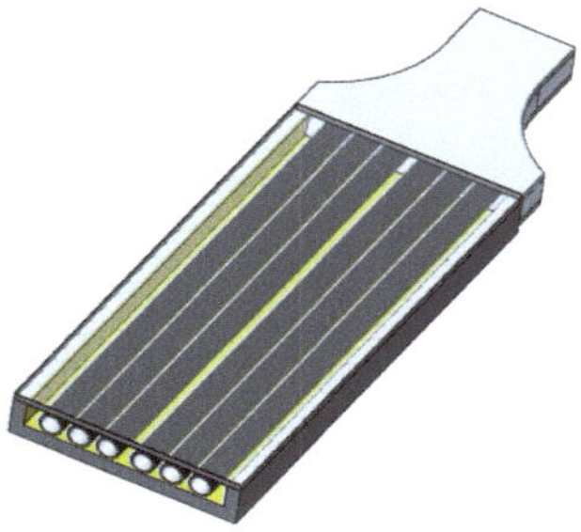

Fig. (12). Solar air heater with tubes.

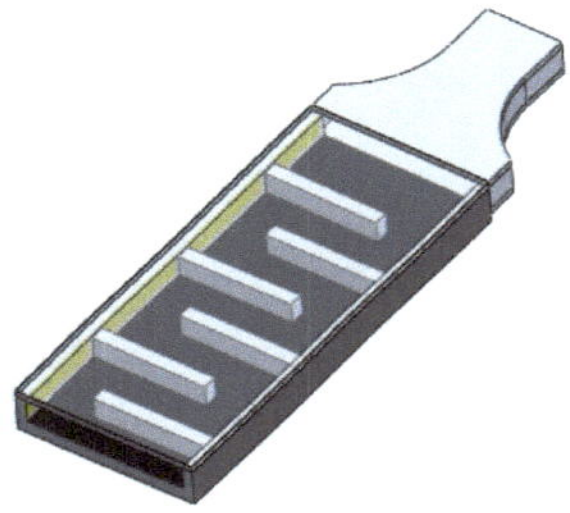

Fig. (13). Solar air heater with some obstacles.

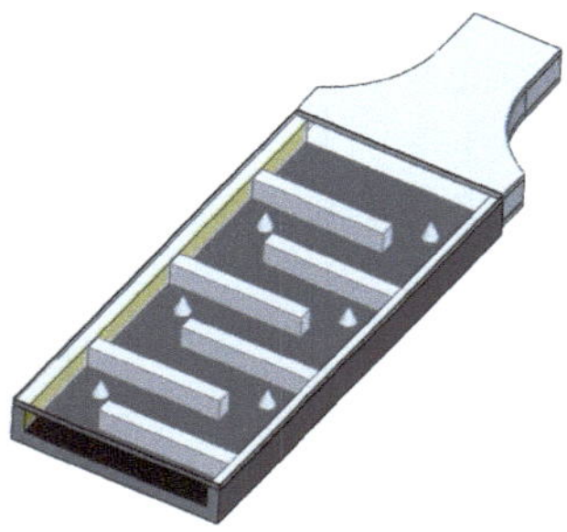

Fig. (14). Solar air heater with many obstacles.

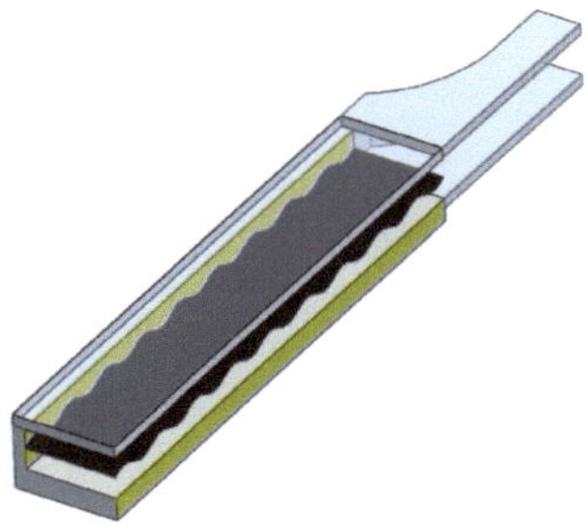

Fig. (15). Solar air heater with double flow.

Solar Hybrid Space Heating Systems

The solar hybrid heating systems are configurations which include the use of solar energy and of another external heat source. The external heat source usually acts as an auxiliary source in order to provide the demanded energy the time periods without available solar energy or storage. The basic idea is to produce hot water and to heat the indoor air with an efficient way. The demanded temperature levels are close to 40-45°C, fact that indicate the use of FPC mainly. In this temperature levels, FPC performs well and their no reason for using a more expensive and complex technology.

The most usual technique is to use Fan coils in order the heat from water to be given to the indoor air. A simple but efficient system is depicted in Fig. (**16**). This system can operate during the day but the use of auxiliary heat is demanded in the night. Fig. (**17**) shows an improved system with extra PCM for storage. This system has been studied by Belmonte *et al.* [17] and they concluded that this design can reduce the auxiliary energy consumption, especially for winter climates.

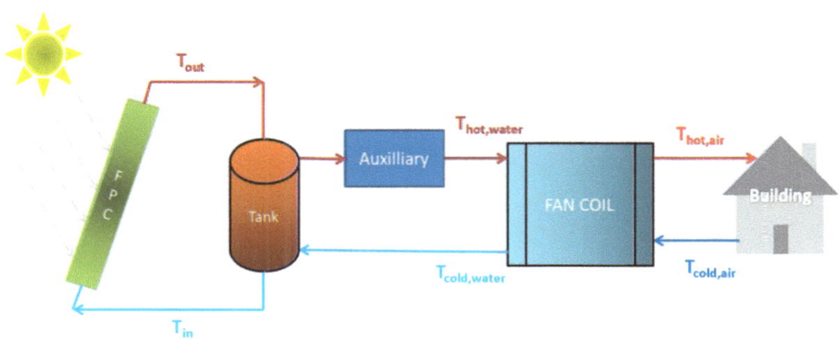

Fig. (16). Solar heating system with fan coils.

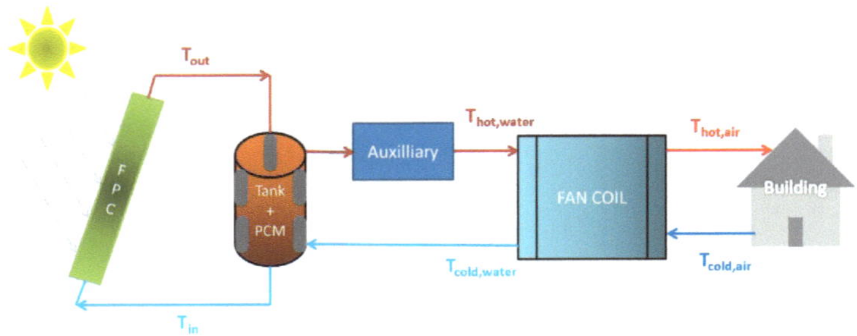

Fig. (17). Solar heating system with fan coils and PCM storage.

Another interesting heating system is the underfloor heating technology, (Fig. **18**). In this system, there are tubes in the ground of the building and hot water flows inside them. The use of solar energy for heating the water is an innovative and renewable way for heating the building. The maximum possible temperature in this system is 45°C for safety reasons in the piping system and the temperature difference between in the hot water circuit is about 5 to 7 K [18].

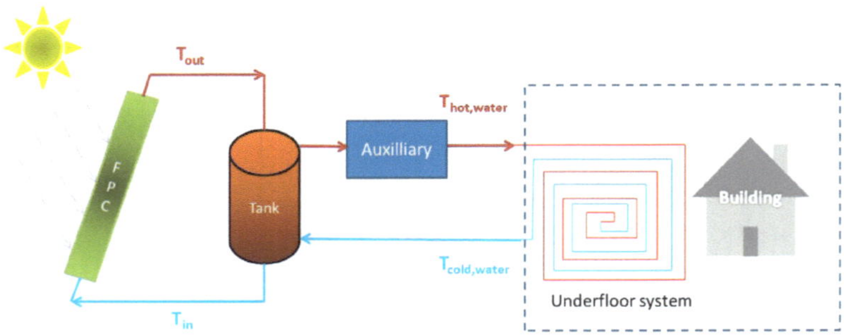

Fig. (18). Underfloor solar heating system.

A realistic goal is to achieve solar coverage (f) close to 70% in order the renewable fraction to be high. A simple way to define the solar fraction is by using the auxiliary energy (Q_{aux}) and the total demanded heating energy (Q_{heat}), as equation 1 shows:

$$f = 1 - \frac{Q_{aux}}{Q_{heat}}, \tag{1}$$

For the system with Fan coils, Fig. (**16**), the solar coverage as a function of the collecting area (A_c) and the storage tank volume (V) is given in Fig. (**19**). This figure corresponds to a usual building in Athens with 100m^2 floor area [19]. The rows in the figure indicate how to size a system with solar coverage of 70% for all the winter period.

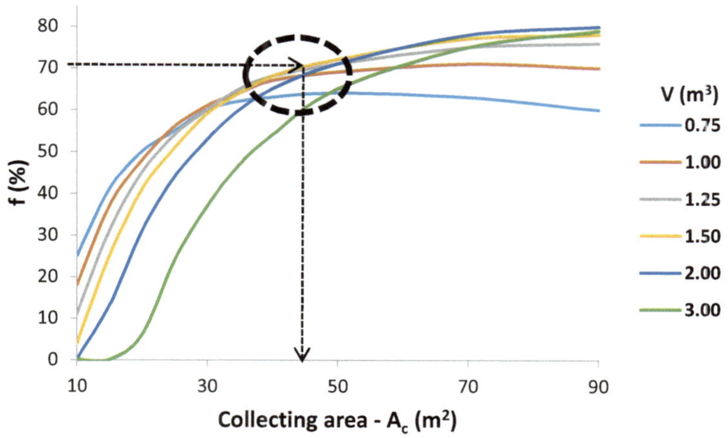

Fig. (19). Solar coverage for the simple system with Fan coils.

Another important parameter in these solar heating systems is the building envelope quality. Fig. (**20**) [18] explains how the insulation thickness influences on the solar coverage for the underfloor heating system of Fig. (**18**). It is obvious that the solar system is suitable for well insulated buildings where the heating needs are lower.

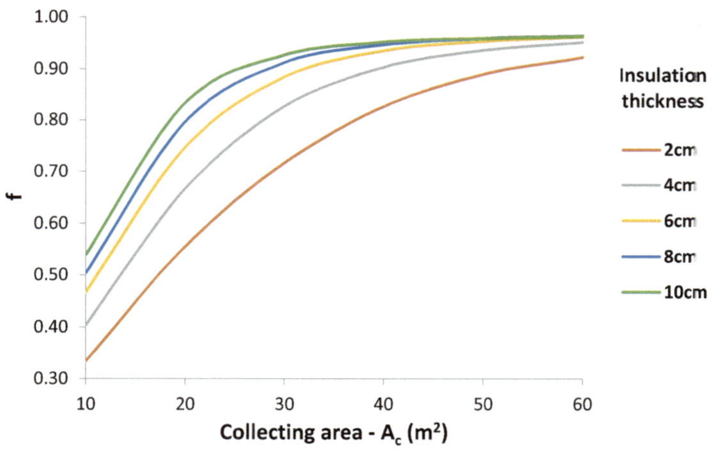

Fig. (20). Solar coverage for various collecting areas and insulation thicknesses of building envelope for the underfloor heating system.

Solar Assisted Systems for Space Heating

The last years, a new idea in solar heating systems has been developed and it gains more and more attention. This idea is based on coupling heat pumps and solar energy in order to produce the space heating demand. These systems can combine FPC, photovoltaic panels (PV) or thermal photovoltaic panels (PVT) with air source or water source heat pumps. Various combinations between these technologies can be made and every configuration presents different advantages and weaknesses.

The solar assisted heat pump (SAHP) is the main technology where a lot of research has been applied. The basic idea of the SAHP is the use of hot water as heat source in order to operate with higher coefficient of performance (COP) and consequently with lower electricity consumption. The COP of a heat pump can be defined as the ratio of the produced heating (Q_{heat}) to the consumed electricity (P_{el}), according to equation 2:

$$COP = \frac{Q_{heat}}{P_{el}}, \tag{2}$$

Water source heat pumps present higher COP than the conventional air source heat pumps. Fig. (**21**) shows the comparison of the COP between an air and a water source heat pump for various heat source temperature levels [20]. It is obvious that the water source heat pump leads to greater COP and it can operate for greater heat source temperature levels. This result proves that the electricity consumption can be reduced with the solar assisted heat pumps. Generally, a conventional heat pump (air source) operates under a COP close to 3, while the water source operates with COP close to 5.

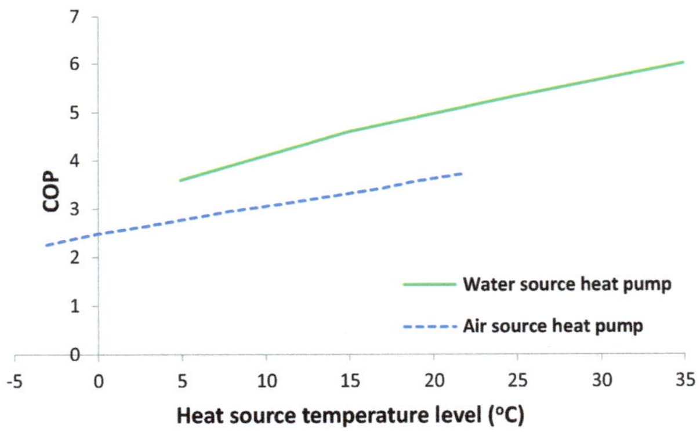

Fig. (21). Comparison of the COP of water and air source heat pump for various heat source temperature levels.

Figs. (**22 - 25**) illustrate innovative solar assisted heat pump heating systems which have been presented by Bellos *et al.* [20] and they have been examined energetically and financially. Fig. (**22**) shows a heating system with an air source heat pump driven by PV. There is inverter and battery in this system, while extra electricity can be taken by the grid. This configuration is partially renewable and it has relative low investment cost.

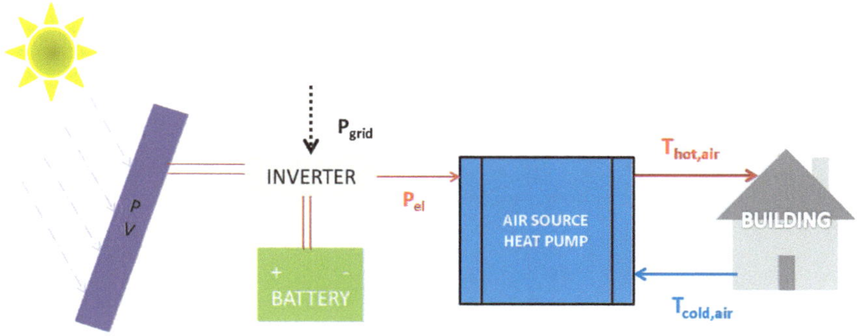

Fig. (22). Air source heat pump heating system with PV.

The next examined system is depicted in Fig. (**23**) and it is a simple solar assisted heat pump space heating system. In this system, FPCs are selected because the hot water has to be produced in low temperature levels. The hot water is stored in a storage tank and it feeds a water source heat pump. This heat pump takes electricity from the grid and this is the weakness of this technology.

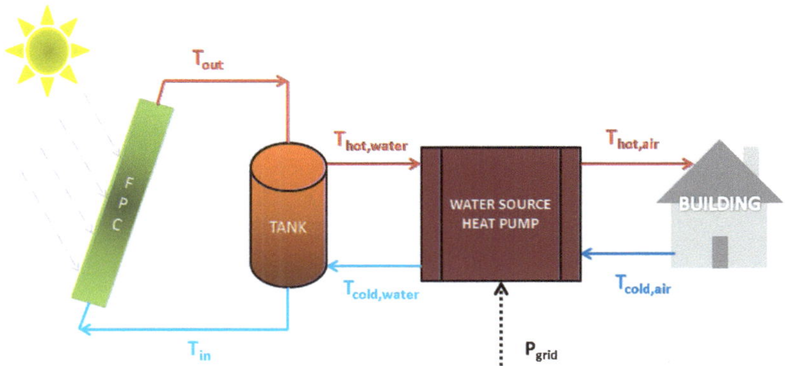

Fig. (23). Water source heat pump heating system with FPC.

In order to improve the previous systems, two new configurations have been suggested. Fig. (**24**) exhibits the use of a PVT with a water source heat pump. This hybrid solar collector is able to give heat and electricity; the two needed energy sources for feeding the water source heat pump. There is a storage tank, batteries and inverter in this system in order to store heat and electricity.

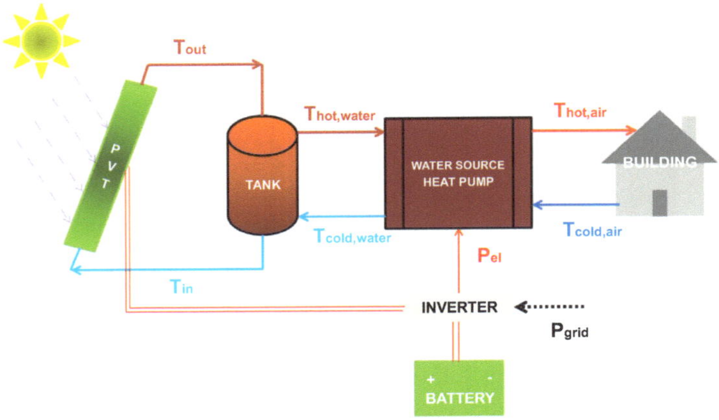

Fig. (24). Water source heat pump heating system with PVT.

The other idea is the use FPC and PV for feeding a water source heat pump, as Fig. (**25**) indicates. This system seems similar to the previous, but the investment cost is greater because the total collecting area is higher in this case. A water storage tank and batteries are also used in this configuration for heat and electricity storage, respectively.

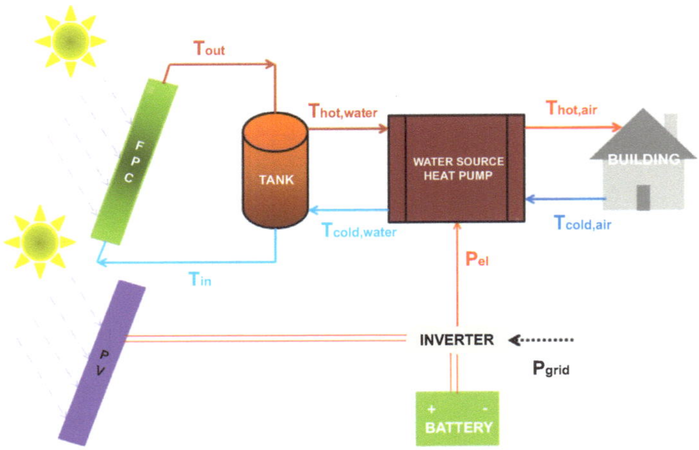

Fig. (25). Water source heat pump heating system with PV and FPC.

These technologies have to be compared in energetic and financial terms. In the energetic comparison, the electricity consumption from the grid (P_{grid}) is the best parameter that has to be investigated. Lower values of this consumption lead to an environmental friendlier technology. Fig. (**26**) shows the grid electrical consumption for a usual building [20]. The results are presented for various electricity prices (K_{el}) and the given combinations have been optimized financially.

It is obvious that the use of PVT leads to the lower energy consumption, especially when the electricity is not expensive. The higher grid consumption is achieved for the simple solar assisted heat pump with only FPC.

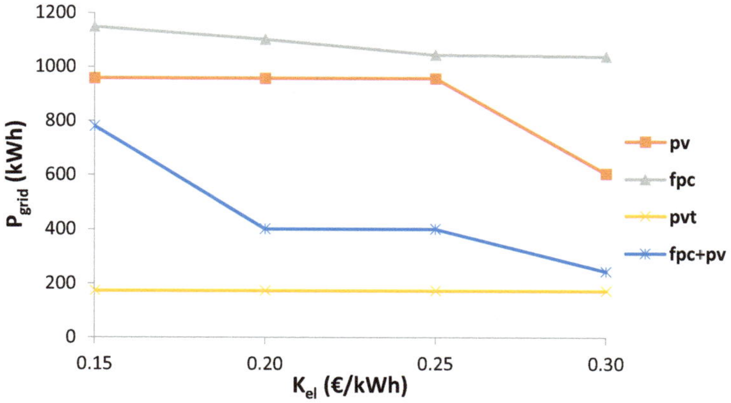

Fig. (26). Energetic comparison of the solar assisted heat pump systems.

The financial comparison of the heating technologies can be easily performed with the total cost index. This financial parameter is the present total cost of the heating, including the investment cost and the operational cost for all the project life. Equation 3 shows the general definition of this index:

$$Total\ Cost = Investment\ Cost + Total\ operational\ cost, \tag{3}$$

The total operational cost includes the operation and maintenance costs (O&M) which are influenced by the grid electricity consumption. In Ref [20], the financial comparison of the above technologies for various electricity prices is presented in Fig. (**27**).

For low electricity prices, the use of PV with an air source heat pump is the most attractive solution, while the use of PVT with a water source heat pump is the best choice when the electricity cost over 0.23 €/kWh. The combination of FPC and PV is the less sustainable choice, as it has been explained.

After the comparison of the presented innovative heating systems, it is useful to examine deeper the literature for similar studies. Sun *et al.* [21], compared a solar assisted heat pump heating system with a conventional air source heat pump heating system and they concluded that the first performs better in all weather conditions. Chargui and Sammouda [22] used TRNSYS in order to simulate a dual source heat pump in a residential house and they finally stated that a higher temperature of hot water improves the performance of the system and COP can reaches values close to 6. Cai *et al*. [23] investigated a novel indirect expansion solar assisted heat pump system for space heating, cooling and domestic hot water production. They performed a detailed parametric analysis in order to explain the impact of the ambient temperature, solar irradiation, and mass flow rate in the system performance. Ahmad *et al*. [24] optimized the control system of a solar assisted heat pump in order to reduce the electricity consumption. Their results proved that the on/off strategy is more beneficial than the linear multi-variable model. Qu *et al*. [25] compared the sensible and latent storage systems in a solar heat pump heating system and they proved that the latent storage with PCM can increase the system performance approximately to 50%. Buker and Riffat [26] examined a novel solar thermal roof for feeding a heat and finally they proved that this investment is feasible with a payback period close to 3 years. The same authors in a recent study [27] stated that many parameters influence on the performance of solar assisted heat pump systems. More specifically, they concluded that the solar collector type, the refrigerant of heat pump, the environmental conditions, the system size and the load characteristics are the parameters that have to be examined in any design.

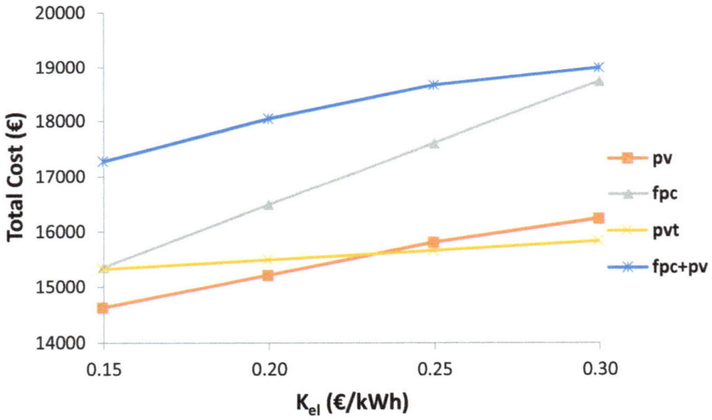

Fig. (27). Financial comparison of the examined systems for various electricity prices.

As a conclusion about the water solar assisted heat pumps, it is useful to state that these technologies perform better than the technologies based on the ambient as heat source [28]. Especially in cold climates, the use of ambient as heat source is very difficult and in these cases the water source heat pumps have to be preferred. Moreover, the use of water source heat pumps with storage can lead to a more steady temperature profile in the indoor air. Furthermore, the use of water source heat pumps has lower needs in electricity consumption due to the increased COP, compared to the conventional heat pumps.

Solar Driven Absorption Heat Pumps

The use of absorption heat pumps for space heating purposes is a promising way for covering the buildings demands. This technology uses heat in low temperature and heating can be produced in a lower temperature levels. The advantage of this technology is that the COP is getting values greater than 1, something that means by putting a heat quantity in the system, a greater can be taken in a lower temperature level. The reason for this result is the utilization of the ambient as heat source.

An important advantage of this technology is that the input heat has to be given in low temperatures, close to 100°C. Thus, renewable energy sources as solar energy and geothermal energy can be coupled with this technology in order to produce high amounts of space heating. More specifically, ETC is the most appropriate solar technology for absorption heat pumps [29] because FPC can operate better in lower temperature levels. On the other hand, concentrating technologies are more expensive and they do not exploit the diffuse solar irradiation which is important in the winter period.

Fig. (**28**) shows a simple modelling of this technology. The heat input is given in the generator of the system, while the useful heating is taken from the condenser and the absorber. The ambient is used as a heat source from the evaporator. The COP in this system can be calculated as:

$$COP = \frac{Q_A + Q_C}{Q_G}, \tag{4}$$

Fig. (**28**) corresponds to a system which operates with LiBr-H_2O as working pair. Also H_2O-NH_3 can be used in order to achieve operation in extremely low ambient conditions, but in this case greater generator temperature level is needed. The system of Fig. (**28**) is able to operate for ambient temperature levels over 0°C, because the water, which acts as refrigerant in the cycle, freezes in lower temperature levels.

The theoretical performance of this system for usual conditions is presented in Fig. (**29**) and Fig. (**30**). These results correspond to ambient temperature equal to 15°C and to heat exchanger efficiency equal to 60%. In Fig. (**29**), the COP is given for various generator temperature levels and heat production temperatures (T_{heat}). Fig. (**30**) shows the system exergetic performance for the same conditions. The exergetic performance (η_{ex}) takes into account the heat production temperature with a great way. This quantity can be calculated according to equation 5:

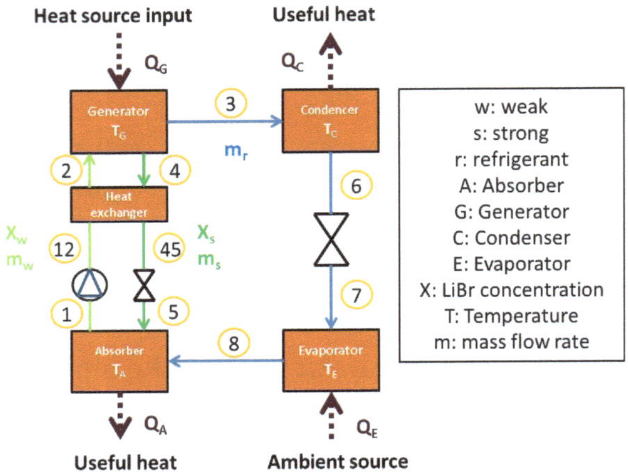

Fig. (28). A simple absorption heat pump modelling.

$$\eta_{ex} = COP \cdot \frac{1 - \dfrac{T_{am}}{T_{heat}}}{1 - \dfrac{T_{am}}{T_G}}, \tag{5}$$

According to Fig. (**29**), the COP is getting greater for higher generator temperature levels, but it takes lower values when the heat is produced in higher temperature levels. The COP is close to 1.7 a satisfying value for this technology.

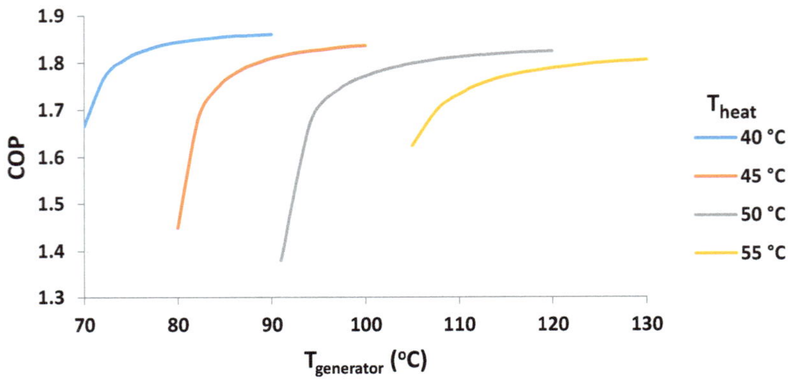

Fig. (29). COP of an absorption heat pump for various generator temperature levels.

Fig. (**30**) illustrates the exergetic performance of the chiller for various generator and heat production temperature levels. In every curve of Fig. (**30**), there is an optimum temperature level which maximizes the exergetic performance of the heat pump. This temperature is the design temperature and the system have to operate close to it in order to achieve the highest heating output with relative low heat input demand. In other words, in this temperature level is the irriversibilities of the system are minimized. Another useful conclusion from the Fig. (**30**) is that greater generator temperature is needed when the heating has to be produced in higher temperature levels. This is a logical result that has to be taken into account in the design of every system. Table **1** includes the results for the optimum operation. It is obvious that the COP and the exergetic efficiency are approximately constant, but the optimum temperature in the generator increases a lot with the increase in the heating temperature.

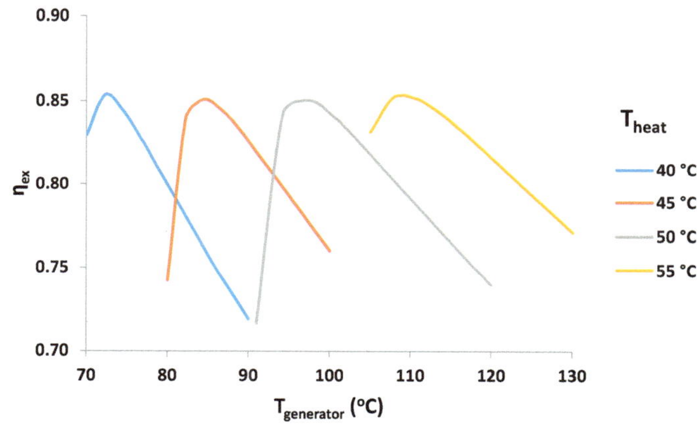

Fig. (30). Exergetic performance of an absorption heat pump for various generator temperature levels.

Table 1. Optimum operation of the absorption chiller for various temperatures of heating production.

T_{heat} (°C)	Optimum T_G (°C)	COP	η_{ex}
40	72	1.771	0.8532
45	84	1.752	0.8508
50	97	1.747	0.8506
55	108	1.702	0.8519

In literature there are numerous studies which examine the use of solar absorption heat pumps and a lot of research has been focused on this area. Wang *et al.* [30] examined a solar assisted absorption heat pump with parabolic trough collectors. They concluded that the mean daily solar fraction can be close to 50%; a very satisfying value. Chaiyat and Kiatsiriroat [31] investigated a 10 kW solar LiBr-H_2O absorption heat transformer. They selected FPC for feeding their system and finally the proved that the operational temperature in the solar circuit has to be close to 70°C. Zhang *et al.* [32] examined a novel cogeneration system which produces electricity and heating. This system is based on the absorption heat pump technology and the final results proved higher performance compared to conventional systems.

NEW PV TECHNOLOGIES FOR BUILDINGS

Concentrated PV

The last years, improvements in the photovoltaic panels have been applied in order to increase their performance. Photovoltaics can be used in the buildings in order to produce the demanded electricity or a part of this. The recent legislation in many countries lets the net-metering and the use of the produced energy for the same building, something that will lead to the increase of the PV installation. The use of concentrators in order to focus more solar energy in the PV is an intelligent way to make PV a more sustainable solution financially.

Many configurations for concentrating PV have been applied. Below some interesting examples are given. Fig. (**31**) gives a simple concentrating PV with a compound parabolic reflector.

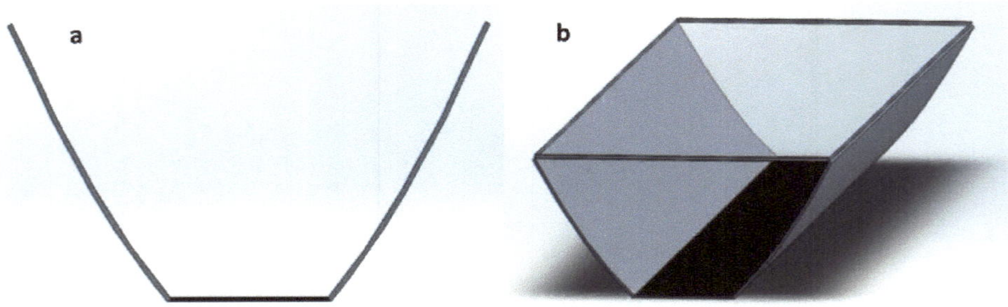

Fig. (31). A simple PV with compound parabolic concentrator a) projection of the collector b) 3-D image.

The previous configuration operates well in a specific range of incident angles. This is a weakness of this system and for improving it, the use of lenses as concentrators has been investigated. The improved system is presented in Fig. (**32**). Su *et al.* [33] proved that the use of lenses increase the range of accepted incident angles but the optical performance is lower in low incident angles. Guiqiang *et al.* [34] stated that the lenses create a more uniform heat flux distribution over the absorber; something very important for PV in order more electricity to be produced.

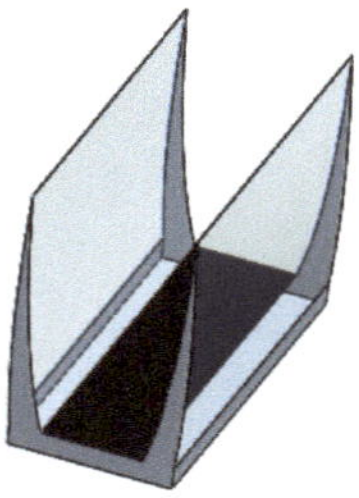

Fig. (32). Concentrated PV with lenses.

The use of three dimensional concentrators, as in Fig. (**33**), is an intelligent solution for improving the optical performance of the solar collector performance for greater range of incident angles [35].

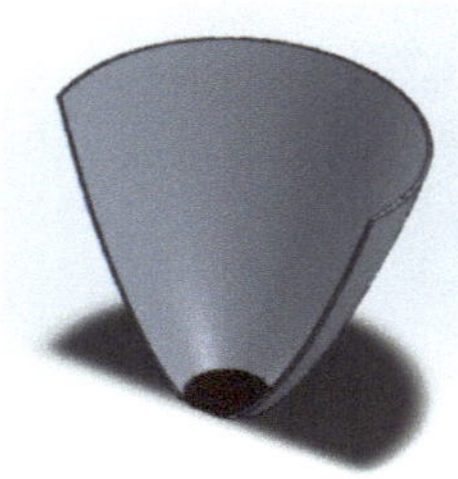

Fig. (33). A usual example of 3-D concentrator.

Abu-Bakar *et al.* [36] presented a high efficient 3-D concentrator which is given in Fig. (**34**).

Fig. (34). An efficient 3-D concentrator.

The use of Fresnel lenses for concentrating the incident solar irradiation to a small region with PV cell is an innovative idea. BenA-tez *et al.* [37] have been examined this idea recently with a configuration similar to Fig. (**35**). The special geometry of the lenses makes the incoming solar irradiation to be focused on a small area leading to high power production.

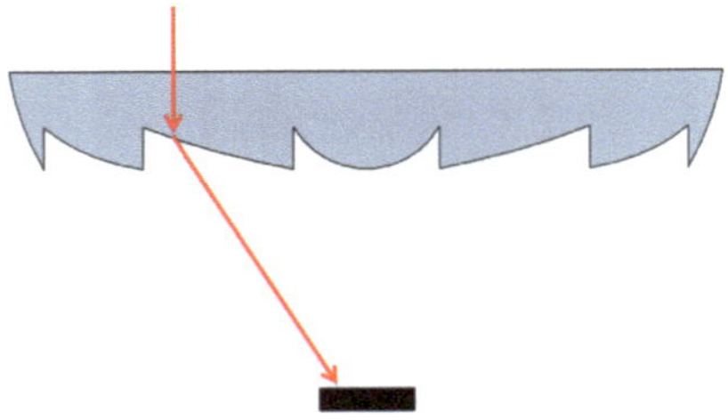

Fig. (35). Fresnel lenses for concentrating PV.

The use of concentrator in PV collectors leads to high temperature levels on the panels, something with negative impact on their performance. On this direction, many techniques have been examined in order to cool the receivers. In a recent study, Micheli *et al.* [38] examined the use of fins in a plate under the cell. A similar configuration of this idea is given in Fig. (**36**). According to this study, a great reduction in the temperature levels can be achieved.

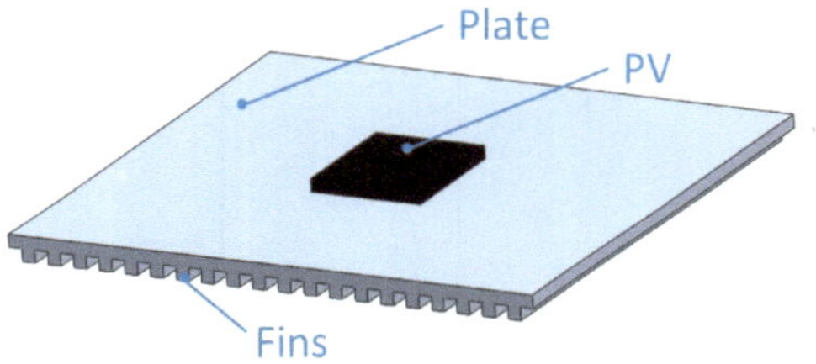

Fig. (36). Fin plate for cooling PV.

Thermal PV

Thermal PV is a promising technology for electricity and heat production simultaneously. The basic idea is to utilize the solar energy which is not exploited by PC cells by heating a flowing working fluid. Moreover, this situation makes the PV cells to be cooled and its efficiency to be increased. Fig. (**37**) illustrates a usual thermal PV with the water tubes under the PV panel.

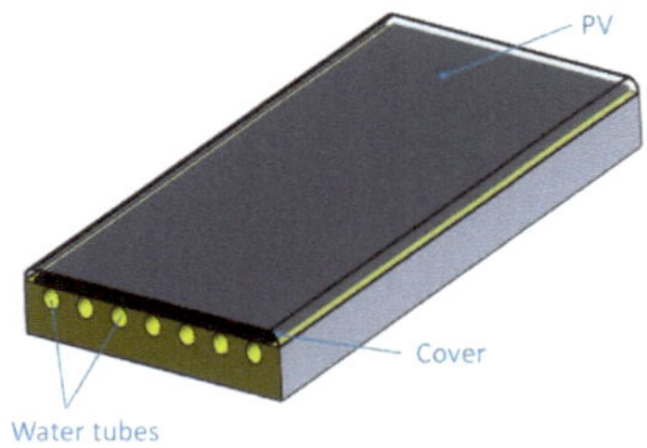

Fig. (37). Thermal photovoltaic collector.

It is essential to explain the thermal performance of a PVT. There are two main outputs (heat and electricity) and thus electrical and thermal efficiency, as well as total efficiency can be defined. Equation 6 is the electrical efficiency determination, equation 7 is the thermal efficiency and equation 8 is the total efficiency.

$$\eta_{el} = \frac{P_{el}}{Q_s}, \tag{6}$$

$$\eta_{th} = \frac{Q_{th}}{Q_s}, \tag{7}$$

$$\eta_{PVT} = \eta_{el} + \eta_{th} = \frac{P_{el} + Q_{th}}{Q_s}, \tag{8}$$

The electrical output (P_{el}), the thermal output (Q_u) and the available solar energy (Q_s) are needed for the above calculations.

For a typical PVT, the various efficiencies vary with the operating conditions, as Fig. (**38**) exhibits. These results have been taken from literature [20] for a usual PVT and they are representative. The parameter [(T_{in}-T_{am})/G_T] is usually used in solar thermal systems and it takes into account the inlet working fluid temperature (T_{in}), the ambient temperature (T_{am}) and the incident solar irradiation (G_T).

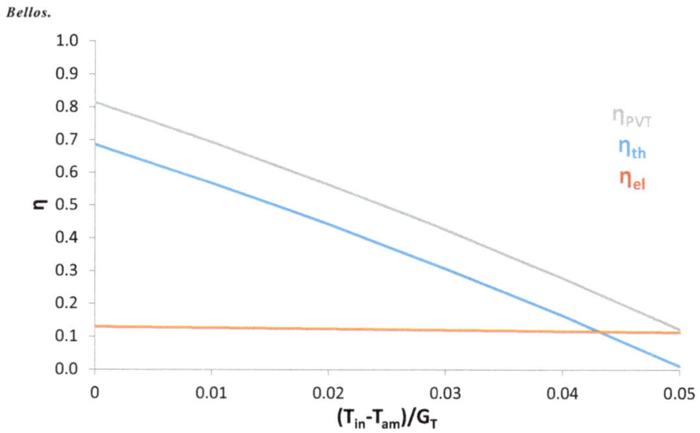

Fig. (38). Electrical, thermal and total efficiency of a typical thermal PV.

It is obvious that the electrical efficiency is approximately constant for higher water inlet temperature levels, while the total and the thermal efficiencies are very influenced by the operating conditions. These results indicate the use of thermal

PV in application with low operation temperature levels. Domestic hot water production, solar assisted heat pumps and similar applications are ideally can be coupled with thermal PV collectors.

The use of air as working fluid in thermal PV is a solution ideal for buildings in order to produce electricity and warm air for space heating purposes. Fig. (**39**) illustrates a simple system with air working fluid.

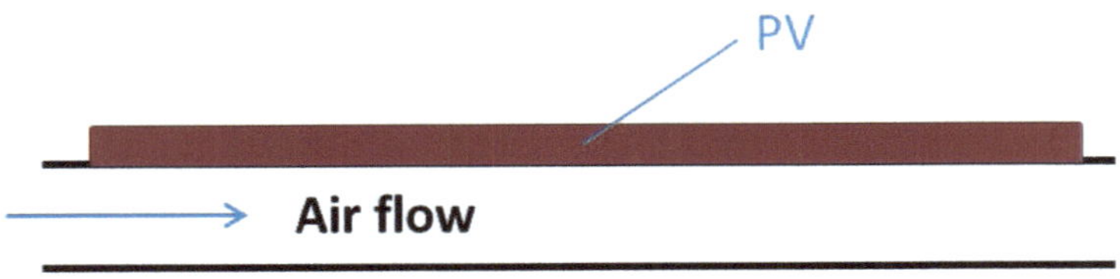

Fig. (39). Thermal PV with air working fluid.

Many researchers the last years have examined the air collectors with PV. Farshchimonfared *et al.* [39] optimized a similar collector coupled in residential building. They stated that it is important to examine the exergetic performance of this system. Equation 9 gives the final exergetic output of this collector (E_u). The useful heat (Q_u), the produced electricity (P_{el}), the electricity consumption in the fan (P_{fan}), as well as the outlet (T_{out}) and inlet (T_{in}) air temperatures have to be taken into account:

$$E_u = Q_u - m \cdot c_p \cdot \ln\left(\frac{T_{out}}{T_{in}}\right) + P_{el} - P_{fan}, \tag{9}$$

The mass flow rate (m) and the specific heat capacity (c_p) have also to be taken into the calculation of the exergy output. The solar available exergy (E_s) is related to the solar the sun temperature (T_s = 5770 K) and the ambient temperature, as equation 10 shows. This equation is the Petela formula [40] which is the most accepted way for estimating the exergy of the undiluted solar irradiation.

$$E_s = Q_s \cdot \left[1 - \frac{4}{3} \cdot \frac{T_{am}}{T_s} + \frac{1}{3} \cdot \left(\frac{T_{am}}{T_s}\right)^4\right], \tag{10}$$

The exergetic performance of this collector (η_{ex}) can be calculated as the ratio of the exergetic output (E_u) and the solar exergy flow (E_s), as equation 11 shows:

$$\eta_{ex} = \frac{E_u}{E_s}, \quad (11)$$

The performance of the collector is fully connected with the mass flow rate. In Ref [39], this specific mass flow rate was examined for values from 0.01 to 0.09 [kg/sm^2] and finally it is prove that this parameter have to be close to 0.03 [kg/sm^2] for optimum operation.

Another important improvement in the thermal PV collectors is the use of concentrators in order to achieve higher electricity and thermal outputs. Abdelhamid *et al.* [41] investigated an innovative thermal PV collector with primary and secondary concentrators. A similar collector is presented by Fig. (**40**). The primary concentrator is a parabolic trough collector. The absorber tube is located inside a tubular cover. The secondary reflector has a compound parabolic shape and it includes PV panels partially. This design leads to heat and electricity production and it is an intelligent configuration for thermal PV systems. Moreover, it is important to be stated that their design is ideal for operation for high temperature levels and for steam production in the case that these collectors are coupled with a steam turbine. The estimated cost was in logical levels and this is a promising technology in the field of solar thermal PV systems.

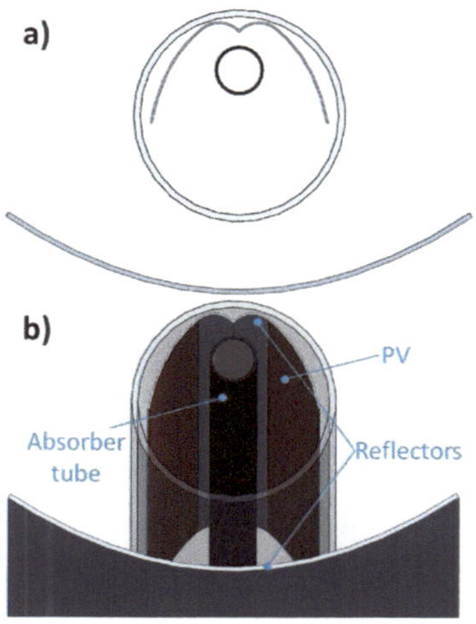

Fig. (40). Concentrating thermal PV with secondary reflector a) 2-D representation b) 3-D representation.

Integrated PV in Walls

The use of photovoltaic panels in the building envelop is a new and promising way for producing electricity in the buildings. This strategy is an important step in order to create green buildings with zero electricity consumption from the grid.

Fig. (**41**) shows a simple case of using PV panels in the external surface of buildings and in the inclined roof. This strategy is well known but the last years gain more and more attention in order the needed environmental goals to be delivered.

At present time, the 2/3 of the residential buildings in Europe is single-family houses with sufficient space on their roof for installing PV panels [42]. This strategy will make easier the creation of zero energy buildings.

Fig. (41). A simple example of using PV panels in on buildings facades.

Evola and Margani [42] investigated the use of PV panels in building facades for various cities in Italy. They concluded that the payback period of this investment can be close to 9 years. Moreover, they stated that many parameters influence on this result as the electrical needs, the building orientation and the number of the floors. Aste *et al.* [43] evaluated the long term use of PV panels on buildings. They examined the system for a 13 year operation and they found a yearly decrease on the efficiency close to 0.3%. This is a low value, compared to the usual assumption of 0.6% per year and this result gives extra advantages in the use of PV in buildings.

Baig *et al.* [44] investigated the use of concentrated PV in the external walls in buildings. They selected compound parabolic concentrators with PV in order to reduce the use of PV panels and simultaneously the investment cost. Finally they proved that the optical performance of their configuration is close to 80%, a high value which leads to efficient concentration. Fig. (**42**) shows a similar system with compound concentrators in a wall. The PV cells are depicted with black shades, while the concentrator with grey. The wall is given with brown shades in this figure. Moreover, Baig *et al.* [45] examined the use of a dielectric based Symmetric Elliptical Hyperboloid (SEH) concentrating element attached to a silicon solar cell for building facades. The concentration ratio of this configuration was about 6; a high value for these systems. The final results proved that this is a promising technology but many improvements can be made.

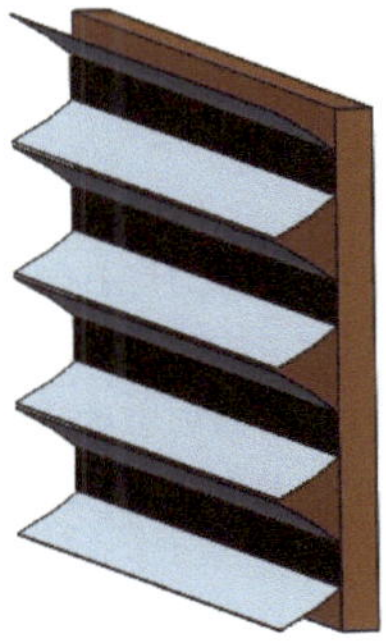

Fig. (42). Concentrated PV (compound parabolic shape) on external walls.

PV with PCM

An important issue in the use of PV is the increase in panel temperature which leads to lower efficiency. Many techniques for cooling the PV have been investigated and in the last years the use of PCM is the most investigated. More specifically, PCM can store high amount of energy, as latent heat storage devices, in the desired temperature levels. Thus, putting PCM in the down side of the PV panels is an easy way for cooling them. Furthermore, the PCM can store energy which can be used in other applications.

The use of PCM in building integrated PV leads to high electrical performance and to better solar energy management for regulating the space heating and cooling loads. Fig. (**43**) shows a configuration that can be used in order to couple PV and PCM in the building envelope.

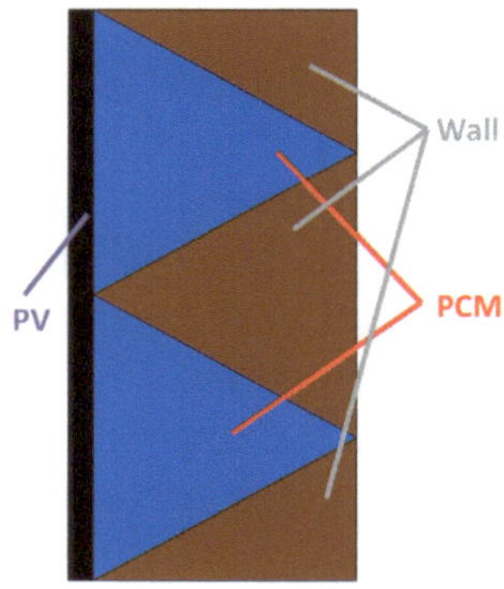

Fig. (43). The use of PCM in building integrated PV.

Huang [46] examined this configuration and proved that the electrical performance of PV panels can be increased up to 30%. Elarga *et al.* [47] examined the use of PCM and PV in double skin facades. More specifically, they proved that this idea can lead to a monthly cooling reduction of 20-30%. This result corresponds to hot climates with high cooling loads. Moreover, the electricity production of the PV panels is about 5 to 8%, according to their results.

Kamela *et al.* [48] investigated the use of building integrated thermal PV with heat pumps and PCM storage. The goal of their study was to create a nearly zero energy building. The final results of this study showed that the mean COP of the heat pump was increased from 2.74 to 3.45, leading to 20% lower electrical consumption during the winter period.

Moreover, Ho *et al.* [49] examined the use of microencapsulated phase change material with PV and they proved high thermal and the electrical performance in the examined system.

Solar Energy Utilization in the Building Envelope

Trombe Wall

Trombe wall is a well-known building structure which is categorized in passive heating systems. This system stores solar energy and manages it properly in order to regulate the building loads. More specifically a Trombe wall is like a solar collector. A massive wall with high thermal mass is located in south direction, in the majority of the cases, and a cover is located between it and the environment. The incident solar irradiation heats the wall and the solar energy is stored as sensible heat in the massive wall. This heat is preserved due to the cover and it enters in the building after some hours. This time lag is important for controlling the thermal comfort conditions in the indoor space.

The air between the cover and the massive wall is a key parameter in the Trombe wall systems. The simpler systems have no holes between the wall and the indoor space; something that leads to high temperature levels in the intermediate air. This configuration can be improved with holes in the massive wall which let the natural ventilation. This is beneficial in the winter period because the hot air can directly heat the indoor space during the day. On the other hand, this is not beneficial in the summer because the extremely hot air will lead to high indoor temperature levels. For this reason, holes between the intermediate space and the ambient are created. This design creates a natural air movement which is natural ventilation, something beneficial in the summer period.

Mechanical ventilation can be used in order the flow rate to be regulated properly.

Moreover, the holes can be closed if it is necessary, something that can be regulated by a suitable control system with thermostats.

Fig. (**44**) depicts the usual Trombe wall without holes which is known as unvented Trombe wall and this configuration is suitable for the winter period. Fig. (**45**) and Fig. (**46**) illustrate the vented Trombe walls. More specifically, Fig. (**45**) shows the internal vented case which is ideal for winter period while Fig. (**46**) depicts the case of ventilation with the ambient which is more suitable for summer period.

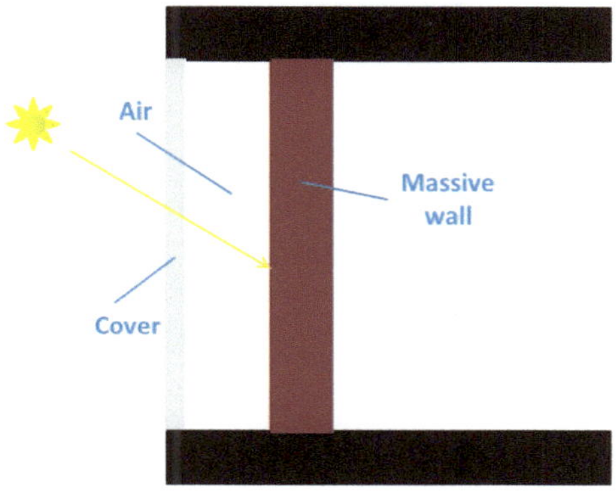

Fig. (44). Unvented Trombe wall.

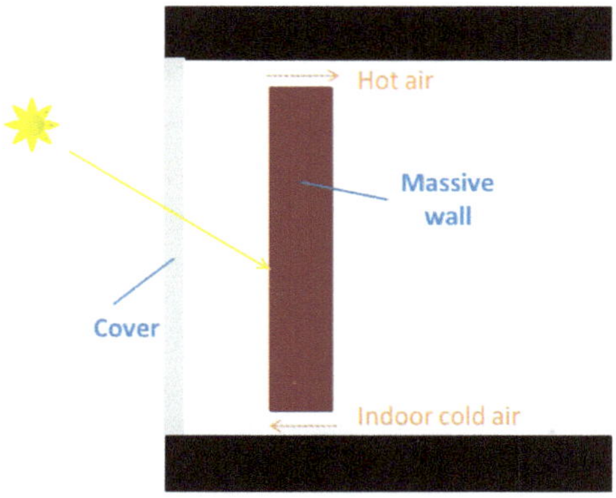

Fig. (45). Vented Trombe wall for winter period.

Many studies in the literature have been performed about Trombe walls and their performance. Abbassi *et al.* [50] examined the use of Trombe wall for Tunisian climate and they concluded that $4m^2$ of this wall can reduce the annually heating auxiliary energy to the half and $8m^2$ can lead to 77% reduction. Bojic *et al.* [51] investigated a Mozart house in Lyon-France and proved that the use of a Trombe wall in the south wall can lead to a yearly reduction on the heating energy consumption close to 20%.

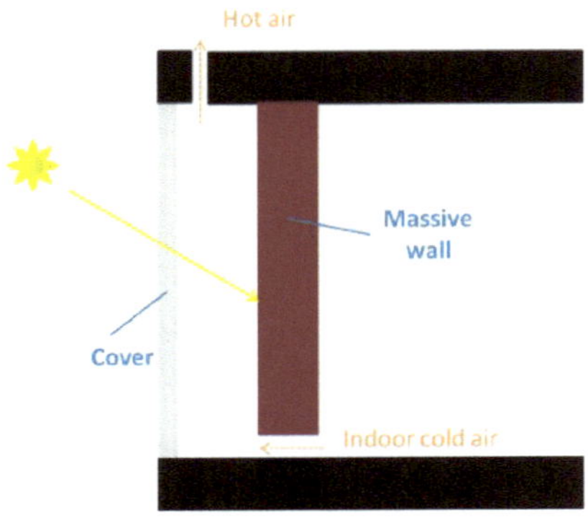

Fig. (46). Vented Trombe wall for summer period.

Abbassi and Dehmani [52] examined the use of internal fins in an unvented Trombe wall in order the heat to be transferred easier to the indoor space. They compared the new design with the conventional and they proved that the existence of fins leads to greater indoor temperature and to lower temperature in the massive wall. Fig. (**47**) presents a similar design with internal fins.

Trombe wall operation is based on the natural convection of the air between the cover and the massive wall. For this reason, many ideas have been applied in order the heat transfer conditions in this area to be improved. Hu *et al.* [53] examined the use of venetian blinds between the cover and the massive mass wall, and they finally proved that the tilt angle of the blinds plays a significant role in the natural convection in the air gap. Moreover, Hong *et al.* [54] studied the use of venetian blinds in Trombe walls and they optimized the examined system. Their results proved that the optimum distance between blinds and cover is 9cm for an air gap of 14cm. Fig. (**48**) shows a simple example of Trombe wall with venetian blinds.

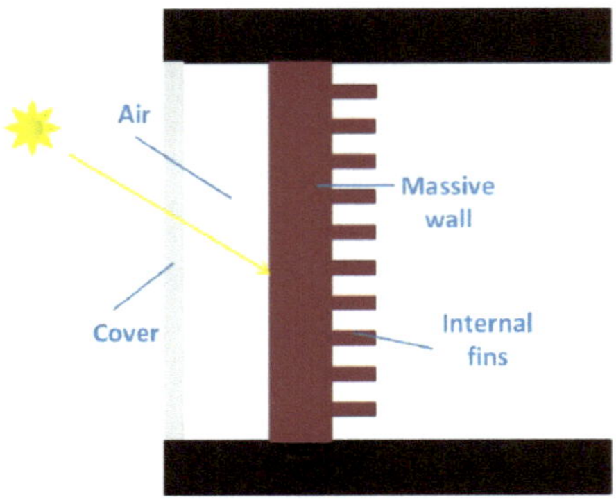

Fig. (47). Unvented Trombe wall with internal fins.

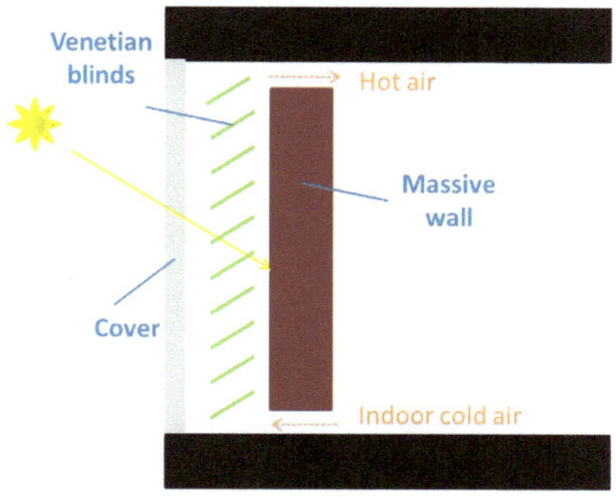

Fig. (48). Vented Trombe wall with Venetian blinds.

Duan *et al.* [55] investigated the use of an insert absorption plate in the middle of the air cavity and not in the massive wall surface. This configuration is shown in Fig. (**49**). They finally concluded that this new design (Fig. **49b**) performs better energetically and exegetically, compared to the conventional Trombe wall (Fig. **49a**).

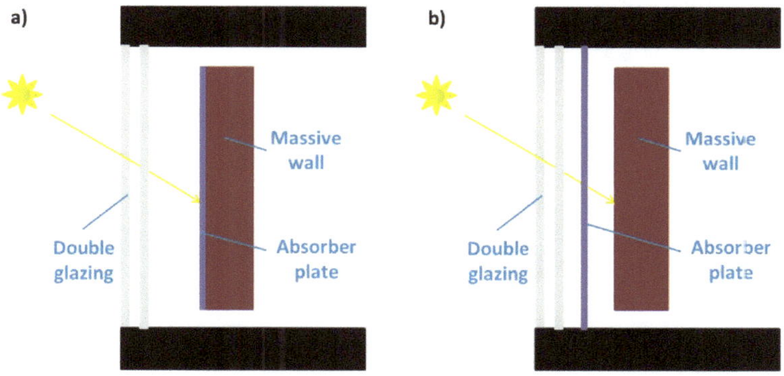

Fig. (49). Comparison of two Trombe walls with absorber plate a) the plate is on the massive wall b) the plate is in the air cavity.

Shen *et al.* [56] compared a classical Trombe wall (Fig. **46**) with a composite Trombe-Michel wall (Fig. **50**) and finally they concluded that the second has better performance in cold and cloudy weather conditions. Rabani *et al.* [57] designed a new Trombe wall which exploits solar irradiation from three directions: East, West and South. This wall covers the half part of the south wall and it is a vented Trombe wall. In the summer it can operates as a solar chimney in order to ventilate the space properly. Finally they concluded that the proposed system can lead to acceptable thermal comfort conditions for the indoor space.

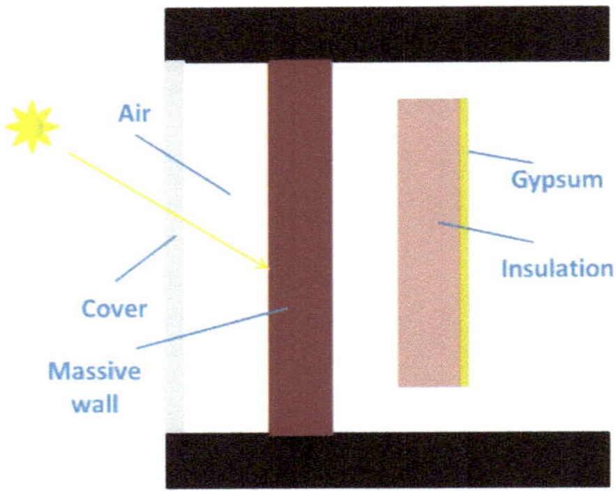

Fig. (50). Composite Trombe-Michel wall.

Zhou and Pang [58] designed and examined an innovative Trombe wall with PCM in the external side of the massive wall. Moreover, they added vortex generators in this side in order to increase the natural ventilation in the air cavity. The final results proved great improvement in the heat transfer conditions in the air gap and greater useful heat for the indoor space. Fig. (**51**) shows a similar configuration of this study.

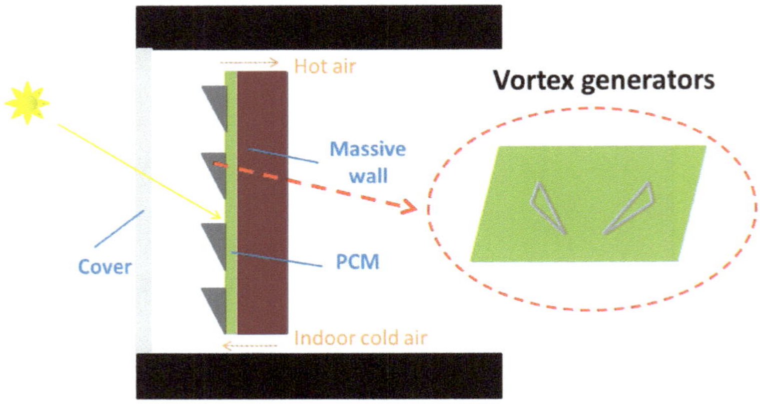

Fig. (51). Trombe wall with PCM an vortex generators.

The use of PV cells in a Trombe wall is another interesting idea. Fig. (**52**) shows a possible design with the PV cells to cover partially the cover. The produced electricity can be used for powering the fan of a forced ventilation system. Moreover, the air ventilation aids the PV to perform better due to their cooling. A similar system has been proposed by Jie *et al.* [59].

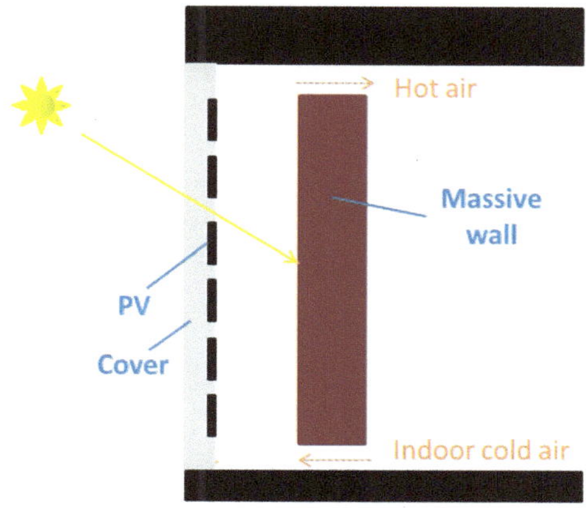

Fig. (52). The use of PV cells in a Trombe wall.

Finally, another interesting idea is the use of an internal window in the unvented Trombe wall, Fig. (**53**). This configuration lets the solar energy to heat directly the indoor space during the day and to improve the indoor lighting. Practically, this design increases the conductivity of the massive wall because the window acts as a thermal bridge; something useful during the day. This is an interesting design with thermal and aesthetic advantages.

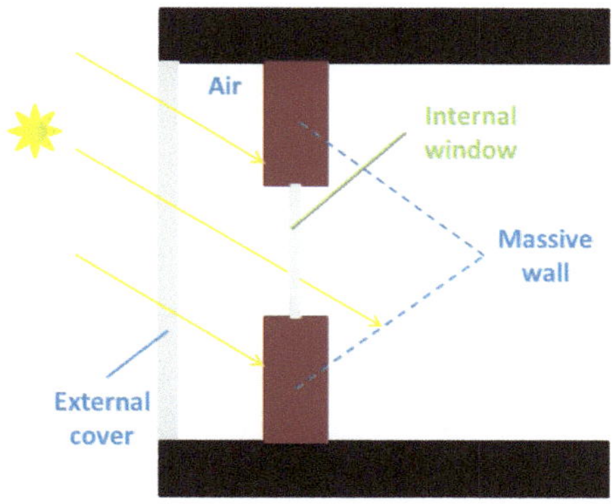

Fig. (53). Trombe wall with internal window in the massive wall.

PCM IN BUILDING ENVELOPE

Phase change materials are integrated in thermal energy systems to increase their efficiency, by storing energy mostly in a latent heat form. Therefore, an ongoing interest in these type of material is expressed over the last 15 years. A considerable amount of reviews on latent heat energy systems and phase change materials (PCM) can be found in the literature [60, 61]. Notably, PCMs are integrated not only in building applications, but also in solar energy heating, heat exchangers, automotive and others.

Latent Heat Storage

This form of energy storage is based on the phase change enthalpy of the material without any change in temperature. The material may undergo solid-liquid (for buildings) phase transition for heat storage and vice-versa for heat dissipation. Latent heat storage technique is attractive over sensible heat storage methods due to its higher energy storage capacity per unit volume and nearly constant temperature during the heat storage. Further PCM materials are available in wide range of temperatures to suit different building applications.

Key Attributes

There is a number of key attributes for a PCM that should be considered:

- Phase change temperature in desired range with sharp melting/freezing point
- High latent heat of fusion (solid to liquid)
- Non-toxic (to humans/animals) & non-carcinogenic
- Commercially available at low cost
- Does not react with and/or act as a solvent for packaging materials
- Landfill disposable and/or waterway disposable
- Biodegradable
- Low/non-flammable (high flash point, low vapor pressure)
- Non-corrosive
- Good stability upon thermal cycling (no super-cooling)
- Limited volumetric expansion/contraction upon freeze/thaw

Organic Phase Change Materials

These include paraffins which are open chain saturated alkanes, fatty acids and vegetable oils. They have high latent heat capacity, are non-reactive, do not undergo phase segregation and super-cooling with melt freeze cycles and have good nucleation property. They however suffer from low thermal conductivities and are flammable in nature. Because of this, salt hydrates are preferred over them in spite of the problems with phase segregation and poor nucleation associated with them.

Paraffins which are the most common building PCM are represented by the chemical formula: C_nH_{2n+2} and they show little or no reactivity at all with most chemical reagents. This material is considered to meet safety and containment constrains, even though the volume increases up to 10% during liquidation. Also, flammability can be an issue as far as the containment is concerned [62]. Plastic containers should be avoided, due to softening. In addition, they present good storage density with respect to mass, little sub-cooling and consistent liquidation and solidification, thus good cycling stability. These characteristics have made parrafins very appealing, despite the low melting enthalpy and thermal conductivity.

Inorganic Phase Change Materials

Inorganic materials are salt hydrates and metallic. They have respectively good thermal conductivity and high latent heat of fusion; they are not expensive and non-flammable. Their main drawback is compatibility with metals, since in some combinations of PCM with metals corrosion can be developed [63] (Table **2**). They require containment; hence, they are inadequate for impregnation into porous building materials.

The most attractive and important PCM materials are salt hydrates, due to their relative high storage density of about 240 kJ/kg, their small volume change during phase transition, and their relative high thermal conductivity of about 0.5W/(mK). Salt hydrates have some disadvantages such as super-cooling, segregation, and corrosion. Concerning Metallic PCMs, they are not within the desired temperature range for building applications.

Table 2. PCM Characteristics.

PCM type	Advantages	Disadvantages
Organic	- Freeze without much undercooling - Ability to melt congruently - Compatibility with conventional material of construction - No segregation - Chemically stable - Safe and non-reactive - Recyclable	- Low thermal conductivity in their solid state - High heat transfer rates are required during the freezing cycle - Volumetric latent heat storage capacity is low - Flammable
Inorganic	- High volumetric latent heat storage capacity - Availability and low cost - Sharp melting point - High thermal conductivity - High heat of fusion - Non-flammable	- Change of volume is very high - Super cooling is major problem in solid ?"liquid transition

Integration of PCMs into Building Elements

A) Means of PCM Containment

PCM can be incorporated into construction materials and elements by direct incorporation, immersion, shape-stabilization and encapsulation.

-- Direct Impregnation

Direct impregnation is the simplest, convenient and economical method in which PCM is directly mixed with gypsum, concrete or other porous materials.

-- Immersion

The immersion technique is an operational approach easily. The construction elements (concrete and brick blocks, wallboards), which are dipped into the liquid PCM, absorb the PCM by capillary action.

-- Shape Stabilization

Shape-stabilized PCM are prepared from a mixture of PCM and a supporting material. The most common supporting materials found in literature are high-density polyethylene and styrene-butadiene.

-- Encapsulation

In this technique, PCM has to be encapsulated before being used into construction elements.

B) Building Applications

The use of PCMs in buildings as thermal storage systems has been of great importance since the second half of the twentieth century. Most frequently, latent heat storage materials are used to stabilize interior building temperatures. The application of PCMs in air conditioning systems reduced room temperature fluctuation by lowering the high temperatures from the external daily temperature and reducing home heating and cooling loads by reducing the electrical power consumption. Also, the use of PCMs to store coolness has been developed for air conditioning applications and reduced the energy consumption, where cold temperature is collected and stored from ambient air during the night, and is released indoors during the hottest hours of the day.

The main applications of PCMs in buildings are when spaces are directed to the sun and require larger thermal storage units to be used as an insulation layer within building envelop components. PCMs were traditionally used to provide a comfortable interior building temperature.

-- PCM Enhanced Wallboards

The wallboards are cheap and widely used in a variety of applications, making them very suitable for PCM encapsulation. However, the principles of latent heat storage can be applied to any appropriate building materials. The idea of improving the thermal comfort of lightweight buildings by integrating PCMs into the building structure has been investigated in various research projects over several decades while most of these attempts applied macro-capsules, microencapsulate PCM or direct immersion processes. From recent studies, it is found that:

1) The concentration of 80% paraffin in PCM was found to be optimum

2) Melting temperatures of shape stabilized PCM in wallboards used in buildings were optimum at the range of 20 to 25°C. Thus the optimal melting temperature of PCM varies with the climate conditions

3) Low thermal conductivity had a little effect on lowering the peak of indoor air temperature.

The efficiency of these elements depends on several factors such as:

1. How the PCM is incorporated in the wallboard
2. The orientation of the wall
3. The climatic conditions
4. The amount of direct solar gains
5. The internal heat gains
6. The color of the surface
7. The ventilation and infiltration rates
8. The thermophysical properties of the chosen PCM mainly the temperature range over which phase-change occurs
9. The latent heat capacity per unit area of the wall.

--PCM Enhanced Cement

The large thermal mass of the concrete walls can be advantageous especially in moderate climates where it can be used to store energy during the day and release it during night time therefore reducing the need for auxiliary cooling and heating. Moreover, the energy storage capacity of concrete can further be enhanced by the incorporation of PCM into the concrete mixtures. Concrete is considered suitable for incorporation of PCM because of the following reasons:

1. They are most widely used construction materials
2. They can be formed into a variety of shapes and sizes
3. They have a larger heat exchange area and smaller heat exchange depth
4. PCM is held by them under capillary and surface tension forces.

--Fenestration

In conventional applications, thermal performance of windows can he improved by the use of absorbing gases filling the gap between the glass sheets, radiation controlling glass surface treatments, and by the application of thermally broken window frames. The other thermal improvement options may incorporate translucent filling materials such as silica aerogel or a semi-transparent PCM. Though, both of these options require serious consideration of optical properties and window functionality. The objective of using PCM in the window glazing or window attachments is to utilize its high latent heat of fusion to reduce the cooling loads by absorbing the solar-generated heat wave before it reaches the indoor space.

--PCM Trombe Wall

As suggested by Fang and Yang [64], passive solar heating of buildings is an area of great interest for renewable energy applications. Several authors have proposed the inclusion of PCMs in solar wall systems to replace masonry big volumes, and many experimental and theoretical tests have been conducted to investigate the reliability of PCMs in this kind of system [65]. The introduction of PCMs in Trombe wall systems could contribute to the development of light, portable, movable and rotating systems fully adapted to the lightweight buildings category. In this new approach, the huge sensible thermal mass of a traditional Trombe wall and the big amount of material could be replaced by the latent heat from the PCMs phase-change processes, and less quantity of material will be necessary. Moghiman *et al.* [66] replaced the classic Trombe wall design by a set of rotating wall segments which can rotate around their vertical shafts. With this configuration, the rotating wall segments are a good absorber during the day and a good radiator during the winter nights. The results showed that, in comparison with classical solar walls, the rotating storage walls can be more efficient, even in cold climates.

Measurement Procedures - Effective Thermal Capacity

In recent years, the application of Phase Change Materials (PCMs) has been investigated in many fields. The key of PCM's diffusion in building applications is the precise information regarding their thermo-physical properties.

Many manufacturers do not provide data on the enthalpy-temperature curve of their products or these data are not suitable for application in a building energy simulation tool. In addition, the capabilities of building simulation tools may not be totally appropriate to simulate every kind of material. Among the experiments that can be conducted for measuring the dependency of the specific heat capacity on temperature are the Differential Scanning Calorimetry (DSC) and the T-history method.

The DSC is the most used laboratory measurement to obtain melting temperature and heat of fusion of PCM samples. However, limitations of the DSC approach are the very small sample size and the strong influence of the test procedure on the results. The T-history method is widely adopted as an alternative to DSC to investigate the thermal behavior of large PCM samples. The T-history method can also be used to evaluate the thermal conductivity of PCMs whose phase change occurs with a clear interface between the two phases. However, thermal conductivity and specific heat cannot be simultaneously determined.

Another similar method is "thermal delay method" [67] which is used for the measurement of the phase change heat H, the temperatures T_l, T_s and the corresponding heat capacities cp_l, cp_s at the ends of the two-phase region of various PCM (Fig. **54**). The most important result of the present measurements is the "oeffective thermal capacity" in terms of the temperature $cp_{eff}(T)$. This function, which represents an equivalent thermal capacity during phase change process, contains much more information than the total phase change heat H, measured by most researchers.

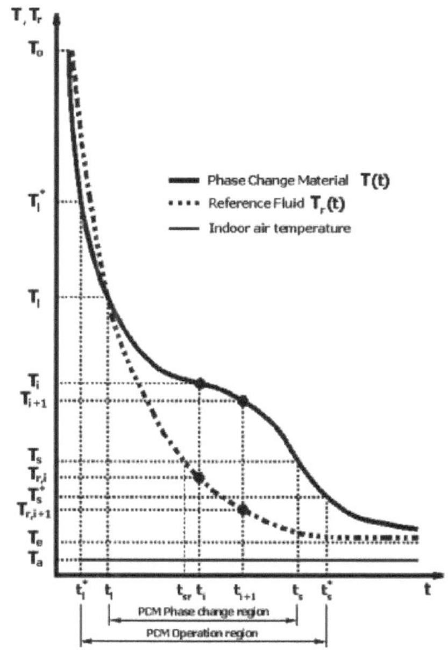

Fig. (54). Schematic representation of PCM and reference fluid cooling curves, illustrating the thermal delay method [67].

NOMENCLATURE

A_c Collecting area, m^2

c_p Specific thermal capacity, J/kg K

c_{pl} Liquid specific thermal capacity, J/kg K

c_{ps} Solid specific thermal capacity, J/kg K

E Exergy flow, W

f Solar coverage, -

G_T Incident solar irradiation on titled surface, W/m^2

H total phase change heat, J/kg

K_{el} Cost of electricity, €/kWh

m_s strong solution mass flow rate, kg/s

m_w weak solution mass flow rate, kg/s

P Power, W

Q Heat rate, W

T Temperature, °C

T_l Liquid temperature, °C

T_s Solid temperature, °C

V Volume, m^3

GREEK SYMBOLS

η Efficiency,-

X_s Strong solution,-

X_w Weak solution,-

SUBSCRIPTS AND SUPERSCRIPTS

A absorber

am ambient

aux auxiliary

C condenser

E evaporator

eff effective

el electrical

ex exergetic

fan fan

G generator
grid grid electricity
heat heat production
in inlet
out outlet
pvt thermal photovoltaic
s sun
th thermal
u useful

ABBREVIATIONS

COP Coefficient of performance
CPC Compound parabolic collector
DHW Domestic hot water
ETC Evacuated tube collector
FPC Flat plate collector
O&M Operation and maintenance
PCM Phase change material
PV Photovoltaic
PVT Thermal photovoltaic
SAHP Solar assisted heat pump
SHE Symmetric Elliptical Hyperboloid

CONSENT FOR PUBLICATION

Not applicable.

CONFLICT OF INTERESTS

The authors declare no conflict of interests.

ACKNOWLEDGEMENT

Declared none.

REFERENCES

[1] K. Wan, D. Li, D. Liu, and J. Lam, "Future trends of building heating and cooling loads and energy consumption in different climates", *Build. Environ.,* vol. 46, pp. 223-234, 2011. [http://dx.doi.org/10.1016/j.buildenv.2010.07.016]

[2] R. Zeng, X. Wang, H. Di, F. Jiang, and Y. Zhang, "New concepts and approach for developing energy efficient buildings: ideal specific heat for building internal thermal mass", *Energy Build.*, vol. 43, pp. 1081-1090, 2011.
[http://dx.doi.org/10.1016/j.enbuild.2010.08.035]

[3] "The European Parliament and the Council of the EU", *Directive 2010/31/EU of the European parliament and of the council of 19 May 2010 on the energy performance of buildings (recast)*, 2010.

[4] "The European Parliament and the Council of the EU", *Directive 2012/27/EU of the European Parliament and of the Council of 25 October 2012 on energy efficiency, amending Directives 2009/125/EC and 2010/30/EU and repealing Directives 2004/8/EC and 2006/32/EC,* 2012.

[5] E. Bellos, C. Tzivanidis, and K.A. Antonopoulos, "Parametric investigation and optimization of an innovative trigeneration system", *Energy Convers. Manage.*, vol. 127, pp. 515-525, 2016.
[http://dx.doi.org/10.1016/j.enconman.2016.09.044]

[6] Y. Tripanagnostopoulos, M. Souliotis, and T.H. Nousia, "CPC type integrated collector storage systems", *Sol. Energy,* vol. 72, no. 4, pp. 327-350, 2002.
[http://dx.doi.org/10.1016/S0038-092X(02)00005-1]

[7] M. Souliotis, P. Quinlan, M. Smyth, Y. Tripanagnostopoulos, A. Zacharopoulos, M. Ramirez, and P. Yianoulis, "Heat retaining integrated collector storage solar water heater with asymmetric CPC reflector", *Sol. Energy,* vol. 85, no. 10, pp. 2474-2487, 2011.
[http://dx.doi.org/10.1016/j.solener.2011.07.005]

[8] E. Bellos, C. Tzivanidis, and K.A. Antonopoulos Optical Performance, *Antonopoulos Optical Performance and Optimization of Two Stationary Compound Parabolic Collectors (CPC), Proceedings of the World Congress on Momentum, Heat and Mass Transfer (MHMT'16), paper no. 101,* 2016 Prague, Czech Republic

[9] A. Regattieri, F. Piana, M. Bortolini, and M. Gamberi, "Emilio Ferrari, □?oInnovative portable solar cooker using the packaging waste of humanitarian supplies", *Renew. Sustain. Energy Rev.,* vol. 57, pp. 319-326, 2016.
[http://dx.doi.org/10.1016/j.rser.2015.12.199]

[10] A. Harmim, M. Merzouk, M. Boukar, and M. Amar, "Performance study of a box-type solar cooker employing an asymmetric compound parabolic concentrator", *Energy,* vol. 47, no. 1, pp. 471-480, 2012.
[http://dx.doi.org/10.1016/j.energy.2012.09.037]

[11] M. Dashtban, and F.F. Tabrizi, "Thermal analysis of a weir-type cascade solar still integrated with PCM storage", *Desalination,* vol. 279, pp. 415-422, 2011.
[http://dx.doi.org/10.1016/j.desal.2011.06.044]

[12] H. Tanaka, "Monthly optimum inclination of glass cover and external reflector of a basin type solar still with internal and external reflector", *Sol. Energy,* vol. 84, pp. 1959-1966, 2010.
[http://dx.doi.org/10.1016/j.solener.2010.07.013]

[13] H.F. Oztop, F. Bayrak, and A. Hepbasli, "Energetic and exergetic aspects of solar air heating (solar collector) systems", *Renew. Sustain. Energy Rev.,* vol. 21, pp. 59-83, 2013.
[http://dx.doi.org/10.1016/j.rser.2012.12.019]

[14] G. Alvarez, J. Arce, L. Lira, and M.R. Heras, "Thermal performance of an air solar collector with an absorber plate made of recyclable aluminum cans", *Sol. Energy,* vol. 77, pp. 107-113, 2004.
[http://dx.doi.org/10.1016/j.solener.2004.02.007]

[15] E.K. Akpinar, and F. Koçyiğit, "Experimental investigation of thermal performance of solar air heater having different obstacles on absorber plates", *Int. Commun. Heat Mass Transf.,* vol. 37, no. 4, pp. 416-421, 2010.
[http://dx.doi.org/10.1016/j.icheatmasstransfer.2009.11.007]

[16] H. Esen, "Experimental energy and exergy analysis of a double-flow solar air heater having different

obstacles on absorber plates", *Build. Environ.*, vol. 43, pp. 1046-1054, 2008.
[http://dx.doi.org/10.1016/j.buildenv.2007.02.016]

[17] J.F. Belmonte, M.A. Izquierdo-Barrientos, A.E. Molina, and J.A. Almendros-IbA□A□ez, "Air-based solar systems for building heating with PCM fluidized bed energy storage", *Energy Build.*, vol. 130, pp. 150-165, 2016.
[http://dx.doi.org/10.1016/j.enbuild.2016.08.041]

[18] C. Tzivanidis, E. Bellos, G. Mitsopoulos, I. Alexopoulos, and K.A. Antonopoulos, "The Impact of Insulation Layer in Various Solar Heating Systems: An Energetic and Financial Evaluation", *International Journal of Mechanical Systems Engineering,* vol. 1, no. 2, p. 110, 2015.
[http://dx.doi.org/10.15344/2455-7412/2015/110]

[19] C. Tzivanidis, E. Bellos, G. Mitsopoulos, K.A. Antonopoulos, and A. Delis, "Energetic and financial evaluation of a solar assisted heat pump heating system with other usual heating systems in Athens", *Appl. Therm. Eng.*, vol. 106, pp. 87-97, 2016.
[http://dx.doi.org/10.1016/j.applthermaleng.2016.06.004]

[20] E. Bellos, C. Tzivanidis, K. Moschos, and K.A. Antonopoulos, "Energetic and financial evaluation of solar assisted heat pump space heating system", *Energy Convers. Manage.*, vol. 120, pp. 306-319, 2016.
[http://dx.doi.org/10.1016/j.enconman.2016.05.004]

[21] X. Sun, Y. Dai, V. Novakovic, J. Wu, and R. Wang, "Performance Comparison of Direct Expansion Solar-assisted Heat Pump and Conventional Air Source Heat Pump for Domestic Hot Water", *Energy Procedia,* vol. 70, pp. 394-401, 2015.
[http://dx.doi.org/10.1016/j.egypro.2015.02.140]

[22] R. Chargui, and H. Sammouda, "Modeling of a residential house coupled with a dual source heat pump using TRNSYS software", *Energy Convers. Manage.*, vol. 81, pp. 384-399, 2014.
[http://dx.doi.org/10.1016/j.enconman.2014.02.040]

[23] J. Cai, J. Ji, Y. Wang, and W. Huang, "Numerical simulation and experimental validation of indirect expansion solar-assisted multi-functional heat pump", *Renew. Energy,* vol. 93, pp. 280-290, 2016.
[http://dx.doi.org/10.1016/j.renene.2016.02.082]

[24] M.W. Ahmad, M. Eftekhari, and T. Steffen, "Abdulhameed Mambo Danjuma, □?oInvestigating the performance of a combined solar system with heat pump for houses", *Energy Build.*, vol. 63, pp. 138-146, 2013.
[http://dx.doi.org/10.1016/j.enbuild.2013.03.055]

[25] S. Qu, F. Ma, R. Ji, D. Wang, and L. Yang, "System design and energy performance of a solar heat pump heating system with dual-tank latent heat storage", *Energy Build.*, vol. 105, no. 15, pp. 294-301, 2015.
[http://dx.doi.org/10.1016/j.enbuild.2015.07.040]

[26] M.S. Buker, and S.B. Riffat, *Build-up and performance test of a novel solar thermal roof for heat pump operation.,* International Journal of Ambient Energy, pp. 1-15, 2015.

[27] M.S. Buker, and S.B. Riffat, "Solar assisted heat pump systems for low temperature water heating applications: A systematic review", *Renew. Sustain. Energy Rev.*, vol. 55, pp. 399-413, 2016.
[http://dx.doi.org/10.1016/j.rser.2015.10.157]

[28] E. Bellos, C. Tzivanidis, A. Delis, and K.A. Antonopoulos, "Antonopoulos, Comparison of a solar driven heat pump heating system with other typical heating systems with TRNSYS", *Proceedings of " The 28th International Conference on Efficiency, Cost, Optimization, Simulation and Environmental Impact of Energy Systems",* 2015 Pau, France

[29] E. Bellos, C. Tzivanidis, and K.A. Antonopoulos, "Exergetic, energetic and financial evaluation of a solar driven absorption cooling system with various collector types", *Appl. Therm. Eng.*, vol. 102, pp. 749-759, 2016.
[http://dx.doi.org/10.1016/j.applthermaleng.2016.04.032]

[30] F. Wang, H. Feng, J. Zhao, W. Li, F. Zhang, and R. Liu, "Performance Assessment of Solar Assisted Absorption Heat Pump System with Parabolic Trough Collectors", *Energy Procedia,* vol. 70, pp. 529-536, 2015.
[http://dx.doi.org/10.1016/j.egypro.2015.02.157]

[31] N. Chaiyat, and T. Kiatsiriroat, "Simulation and experimental study of solar-absorption heat transformer integrating with two-stage high temperature vapor compression heat pump", *Case Studies in Thermal Engineering,* vol. 4, pp. 166-174, 2014.
[http://dx.doi.org/10.1016/j.csite.2014.09.007]

[32] H. Zhang, Z. Li, and H. Zhao, "Thermodynamic performance analysis of a novel electricity-heating cogeneration system (EHCS) based on absorption heat pump applied in the coal-fired power plant", *Energy Convers. Manage.,* vol. 105, pp. 1125-1137, 2015.
[http://dx.doi.org/10.1016/j.enconman.2015.08.075]

[33] Y. Su, S.B. Riffat, and G. Pei, "Comparative study on annual solar energy collection of a novel lens-walled compound parabolic concentrator (lens-walled CPC)", *Sustainable Cities and Society,* vol. 4, pp. 35-40, 2012.
[http://dx.doi.org/10.1016/j.scs.2012.05.001]

[34] L. Guiqiang, P. Gang, S. Yuehong, J. Jie, and S.B. Riffat, "Experiment and simulation study on the flux distribution of lens-walled compound parabolic concentrator compared with mirror compound parabolic concentrator", *Energy,* vol. 58, no. 1, pp. 398-403, 2013.
[http://dx.doi.org/10.1016/j.energy.2013.06.027]

[35] K. Shanks, S. Senthilarasu, and T.K. Mallick, High-concentration optics for photovoltaic applications.*High concentrator photovoltaics fundamentals, engineering and power plants.,* P. PA(c)rez-Higueras, E.F. Fernández, Eds., 1st ed. Springer In-ternational Publishing: Cham, 2015, pp. 85-11.
[http://dx.doi.org/10.1007/978-3-319-15039-0_4]

[36] S.H. Abu-Bakar, F. Muhammad-Sukki, R. Ramirez-Iniguez, T.K. Mallick, M. Abu Bakar, S.H.M. Yasin, and R. Abdul Rahim, "Rotationally asymmetrical compound parabolic concentrator for concentrating photovoltaic applications", *Appl. Energy,* vol. 136, pp. 363-372, 2014.
[http://dx.doi.org/10.1016/j.apenergy.2014.09.053]

[37] P. BenA-tez, J.C. Miñano, P. Zamora, R. Mohedano, A. Cvetkovic, M. Buljan, and M. Julio Chaves, "Hernández, High performance Fresnel-based photovoltaic concentrator", *Opt. Express,* vol. 18, pp. A25-A40, 2010.
[http://dx.doi.org/10.1364/OE.18.000A25]

[38] L. Micheli, K.S. Reddy, and T.K. Mallick, "Plate micro-fins in natural convection: an opportunity for passive concentrating photovoltaic cooling", *Energy Procedia,* vol. 82, pp. 301-308, 2015.
[http://dx.doi.org/10.1016/j.egypro.2015.12.037]

[39] M. Farshchimonfared, J.I. Bilbao, and A.B. Sproul, "Full optimisation and sensitivity analysis of a photovoltaic–thermal (PV/T) air system linked to a typical residential building", *Sol. Energy,* vol. 136, pp. 15-22, 2016.
[http://dx.doi.org/10.1016/j.solener.2016.06.048]

[40] R. Petela, "Exergy analysis of solar radiation, (Chapter 2)", In: *Solar Thermal Sciences and Engineering Applications* CRC Press, Taylor & Francis Group, 2013.

[41] M. Abdelhamid, B.K. Widyolar, L. Jiang, R. Winston, E. Yablonovitch, G. Scranton, D. Cygan, H. Abbasi, and A. Kozlov, "Novel double-stage high-concentrated solar hybrid photovoltaic/thermal (PV/T) collector with nonimaging optics and GaAs solar cells reflector", *Appl. Energy,* vol. 182, pp. 68-79, 2016.
[http://dx.doi.org/10.1016/j.apenergy.2016.07.127]

[42] G. Evola, and G. Margani, "Renovation of apartment blocks with BIPV: Energy and economic evaluation in temperate climate", *Energy Build.,* vol. 130, pp. 794-810, 2016.

[http://dx.doi.org/10.1016/j.enbuild.2016.08.085]

[43] N. Aste, C. Del Pero, and F. Leonforte, "The first Italian BIPV project: Case study and long-term performance analysis", *Sol. Energy,* vol. 134, pp. 340-352, 2016. [http://dx.doi.org/10.1016/j.solener.2016.05.010]

[44] H. Baig, E.F. FernA□ndez, and T.K. Mallick, "Influence of spectrum and latitude on the annual optical performance of a dielectric based BICPV system", *Sol. Energy,* vol. 124, pp. 268-277, 2016. [http://dx.doi.org/10.1016/j.solener.2015.11.043]

[45] H. Baig, N. Sellami, and T.K. Mallick, "Performance modeling and testing of a Building Integrated Concentrating Photovoltaic (BICPV) system", *Sol. Energy Mater. Sol. Cells,* vol. 134, pp. 29-44, 2015. [http://dx.doi.org/10.1016/j.solmat.2014.11.019]

[46] M.J. Huang, "The effect of using two PCMs on the thermal regulation performance of BIPV systems", *Sol. Energy Mater. Sol. Cells,* vol. 95, pp. 957-963, 2011. [http://dx.doi.org/10.1016/j.solmat.2010.11.032]

[47] H. Elarga, F. Goia, A. Zarrella, A. Dal Monte, and E. Benini, "Thermal and electrical performance of an integrated PV-PCM system in double skin façade: A numerical study", *Sol. Energy,* vol. 136, pp. 112-124, 2016. [http://dx.doi.org/10.1016/j.solener.2016.06.074]

[48] R. Kamela, N. Ekramia, P. Dasha, A. Funga, and G. Hailub, "BIPV/T+ASHP: Tech-nologies for NZEBs", *Energy Procedia,* vol. 78, pp. 424-429, 2015. [http://dx.doi.org/10.1016/j.egypro.2015.11.687]

[49] C.J. Ho, A.O. Tanuwijava, and C-M. Lai, "Thermal and electrical performance of a BIPV integrated with a microencapsulated phase change material layer", *Energy Build.,* vol. 50, pp. 331-338, 2012. [http://dx.doi.org/10.1016/j.enbuild.2012.04.003]

[50] F. Abbassi, N. Dimassi, and L. Dehmani, "Energetic study of a Trombe wall system under different Tunisian building configurations", *Energy Build.,* vol. 80, pp. 302-308, 2014. [http://dx.doi.org/10.1016/j.enbuild.2014.05.036]

[51] M. Bojić, K. Johannes, and F. Kuznik, "Optimizing energy and environmental performance of passive Trombe wall", *Energy Build.,* vol. 70, pp. 289-286, 2014. [http://dx.doi.org/10.1016/j.enbuild.2013.11.062]

[52] F. Abbassi, and L. Dehmani, "Experimental and numerical study on thermal performance of an unvented Trombe wall associated with internal thermal fins", *Energy Build.,* vol. 15, pp. 119-128, 2016.

[53] Z. Hu, W. He, X. Hong, J. Ji, and Z. Shen, "Numerical analysis on the cooling performance of a ventilated Trombe wall combined with venetian blinds in an office building", *Energy Build.,* vol. 126, pp. 14-27, 2016. [http://dx.doi.org/10.1016/j.enbuild.2016.05.016]

[54] X. Hong, W. He, Z. Hu, C. Wang, and J. Ji, "Three-dimensional simulation on the thermal performance of a novel Trombe wall with venetian blind structure", *Energy Build.,* vol. 89, pp. 32-38, 2015. [http://dx.doi.org/10.1016/j.enbuild.2014.12.014]

[55] S. Duan, C. Jing, and Z. Zhao, "Energy and exergy analysis of different Trombe walls", *Energy Build.,* vol. 126, pp. 517-523, 2016. [http://dx.doi.org/10.1016/j.enbuild.2016.04.052]

[56] J. Shen, S. Lassue, L. Zalewski, and D. Huang, "Numerical study on thermal behavior of classical or composite Trombe solar walls", *Energy and Buildings,* vol. vol. 39, pp. 962-974, 2007. [http://dx.doi.org/10.1016/j.enbuild.2006.11.003]

[57] M. Rabani, V. Kalantar, A.A. Dehghan, and A.K. Faghih, "Experimental study of the heating performance of a Trombe wall with a new design", *Sol. Energy,* vol. 118, pp. 359-374, 2015.

[http://dx.doi.org/10.1016/j.solener.2015.06.002]

[58] G. Zhou, and M. Pang, "Experimental investigations on thermal performance of phase change material □?" Trombe wall system enhanced by delta winglet vortex generators", *Energy,* vol. 93, no. 1, pp. 758-769, 2015.
[http://dx.doi.org/10.1016/j.energy.2015.09.096]

[59] J. Jie, Y. Hua, P. Gang, J. Bin, and H. Wei, "Study of PV-Trombe wall assisted with DC fan", *Build. Environ.,* vol. 42, pp. 3529-3539, 2007.
[http://dx.doi.org/10.1016/j.buildenv.2006.10.038]

[60] B. Zalba, J.M. Marin, L.F. Cabeza, and H. Mehling, "Review on thermal energy storage with phase change: materials, heat transfer analysis and applications", *Appl. Therm. Eng.,* vol. 23, pp. 251-283, 2003.
[http://dx.doi.org/10.1016/S1359-4311(02)00192-8]

[61] A. Sharma, V.V. Tyagi, C.R. Chen, and D. Buddhi, "Review on thermal energy storage with phase change materials and applications", *Renew. Sustain. Energy Rev.,* vol. 13, pp. 318-345, 2009.
[http://dx.doi.org/10.1016/j.rser.2007.10.005]

[62] F. Kuzriik, D. David, K. Johannes, and J-J. Roux, "A review on phase change materials integrated in building walls", *Renew. Sustain. Energy Rev.,* vol. 15, pp. 379-391, 2011.
[http://dx.doi.org/10.1016/j.rser.2010.08.019]

[63] C.V. Konstantinidou, *A. Novoselac, Integration of Thermal Energy Storage in Buildings.* Univ.: Austin, Tex., 2010.

[64] X. Fang, and T. Yang, "Regression methodology for sensitivity analysis of solar heating walls", *Appl. Therm. Eng.,* vol. 28, no. 17-18, pp. 2289-2294, 2008.
[http://dx.doi.org/10.1016/j.applthermaleng.2008.01.009]

[65] L.F. Cabeza, M. Medrano, C. CastellA3n, A. Castell, C. SolA(c), J. Roca, and M. NoguA(c)s, "Thermal energy storage with phase change materials in building envelopes", *Contrib. Sci.,* vol. 3, no. 4, pp. 501-510, 2007.

[66] M. Moghiman, M. Hatami, and M. Boghrati, "Improvement the winter space heating by the effect of rotating thermal wall storage", *Proceedings of ECOS,* 2011 Novi Sad, Serbia.

[67] E.D. Kravvaritis, K.A. Antonopoulos, and C. Tzivanidis, "Experimental determination of the effective thermal capacity function and other thermal properties for various phase change materials using the thermal delay method", *Appl. Energy,* vol. 88, no. 12, pp. 4459-4469, 2011.
[http://dx.doi.org/10.1016/j.apenergy.2011.05.032]

CHAPTER 4

Applications of Bioenergy - Modeling of Anaerobic Digestion

Emmanouil D. Rogdakis* and **Panagiotis I. Bitsikas**

Laboratory of Applied Thermodynamics, School of Mechanical Engineering, National Technical University of Athens, Athens, Greece

Abstract: Due to the necessity for the reduction of the utilization of fossil fuels, the share of renewable energy sources is expected to further increase in the near future. Energy from biomass, alternatively defined as bioenergy, already provides the majority of the renewable energy worldwide. In this chapter, the use of solid, liquid and gaseous biomass as an energy source is presented, with particular emphasis being given to biogas production by anaerobic digestion.

The chapter is divided into six parts: In the first, a brief description of the global energy consumption and the expected trends are given. In the second part, the advantages and disadvantages associated with the use of biomass as an energy source are discussed. Furthermore, the sources of biomass (forestry, agriculture residues and energy crops, waste), along with its properties are investigated. In the third part, potential fields for the utilization of the three forms of biomass (solid, liquid and gaseous) for energy production are described, along with ways for biomass upgrading.

In the fourth part, the process of Anaerobic Digestion is described in detail. Anaerobic Digestion is a popular method of biological treatment that is suitable for many types of wastes (sewage sludge produced by wastewater treatment, animal waste, the organic matter of municipal solid waste) and leads to biogas production. Then, a common model for the modeling of the process is presented. ADM1 (Anaerobic Digestion Model No. 1) is applied on two wastewater treatment plants operating in Stockholm and in Athens. In the fifth part, the factors that mostly affect Anaerobic Digestion are discussed and the effect of selected parameters (temperature, digester volume, inflow rate and number of digestion stages) is practically examined with the use of ADM1. In the sixth part, the modification of ADM1 to enable its use on the treatment of olive-mill waste (wastewater and solid waste) is presented. The model is then applied to the combined treatment of wastewater and solid waste of a small rural olive-mill. Finally, in the Appendices, the suggested values of ADM1 parameters and the effect of selected parameters of the model in biogas quantity and energy content are presented, along with the suggested modification of the model parameters for its application on olive-mill waste.

* **Corresponding Author Emmanouil D. Rogdakis:** Laboratory of Applied Thermodynamics, School of Mechanical Engineering, National Technical University of Athens, Athens, Greece; Tel: +301107713966; Email: rogdemma@central.ntua.gr

Emmanuel D. Rogdakis & Irene P. Koronaki (Eds.)

Keywords: Bioenergy, Biogas, Energy Production, CHP, Power, Heating, Waste Treatment, Anaerobic Digestion, Wastewater.

GLOBAL ENERGY CONSUMPTION AND SUPPLY

World has been facing a huge increase in energy consumption during the last 200 years. The extreme rise in energy demand has been covered almost completely by the use of fossil fuels. Biomass, hydro-electric power and nuclear energy account for a small share of energy production [1]. However, global warming emissions resulting from energy production are a major environmental problem that is going to become more intense. In order to limit the rise of globe's temperature to 2 °C, carbon emissions must be reduced by 75% in industrial countries by 2050. This will be achieved mainly by the reduction of the use of coal and the development of renewable energy sources.

Coal has faced the largest growth among all energy sources within 2000 and 2012, followed by oil and natural gas. Renewable energy was also developed at a great extent compared to the previous years (Fig. **1**)[2*]. The main part of the energy is used in residential activities and services (36%) while industry and transportation account for 27.8% and 27.3%, respectively [2]. For 2015, the three biggest energy consumers were China, USA and India, followed by Russia, Japan, Germany, Brazil, South Korea, Canada and France [3].

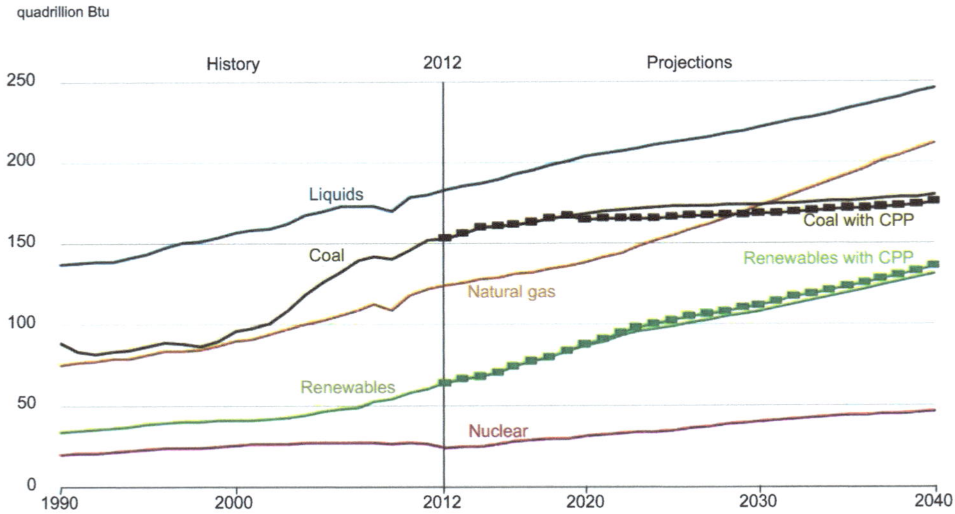

Fig. (1). World energy consumption by energy source 1990-2040 [4].

However, the highest energy consumption per capita has been seen in USA, Canada and countries of Northern Europe like Norway and Sweden. The average power consumption per capita of China is about 4 times lower to that of the USA, while the power consumption per capita in India is among the lowest globally [5]. The low amount of energy consumption per person in developing countries of high population implies that energy demand is going to further increase in the future. Indeed, a 48% increase in global energy demand is expected to occur within 2012 and 2040 [4]. Member countries of the Organization of Economic Co-Operation and Development (OECD)[***3] will account for a small part of this increase, as their energy demands will rise by about 18%. The main part of this increase will from non-OECD members, as their demand of energy is expected to increase by 70%. Asia is expected to show the greatest increase in absolute values, while the energy demand in Middle East and Africa is expected to double.

Renewable energy sources will face the greatest development within now and 2040, followed by nuclear energy and gas. On the contrary, coal is expected to face the lowest growth among all. Except of coal, the share of liquid fuels and petroleum is also expected to drop (Fig. **1**).

BIOMASS AS SOURCE OF ENERGY

Biomass can be defined as "*the biodegradable fraction of products, waste and residues from biological origin from agriculture, including vegetal and animal substances, forestry and related industries including fisheries and aquaculture, as well as the biodegradable fraction of industrial and municipal waste*" [6]. It includes all forms of organic matter that can be used as a source of energy, such as trees, crops and plants, as well as effluents, sludge, manure, industrial byproducts and the organic fraction of municipal solid waste [7].

The carbon dioxide from the atmosphere, along with water absorbed by the plants roots produce carbohydrates or sugars in the photosynthetic process. The carbohydrates and sugars finally form biomass. When biomass undergoes combustion, oxygen from the atmosphere is combined with the carbon in biomass to produce CO_2 and water. The process is therefore cyclic because the carbon dioxide is then available to produce new biomass. This can explain why bioenergy is potentially considered as carbon-neutral and biomass is regarded as a renewable energy source [7].

Bioenergy already provides the majority of renewable energy worldwide and is considered to have the potential to provide a large fraction of world energy demand over the next century [8]. Biomass energy systems can be based on a wide range of feedstock. This feedstock can be used to provide heat and electricity or to power vehicles, either by directly use or after a conversion process. In order to improve the quality of the produced energy, biomass can be upgraded to a fuel with properties close to those of fossil fuels. Alternatively, the physical form in which biomass is used can be modified [9].

Advantages and Disadvantages from the Use of Biomass as Energy Source

As it has already been explained, biomass is regarded as a renewable energy source. In addition to that, advantages occurring from the use of bioenergy that are related to both the economy and the environment are listed below.

- Biomass is widely available all over the world and will also be in the future, in the form of either organic matter from plants and animals, or as waste. The United Nation Environmental Program estimates that biomass provides renewable energy to 1.6 billion people in developing countries [10].
- Biomass is usually a home-grown energy source. As a result, the energy security of a country is enhanced by the utilization of bioenergy [9].
- The use of biomass contributes to the reduction of emissions that account for the climate change [9]. The utilization of biomass in energy applications reduces CO_2 emissions from 55% to 98% compared to fossil fuels, even if transporting of biomass for long distances is needed [11].
- Fuels derived from biomass contain less sulfur, while properly designed systems reduce atmospheric pollutants and improve air quality [9].
- Biomass can be used in all forms, solid, liquid and gaseous. Moreover, it can be stored at times of low demand and provide energy when needed.
- The substances used as biomass for energy production are usually cheap.
- Energy production from biodegradable waste is an efficient way to reduce hazards associated with waste landfilling or bad treatment.
- Bioenergy production can generate jobs and improve local economy and rural development [11].

However, there are serious concerns for bioenergy production associated with land overuse and environmental issues, as listed below:

- Combustion of biomass is not totally clean, as it leads to the production of substances and gases like NO_x, soot, ash, CO and CO_2. Moreover, further emissions of CO_2 may occur during transport of wood or biofuels.
- Harvesting crops for energy production competes directly with food production [9]. Moreover, cultivation of energy crops is associated with the overuse of water resources and impacts of intensive agriculture [10].
- The utilization of wood biomass for energy production on large scale may lead to deforestation, affecting thus the natural habitats and biodiversity. Moreover, biomass coming from harvesting of old forests is associated with the release of large amounts of CO_2 that have been stored for centuries in vegetation and the surrounding soil.

Another source of concern and dispute is that sometimes the energy gain from biomass is small or the production of bioenergy is expensive:

- Biomass is still not cost competitive compared to fossil fuels [11]. Although it is a cheap substance, the process needed for energy production is often expensive.
- The treatment of biomass prior to energy production may need a considerable amount of energy. As a result, the energy gain may be low, especially when biomass must be transported to the energy plants.
- Some crops that are used to produce biofuels are seasonal and, as a result, there are not available during the whole year.
- The calorific value of biomass is still low compared to coal and oil. Moreover, biofuels such as biodiesel are still inefficient compared to gasoline [10].

Bioenergy is believed to be able to play an important role in the near and long-term future [11]. Biomass holds the potential to meet up one third of the projected global energy demand in 2050, with all the constraints relevant to land usage and the environment being taken into account [12]. Moreover, it is anticipated that if energy from biomass is produced in a larger scale and due to the increasing experience, processing and production costs can be reduced by an amount of 15-40% until 2020.

SOURCES AND PROPERTIES OF BIOMASS

Biomass to be used for energy production originates from three general types of **sources**: wood and forestry by-products, energy crops and agriculture residues and waste. Below, each of the three sources will be briefly described.

Wood and Forest Residues

Typical types of wood biomass are harvesting residues such as tops and branches or industry residues like dust, bark, wood chips or liquor [10, 13]. Biomass originating from forestry products and by-products is greatly used for heat and electricity production, either in residential activities or in the industry [13].

The moisture content of biomass to undergo combustion must be lower to 15%. However, solid biomass sources are of higher content. For example, the moisture content of fresh forest wood chips and sawdust may reach 40% and 60%, respectively [7]. Direct combustion of wet biomass reduces boiler efficiency and causes higher gas emissions, so solid biomass may often have to undergo drying pre-treatment [14]. A potential solution is densification, where the material is dried and compresses, acquiring thus uniform properties. Densified products can be found as briquettes or pellets [15].

Biomass originating from forestry has a low heating value. For example, the energy content of pellets is about 18 MJ/kg, lower than the half of the lower calorific value of oil (42 MJ/kg). Its main advantage is that it is abundant in many areas of the world (*e.g.* Scandinavia). Especially in the case of residential heating, where relatively low amount of energy is needed, wood and relevant by products can be used without extensive pretreatment.

Energy Crops and Agriculture Residues

Agricultural residues comprise mainly straw, leaves, fruits and stalk from grass species, such as wheat, maize, barley, and rye [16]. Livestock waste such as manure is also considered to be a kind of agricultural residues. Depending on their type, agricultural residues are able to produce energy by direct combustion, lead to biofuel production or be treated for biogas production.

In the case of **energy crops**, oil seeds are cultivated to produce methylesters and sugar or starch crops are used to produce ethanol [7]. Other crops like willow, poplar and Eucalyptus are sources of solid fuel, while plants such as the giant reed, Cynara and Miscanthus can provide biomass suitable for direct combustion or thermochemical conversion [9].

In 2012, agricultural residues and by-products along with energy crops accounted for about 14% of total biomass supply within the EU [10]. However, an imbalance exists, as the utilization of energy crops is still undeveloped [11]. The utilization of energy crops is still associated with technical difficulties related to the harvesting and storing of the material grown, as well as the high cost of production [9]. Except of the use of land, some types of energy crops may grow once every 3-5 years, so their production is not profitable. However, the production of energy crops is expected to rapidly increase and be able to meet a major part of the increasing bioenergy demand in the near future [10].

Waste

As they can be produced everywhere, waste can be a great source of energy. At the same time, waste disposal and treatment is a major global issue. In 2011, the average amount of municipal solid waste (MSW) alone per capita within Europe was 500 kg of waste per person [17]. Other common types of waste sources are waste coming from agriculture and farming, waste produced by industry and industrial or municipal wastewater.

Landfills are an easy way of waste disposal. From the organic fraction of waste placed at the landfill, a mix of different gases, defined as **landfill gas** is produced. It is created by the action of waste degrading microorganisms and is consisted of methane and carbon dioxide, while trace amounts of volatile organic compounds and hydrocarbons may be present. However, landfill gas can be hazardous if it is not treated. For example, about 62 m^3 of methane are produced from 1 metric ton of MSW placed at a landfill. Those emissions have more than twice the global warming potential than the 1 metric ton of carbon dioxide that would have been produced if the same amount of waste were combusted. However, if landfill gas is collected and burnt, energy is produced and emissions are reduced [18].

Generally, the disposal of waste in landfills is not desirable as poor landfilling pose a threat to the environment and risk on the health of humans living close to them. The non-organic waste can be recycled and get processed into reusable objects. However, recycling is impossible if the material is dirty or contaminated, or consisted of different parts that cannot be separated. In that case, waste can be turned into energy by **thermal treatment**.

Thermal treatment is considered as a good solution for residual waste. It can generally be described as the process of generating electricity and heat from the primary treatment of waste and is greatly developed in many countries, especially in Central and Northern Europe [17]. Direct combustion (incineration) with energy recovery is the most common method of Waste-to-Energy (Wte). About half (50%) of the energy produced by thermal treatment is considered to be renewable. However, it is associated with environmental concerns, such as the potential of pollutants like fine particulate, heavy metals, trace dioxin, acid gas and NO_x entering the atmosphere. Environmental measures in modern incinerators include the reduction of the original waste volume by about 95 %, the strict control of pollutants concentration and the reuse of bottom ash in civil works. A modern Wte plant can save up to 100-450 kg CO_2 per ton of waste processed. Another method suitable for MSW is Refused Derived Fuel (RDF). Its main difference to incineration is that RDF includes treatment prior to waste burning. Recyclable and noncombustible fraction is removed from MSW and the remaining waste is shred to form a combustible material [19].

In general, the energy content of solid waste is low. The calorific value of Municipal Solid Waste is about 9-10 MJ/kg, while by the use RDF the energy content is increased and may reach 14 MJ/kg.

The most suitable method for the treatment of the waste organic fraction is **biological treatment**. Biodegradable waste includes sewage sludge from municipal or industrial wastewater treatment, residues from parks and gardens, food waste from industry or households, biodegradable waste placed in landfills, as well as the organic fraction of the municipal solid waste [7]. The treatment of biodegradable waste leads to energy production in the form of biogas, a gas consisted mainly of methane and carbon dioxide. The energy content of biogas may reach 21-22 MJ/kg.

Maybe the most popular method for biological treatment is anaerobic digestion.

The way of biogas production from Anaerobic Digestion is based on the same concept with the production of landfill gas, as during Anaerobic Digestion organic material is also broken down in several steps. In the case of Anaerobic Digestion, there is an extra by-product along with biogas, called **digestate**. Digestate is slurry of solid fraction consisted of what is left on the treated substrate and contains all the remaining inorganic compounds such as nutrients. As a result, it is an excellent natural fertilizer [20]. Except of its use as fertilizer, digestate can also be further processed into compost [21].

Forestry is by far the prevailing source of bioenergy, although its share is expected to decrease in the near future. According to Table **1**, the utilization of biodegradable waste and agricultural residues is expected to significantly have been increased by 2020 within the European Union (EU). Nevertheless, forestry will continue accounting for more than half the energy produced from biomass [10].

Table 1. Bioenergy production in 2012 and 2020 (expected) within the European Union [10].

Year		Forestry	Agricultural Residues – Crops	Biodegradable Waste	Total
2012	Energy (Mtoe)	71	13.2	10.8	95
	Share (%)	74.7	13.9	11.4	100
2020	Energy (Mtoe)	73.6	41.7	16.7	132
	Share (%)	55.8	31.6	12.7	100

It is clear that the total bioenergy production is expected to increase in the near future. In Europe, an increase of 40% between 2012 and 2020 is expected. In addition to the energy produced by biodegradable waste, an additional amount of renewable energy is produced by the thermal treatment of residual waste. In 2011, about 10.44 Mtoe of energy were produced by Waste-to-Energy plants, half of which is renewable. Thermal treatment of MSW is also developing as both the number and the treatment capacity of plants is increasing [17].

Properties of Biomass

The energy content of biomass is not standard, but varies according to specific factors. A decisive factor in the case of solid biomass is the moisture content. For example, the energy content of industrial softwood chips of 20% moisture is 15.2 MJ/kg, while for the same kind of woodchips of 50% moisture, the lower heating value drops to 9.5 MJ/kg [7]. The important factor for gaseous biomass is the gas composition and, more specifically, the methane content in gas. The lower heating value of biogas is directly proportional to its methane rate. The composition of biogas is not standard. In Table **2**, the range of concentration (Vol %) of biogas components is listed. Biogas produced by Anaerobic Digestion is compared with Natural Gas and Landfill Gas [21].

It can be seen than the composition of biogas is quite similar to the composition of landfill gas. Landfill gas is a type of biogas that contains higher quantity of oxygen, as a considerable amount of air may suck into landfill gas during its collection. Composition of biogas is quite different to that of natural gas, as natural gas contains significantly more methane and almost no carbon dioxide. Based on the methane content of each gas, biogas from Anaerobic Digestion is expected to be of lower calorific value compared to natural gas and of higher compared to Landfill Gas.

Table 2. Composition of biogas, natural gas and landfill gas [21].

Constituent	Natural Gas	Biogas	Landfill Gas
CH_4	91	55-70	45-58
C_2H_6	5.1	0	0
C_3H_8	1.8	0	0
C_4H_{10}	0.9	0	0
C_5H_{12}	0.3	0	0
CO_2	0.61	30-45	32-45
N_2	0.32	0-2	0-3
Volatile Organic Compounds (VOC)	0	0	0.25-0.5
H_2	0	Trace	<1
H_2S	1	round 500	10-200
NH_3	0	round 100	0
CO	0	Trace	Trace

In Table **3**, the Lower Heating Values (LHV) of different forms of solid and gaseous biomass are presented and compared with commonly used fossil fuels and alternative fuels such as ethanol and hydrogen. Three different compositions of biogas are taken into account, corresponding to low, medium and high methane content.

Table 3. Lower heating values of biogas and other fuels [7, 22, 23].

			LHV (MJ/kg)		
Solid Fuels		**Gaseous and Liquid Fuels**		**Gas from Waste Treatment**	
Coal	25	Crude Oil	42.7	Biogas, 50% CH_4	13.3

(Table 3) cont.....

			LHV (MJ/kg)		
Solid Fuels		**Gaseous and Liquid Fuels**		**Gas from Waste Treatment**	
Forest residue	15.4	Gasoline	43.5	Biogas, 60% CH_4	17.6
Pellets	17.5	Ethanol	26.7	Biogas, 70% CH_4	23.0
MSW	9-10	Low-sulfur diesel	42.6	Landfill gas, 55% CH_4	15.6
Pile wood, 50% moist.	9.5	Natural Gas	43.6-49.5	Methane	50
RDF	14	Hydrogen	142	Syngas	3-14

It is clear that biomass (both solid and gaseous) is of significantly lower energy content if they are to be compared with commonly used fossil fuels. The energy content of biomass originated from forestry and waste can be increased by densification. Gaseous biomass is of higher energy content compared to solid biomass, but only biogas rich in methane is of heating value similar a fossil fuel of low energy content such as coal. Moreover, biomass is of lower energy content compared to alternative fuels such as ethanol.

Biomass cannot be competitive to solid fuels in applications requiring high energy density, such as vehicle propulsion. However, there are fields such as the production of heat and power for residential or industrial purposes that biomass can be greatly used. The main reasons for biomass utilization despite its low energy content are either its abundance, in the case of wood and forestry by-products, or the fact that the production of bioenergy is associated with waste treatment, as it happens in the case of anaerobic digestion, landfill gas or even the thermal treatment of MSW.

Because of the lower energy content of biomass, the most cost-effective option for large-scale electricity and heat production from any form of biomass is the co-firing with fossil fuel. This approach makes use of the existing infrastructure, requiring only minor investments in biomass pre-treatment and feed-in systems. It is also benefited by the higher conversion efficiency of the large-scale-coal plants. Coal-fired plants that are near the end of their lifetime are often converted to operate entirely on biomass [10].

APPLICATIONS OF BIOENERGY

The utilization of biomass for energy production is continuously increasing, as it can be used in a variety of applications. In the case of Europe, the majority of bioenergy (75%) is used for heating purposes during 2013, while electricity and transportation accounted for 12.9% and 12.5%, respectively. Transportation has been showing the most rapid increase until now and is expected to continue developing at higher rate in the near future compared to heating and electricity.

Biomass in solid form (forestry, animal waste, plant matter, residues) is used mainly for generation of heat and electricity, along with biomass in gaseous form (landfill gas and biogas produced by the biological treatment of organic waste). Liquid biomass (biofuels) is used as a fuel in transportation sector. Of course, there are many alternative ways of utilization that may require additional treatment, such as the transformation of solid biomass to its gaseous phase or the upgrade of biogas to biomethane.

The utilization of bioenergy has been greatly developed during the last 15 years. Generally, solid biomass is the prevailing form at which biomass is utilized for energy production. According to the respective reports of the European Observatory for 2013 and 2014, 77.4% of the total bioenergy that was consumed within the EU in 2013 came from the utilization of solid biomass. The shares of liquid biomass (biofuels) and gaseous biomass (biogas) were 11.2% and 11.2% respectively [24 - 26].

The largest amount of the globally utilized biomass is burnt for purposes such as cooking, space heating, industrial process heat production and generation of electricity [9]. Biomass is greatly utilized in **residential** activities mainly for heating, while in terms of **industrial** use, generation of electric power used to be the main purpose of biomass utilization.

However, the power efficiency of a biomass-fired power plant is low, at a range of 10-25% [9]. If the rejected heat is used, the total efficiency is significantly increased. In a combined heat and power (CHP or cogeneration) unit, waste heat recovered from the engine cooling system or the exhaust gases of the power production unit is used to heat water or low-pressure steam and thus, satisfy the heat demands of the industrial processes in the facility, either completely or at a great extent [19]. Excess heat can also be used to drive an absorption cooling system. Then, cooling is additionally produced (CCHP, trigeneration). The recovered heat may also be utilized for generation of additional power if required, or for wet biomass drying. A typical ratio of electricity to heat production for a CHP plant is 1:2 to 1:3. The total efficiency of small CHP plants has a value of 70 %, while the total efficiency of a big plant may reach 90%.

A portion of heat recovered in large CHP plants, as well as in waste incinerators or industrial processes can be used in homes or remote industries through district heating [27]. Heat is distributed by the recirculation of hot water or low-pressure steam through underground piping. The heat delivered can also be used for air-conditioning, if a heat pump or an absorption system is added. Biomass-fired district heating is regarded as cost-effective and environmentally friendly. In 2012, Rezaie and Rosen concluded that biomass such as wood chips and wood waste for district heating and cooling is cheaper and more effective in terms of greenhouse gases emissions compared to fossil fuels [28].

Biomass CHP plants for district heating are widespread in Northern and Eastern member states of the EU [10]. For example, in Sweden about 60% of the heat demand of buildings is satisfied by district heating and 50% of the district heating demands are covered by biomass. Ericson and Werner found out that the utilization of biomass for district heating reduced CO_2 emissions by 90% compared to 1970, when only fossil fuels were used. In terms of heat production, it is preferred that new plants suitable only for biomass are used instead of converted fossil-fuel plants [27].

Solid Biomass

Solid biomass is the predominant source of bioenergy globally. It can be found in many forms and is mainly used for the production of heat and power. Solid biomass can also get transformed to gaseous biomass without combustion.

Generation of Heat and Power

In terms of **residential activities**, garden trimmings, wood, pellets or residues are mainly used for the production of heat by boilers, inserts or stoves. Although biomass was initially used for household heating, power is also produced with the use of small-scale CHP plants, especially in rural areas. Currently, biomass is largely used for cooking and lighting, especially in rural areas of developing countries. In 2009, Champier investigated the use of biomass-fired stoves for heat and electricity production [29]. The use of bioenergy on 100 kW_e CHP systems in order to satisfy the heat and power demands of a rural village was investigated by Kimming *et al.* [30]. They concluded that the production of biogas from ley and its combustion in a microturbine was a better solution than the utilization of fossil fuel, as the emissions were reduced for the production of the same amount of energy.

In Europe, 89.1 Mtoe of bioenergy from solid biomass were consumed during 2014. At the same time, 84.8 TWh (7.29 Mtoe) of electricity from solid biomass were produced, along with 9.08 Mtoe of heat. Generally, heat is produced regionally in many plants that are relatively small, but CHP and power-producing plants are bigger in order to increase efficiency [13]. More than 62% of the total electricity was produced by CHP plants. District heating accounts for 13% of the total heat consumed in the EU. It is interesting that in northern countries like Sweden and Finland, power was produced totally by CHP plants. Moreover, district heating was largely used, representing 30.7% and 27.3% of the total amount of heat consumed, respectively.

Some of the biggest solid biomass-fired power plants in Europe are converted old coal-fired plants. For example, two plants of this type in the United Kingdom (Drax and Ironbridge) are currently the biggest solid biomass plants in Europe, utilizing pellets and accounting for 1260 MW and 740 MW, respectively [24]. Electricity production was also boosted by the expansion of biomass cogeneration. The co-firing of biomass with fossil fuels is a field of interest and research during the last years, in order to define the most suitable combination of biomass and fossil fuel, as well as the most efficient biomass ratio. In 2013, Pantaleo *et al.* assessed thermally and economically a 100 kW_e CHP plant fired by solid biomass and natural gas. After testing, they concluded that the best scenario in terms of profitability and energy saving is that the rate of biomass in the fuel mix is equal to 70% [31].

Transformation to Gaseous Biomass

Solid biomass can be transformed to gaseous phase in order to generate higher value products. As a result, the utilization of solid biomass can lead to energy production without direct combustion. The two methods for phase transformation of solid biomass are **pyrolysis** and **gasification**.

Pyrolysis is the thermo-chemical decomposition of organic material that occurs at temperatures higher to 400 °C, without the presence of oxygen [32]. The process involves the simultaneous and irresistible change of chemical composition and physical phase. The main product of pyrolysis is called synthesis gas or **syngas**, while the byproducts include unconverted carbon, charcoal and ash.

Through **gasification**, organic materials containing carbon are converted into syngas at temperatures of around 800 °C or more. The material is reacted with a controlled amount of oxygen or steam. Partial oxidation can be also carried out using air, but the gas produced is of lower quality. **Syngas** is a fuel gas mixture consisting mainly of carbon monoxide (30-60%), hydrogen (25-30%), carbon dioxide (5-15%), methane (0-5%) and water vapor. Syngas produced by air oxidation contains high amount of nitrogen and, therefore, its calorific value is low (3-5 MJ/Nm^3). Opposite to that, the lower heating value of syngas produced by oxygen or steam oxidation varies between 5-12 MJ/Nm^3, depending on the way of production [33]. If syngas is upgraded, its energy content will have a value of 12-14 MJ/Nm^3 and may increase even to 20 MJ/Nm^3 [34, 35].

As the Lower Heating Value (LHV) of natural gas lies between 33-36 MJ/Nm^3, it is obvious that syngas produced by oxygen/steam gasification or pyrolysis is, in most of the cases, a product of medium calorific value. Syngas is combustible and can be directly burnt in a CHP unit. In 2014, Lee *et al.* successfully coupled a downdraft biomass gasification system with a spark-ignited Internal Combustion Engine, using pine, red oak and horse manure as gasification feedstock. The efficiency of gasifier was between 81-85%, while the total efficiency was 20-23% [36]. Moreover, Huang *et al.* modeled a biomass fired trigeneration system consisted by an IC Engine integrated with biomass gasification for district heating and cooling in selected buildings [19].

However, syngas produced by steam or oxygen gasification is mainly used as an intermediate resource for the production of fuels such as synthetic natural gas, hydrogen, biodiesel and synthetic petroleum, or as an intermediate for the production of other chemicals. Gassner and Marechal converted the 66-75% of the dry wood calorific content to Synthetic Natural Gas, also producing power and industrial heat [37]. Moreover, Balat and Kirtay concluded that biomass gasification is a possible system for large-scale production of renewable hydrogen [38]. Because of its low energy density, syngas produced by air gasification cannot be distributed through pipelines and is only suitable for low power and heat production through combustion.

Some of gasification technologies are relatively simple, such the downstream and upstream gasification. Meanwhile, other gasification methods are associated with complexity and higher costs, but also with the production of better quality syngas. Typical examples of those are the circulating fluidized bed and the twin fluidized bed [33]. An option that leads to the production of high-quality syngas is the gasification with the utilization of plasma torch or plasma arc, where carbon electrodes, copper and other metals are used to initiate temperatures higher to 2000 °C [32]. Moreover, high-quality syngas can be produced by the coupling of biomass gasification with in situ CO_2 absorption in a dual fluidized bed system, in a method called Absorption Enhanced Reforming [23]. In 2011, Goeransson *et al.* produced high-quality syngas to be used for automotive fuel synthesis of 12-20 MJ/Nm^3 energy content in a dual fluidized bed gasifier by steam oxidation [35]. Finally in 2010, Rapagna *et al.* increased gas and hydrogen yield by 69% and 130% respectively, by the use of a gas cleaning filter in a bench-scale fluidized biomass gasification plant operating at 800 °C [39].

Biogas

Biogas is often produced in landfills (landfill gas) or in large scale digesters in industrial countries for sewage sludge treatment and stabilization. But the most common source of biogas production are waste (both farming and municipal) and crops, in decentralized agricultural plants, MSW methanisation plants or centralized co-digestion plants. Biogas from agricultural and waste sources accounted for almost 70% of total biogas production within the EU in 2013.

Heat and Power Production

Biogas is widely used in household applications to provide energy for cooking and lighting. Thousands of agricultural biogas plants are currently under operation in Europe, as well as in North America and developing countries. For example, more than 2 million small rural family-owned biogas plants exist in China, India and the countries of South-East Asia, where biogas from Anaerobic Digestion provides the energy needed by a family on a daily basis for cooking, heating and lighting [20].

Biogas is also used in industrial activities. The combustion of biogas in a boiler to produce steam is an established and reliable technology [20]. As boilers do not have a high gas quality requirement, no extensive treatment of biogas prior to its combustion is needed. In most of the cases, a small portion of the produced biogas is burnt for production of heat and the main part of the gas is used in a CHP unit for Combined Heat and Power production [40]. As biogas is a gaseous fuel, power can be generated by the use of both internal and external combustion engines. At the moment, more than 90% of biogas plants in Europe produce heat along with power. Within the European Union, electricity production is the main form of biogas energy recovery, accounting for 4.5 Mtoe in 2013. About 65% of the total electricity production came from CHP plants. At the same time, heat production from biogas reached 2.44 Mtoe. Almost 18% of those were sold to district heating networks [25].

It is desirable that the site of biogas production is close to the site of its utilization. In case biogas cannot be used at the site of its production, it can be conveyed through pipelines to a CHP plant [40]. If the power produced exceeds the demands of the plant, surplus electricity is sold to the grid.

Upgrade to Biomethane

Biogas can be upgraded to biomethane by the removal of impurities, along with quantities of CO_2, H_2 and hydrocarbons. The content of methane in the upgraded biomass must be higher to 95%. Biomethane can be used in a CHP plant at the same way as biogas, but with significantly higher energy production per unit of mass burnt. Moreover, biomethane is often injected into natural gas grid to substitute natural gas [40]. If the facility of biogas production and upgrade in sited close to natural gas grid, biomethane can be stored at low cost and utilized when required.

Biomethane can also be used as renewable transport fuel in vehicles designed on to run on compressed natural gas (CNG) or liquefied natural gas (LNG), either exclusively or alongside natural gas at any ratio. Before its use as fuel, biomethane has to be compressed, as it is stored inside the vehicle at a pressure of 200-250 bar. The use of biomethane as a fuel offers significant advantages, as it is highly efficient and can use the existent infrastructure of natural gas filling stations. Moreover, biomethane is one of the cleanest fuels with little impact on environment and human health [40]. Its use reduces NO_x and hydrocarbons emissions for up to 80% compared to petrol and diesel, while the emissions of particles and soot are also drastically reduced [20].

Biomethane production is increasing within the EU. At the end of 2014, at least 258 biomethane plants were under operation in 12 countries. More than half of those (151 plants) are located in Germany, while biomethane production has also been developed in Sweden, Netherlands and Austria.

Biofuels

Biofuels are liquid fuels produced by biomass sources. They can be divided in primary and secondary ones. The primary biofuels are produced directly by food crops enriched with sugars, starch and oils [41].

The most common **primary** biofuels produced on a commercial basis from conventional agricultural crops are **bioethanol** and **biodiesel** [11]. **Bioethanol** is mainly produced from corn, sugar cane and sugar beets, but it can also be made from wood, paper waste and grown grasses [16]. Bioethanol is very scarcely utilized directly in vehicles and its most common use is as a petrol additive in Spark Ignition Engines [42].

Biodiesel is produced from vegetable oil extracted from seeds or from animal oil by the reaction of oil with methanol or ethanol in a catalytic process. Glycerol is also provided as a by-product [9]. Biodiesel is mainly utilized as an additive to diesel fuel in Diesel Engines [42].

The use of biofuels is developing and is expected to increase more in the future. In Europe, about 13.1 Mtoe of bioenergy were utilized in transport during 2013 within the EU. This amount is expected to have been increased by about 120% by 2020. Biodiesel accounts for about 80% of the totally consumed biofuels, while bioethanol accounts for 19% and biogas and vegetable oil account for the rest 1% [13].

Although the production of primary **biofuels** is relatively simple and cheap, it directly competes with food industry for the use of lands. Moreover, only a very small part of cultivated biomass is currently used for biofuel production. On the contrary, **secondary** biofuels are produced by lignocellulosic biomass, which is highly present in plants and trees. Biomass is then enriched by the extraction of the most energy-intensive substances and, finally, biofuels are produced. The advantages associated with second generation biofuels, are the significantly larger area of land that can be used for the growth of suitable biomass as well as the low impact on food sector. On the other hand, the treatment techniques that lead to biomass enrichment are still difficult [41].

Biofuels in the Future

There are intense efforts over new methods for biofuel production. In 2016, Liu and Wu proposed a two-stage continuous system that is able to produce biohydrogen and biomethane during **wastewater treatment** by the intermediate conversion of biomass to sugars. It was concluded that the use of agricultural residues and waste, liquid waste of food industry and industrial wastewater for the production of bioenergy in the form of hydrogen, methane and ethanol is feasible [43].

Another interesting field is the production of biohydrogen. Hydrogen has by far the highest energy density among all the common fuels and is additionally a clean energy carrier. It can be produced from compounds containing H_2, such as natural gas, coal, heavy oil and water. However, production of hydrogen from fossil fuels requires high amount of energy and causes greenhouse gases emissions. This makes the development of hydrogen production from biomass desirable. Unfortunately, production of hydrogen from biomass by the use of microorganisms is still limited and results in low-yield production. Rahman *et al.* (2016) integrated biohydrogen production and separation with the applications of **fuel cells** for electricity production. Biohydrogen was produced by dark fermentation, a method utilizing carbohydrates such as glucose for the energy and carbon sources, and was purified by the use of membranes. Finally, polymer electrolyte membrane fuel cells (PEMFC) were preferred [22].

Finally, research is already being conducted over the production of third generation biofuels. During 2016, Voloshin *et al.* studied the possibility of production of biofuels with algal as raw material [41]. Algal is an attractive potential, as it does not compete with fuel crops for food or biofuel production, does not require extensive treatment and can grow in conditions unsuitable for other plants. Photobioreactors and open-air systems can be used for the production of biodiesel, bioalcohol and biohydrogen. However, the production of third generation biofuel is still at its development stage.

Pretreatment of Biomass

Biomass often needs treatment before its use, in order that its potential for energy production is increased. Moreover, treatment helps on reducing greenhouse gases.

Solid biomass can be treated prior to its combustion without undergoing phase transformation. It has already been mentioned that pretreatment is required for wet solid biomass in order to reduce its moisture content lower to 15%. A possible solution for wet wood or relevant by-products is densification. The densified material is of moisture content lower to 10%, it has increased energy content and its combustion is more efficient [15]. Drying is another method for the reduction of moisture content. Li *et al.*, managed to reduce moisture content of solid biomass feedstock by 5-15 times by the utilization of low grade waste heat from a 100 MW industrial plant. The dried biomass was efficiently used as fuel in a 40 MW power plant [14].

Torrefaction is a method of solid biomass thermal pretreatment under anoxic conditions. Phanphanich and Mani improved the fuel characteristics of wood chips by torrefaction at 250-300 °C. They found that gross heating value of chips increased with treatment temperature. When temperature was set to 300 °C, wood chips obtained properties near to those of coal [44].

Except of solid biomass, pretreatment is often applied to biogas immediately after the stage of Anaerobic Digestion in order to improve the quality of gas. This stage of treatment includes the removal of substances such as sulfur, ammonia and water vapor, in order to prevent corrosion and wear of the mechanical equipment [39]. Also, chemical treatment for the removal of H_2S, NH_3 and trace elements is applied [45].

Co-generation Systems Used for Energy Production from Biomass

In a co-generation system fired by solid biomass, power is produced by an External Combustion System such as a boiler coupled with a Steam Turbine, while Stirling Engine is another option. If liquid or gaseous biomass is used as a fuel, an Internal Combustion Engine can be used. However, external combustion systems are more broadly used, as not only the main part of bioenergy is produced by solid biomass but also such systems are utilized in plenty of cases where energy is produced from biogas.

An external combustion biomass-fired cogeneration system is consisted of a boiler or a similar system through which heat is provided by the combustion of biomass, the unit for power generation (expander and alternator for power production), the cooling system (*e.g.* a condenser) and a heat exchanger for heat recovery. For the additional production of cooling, a single-effect absorption chiller must be added [46].

The Use of Organic Rankine Cycle (ORC)

Organic Ranking Cycle (ORC) describes closely the process by which steam-operated engines generate power. It usually gets energy from low-temperature heat sources such as geothermal and industrial waste heat. ORC technology allows the use of heat sources at a temperature as low as 100 °C. As a result, it can be integrated in biomass applications such as a biomass-fired micro-CHP system or waste heat recovery for biogas digestion plant [47].

An ORC is consisted of a boiler that provides the heat required, evaporator, expander, condenser and alternator that converts work to power. It is associated with reduced cost and complexity, as it operates at lower temperature and pressure than the conventional steam process [48]. Biomass-fuelled ORC plants coupled with CHP have been successfully designed and improved for power production in the range of 200-2000 kW_e [49].

In 2011, Al-Sulaiman *et al.* used a biomass-fired ORC for combined cooling, heating and power production, concluding that the use of cycle for trigeneration instead of only power production increases efficiency and decreases carbon emissions [46]. In 2012, Qiu *et al.* investigated experimentally an ORC-based micro-CHP system fired by pellets, reaching about 95% thermal efficiency and producing heat for domestic water and underfloor heating [48]. During the same year, Taijan *et al.* tried to develop a biomass-fired ORC CHP system with heat storage [50]. Finally, Huang *et al.* concluded in 2013 that the operation of a 150 kW_e ORC getting energy from biomass direct combustion or biomass gasification is viable if heat is recovered and used along with power [51].

Use of Stirling Engines for Bioenergy Production

Stirling Engines (SE) are considered to be the best technology for CHP generation based on solid and gaseous biomass fuels for electric power output up to 150 kW_e [52]. In general, Stirling engines are based on a closed cycle, where the working gas is alternatively compressed in a cold cylinder volume and expanded in a hot cylinder volume. The heat input from the combustion of fuel is transferred to the working gas through a hot heat exchanger (**heater**) at high temperature, typically between 650 and 750 °C. The two pistons are driven by heat and, as a result, drive the shaft for power production. The amount of heat that is not converted into work on the shaft is rejected to the cooling water in a cold heat exchanger (**cooler**) at 25-75 °C. Between heater and cooler, an intermediate heat exchanger called **regenerator** is placed. It contains a porous material and is used for transferring heat from and to the working medium flowing between the heating and cooling space (Fig. **2**) [53].

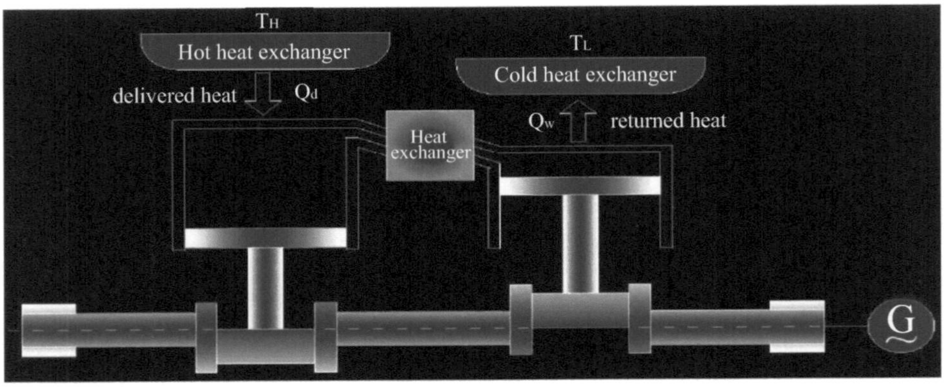

Fig. (2). A-type Stirling engine [53].

In Stirling Engines, heat is supplied to the working medium through the heating of the heater's external wall. So it is possible to supply the engine's primary energy practically from any source that can deliver flue gases with low or no particles and ash. Heat can be provided to a Stirling Engine by the direct combustion of fuels such as coal, natural gas, oil derivatives, wood and its by-products, waste products, biogas or bio-oil. Other sources of heat include solar radiation, nuclear power and gasified biomass. The efficiency of a Stirling Engine is mainly dependant on temperature difference between the heater and the cooler, and, in general, it does not exceed 30% [53].

The main advantage associated with the operation of Stirling Engines is their ability to use a variety of fuels [53]. Other advantages include the low production of noise, their little need of maintenance and their ability for long term operation. Moreover, the closed cycle of Stirling Engines enables the use of a working gas that is better suited for heat transfer to and from the cycle than air, such as hydrogen or helium. Finally, the combustion related problems are reduced with the use of a Stirling Engine, as its combustion system can be based on proven furnace technology [54].

A typical application of Stirling Engines is their use as a CHP biomass combustion plant. (Fig. **3**) The total efficiency of a SE for combined heat and power production can be up to 90% [53]. As the efficiency is highly dependent on the temperature of heater, combustion air must be preheated by the flue gas leaving the engine's heater. The temperature of the combustion air is raised to 500-600 °C, resulting in high temperatures in the combustion chamber [54].

Efforts in design of small-scale biomass-fired CHP plants were intensified after 2000. In 2003, a collaboration between the Technical University of Denmark and Austrian manufacturers led to the development of a pilot small-CHP plant of electric power output of 35 kW_e and thermal output 200 kW_{th} in Austria. The plant was based on a hermetic four cylinder Stirling engine and utilized woodchips fuels [54]. One year after, a small-scale CHP plant based on a 75 kW_{el} hermetic eight cylinder Stirling engine was developed, reaching thermal power of 475 kW and overall efficiency of 86% [52]. As the reliability of Stirling Engines was proven after several hours of operation, a hot gas heat exchanger and a new cleaning system were developed, in order to optimize the performance of the engine [55].

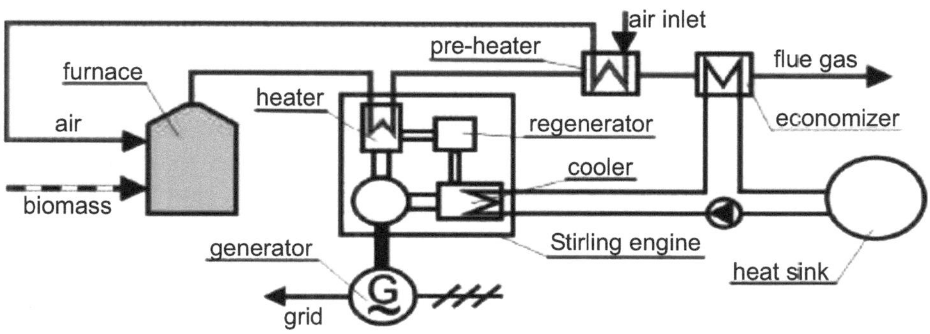

Fig. (3). CHP biomass combustion plant with Stirling [53].

In addition to this work, several efforts have been made in order that biomass-fired Stirling Engines are efficiently utilized for energy production. In 2012, Szewczyk and Trzepiecinski coupled a SE with a direct combustion boiler to cover the energy demands of small farms in rural areas, achieving a fuel economy of 35% and reducing CO_2 emissions [53]. Two years later, Koelling *et al.* developed a high temperature heat exchanger for the coupling of biomass boiler and Stirling Engine of 4.5 kW_e capacity [56]. A 55 kW_e Stirling engine CHP system utilizing pulverized wood powder was developed by Nishiyama *et al.* [57]. Finally, Maraver *et al.* concluded that Stirling Engines are the best alternative to be utilized as a Prime Mover in a biomass-fired CCHP (Combined Cooling, Heat and Power) small scale plant [58].

In terms of the utilization of gaseous biomass as a heat source, Cotana *et al.* investigated the potential of ORC turbines and Stirling engines to produce electricity from gasified poultry waste [49]. Rokni investigated the case of a small-scale integrated gasification solid oxide fuel cell (SOFC) and SE for combined heat and 120 kW_e power, with woodchips being used as gasification feedstock. The cost of SOFC and Stirling engine technology was found to be competitive compared to renewable systems of the same size [59, 60].

Finally, efforts are done in order to develop Stirling Engines that will be able to operate efficiently while getting energy from a low-temperature heat source. Dellali *et al.* worked on the development of a SE that will be able to utilize waste heat at low temperatures [61]. Toda and Matsubara worked on developing low temperature SE that can generate about 1 kW_e by using a heat source of about 100 °C [62].

INTRODUCTION TO ANAEROBIC DIGESTION

Anaerobic Digestion (AD) is a process of **naturally occurring decomposition and decay**, where organic matter is **broken down** by microorganisms to its simpler chemical components **in the absence of oxygen** in order to produce **methane and carbon dioxide**. Anaerobic digestion is used for both industrial and domestic purposes and results to both biogas production and waste management. This process can be applied to different types of organic materials. Actually, wood is the only material containing organic fraction that cannot undergo anaerobic digestion. The most common materials to which anaerobic digestion is applied are listed below [39]:

- Animal waste, such as manure and slurry from cattle, pig and poultry, agricultural waste and crop waste [21].
- Sewage sludge, as a by-product of municipal and industrial wastewater treatment [39].
- Energy crops: Typical crops are maize or sweet sorghum [39].
- Green wastes, such as garden and parks waste. They are generally shredded, in order that a more homogeneous feedstock is obtained [21].
- By-products of the food industry, food waste, used kitchen oil.
- Organic fractions of other industrial waste, such as wastes from paper industry, chemical and pharmaceutical waste [39].
- The organic fractions of household waste and Municipal Solid Waste.

Prior to the beginning of Anaerobic Digestion, pre-treatment of feedstock is essential. The non-biodegradable materials are removed, providing a uniform small particle size feedstock for efficient digestion.

The anaerobic digestion of two or more biodegradable substrates is a common practice. It is applied in case that one type of feedstock cannot be anaerobically digested. The simultaneous digestion of multiple substrates is defined as **co-digestion**. Co-digestion may also apply to increase biogas production. For example, the mix of maize and manure is the most commonly used feedstock for decentralized agricultural biogas plants [39].

Anaerobic Digestion consists of 4 steps, namely hydrolysis, acidogenesis, acetogenesis and methanogenesis [63]. The steps of Anaerobic Digestion are described later, along with the presentation of ADM1.

Advantages and Disadvantages of Anaerobic Digestion

The process of anaerobic digestion poses significant advantages in terms of environment protection and energy production.

- It utilizes materials that are widely available [64]. Moreover, it is a way of treating waste that, otherwise would be conveyed to sea, deposited at landfills or burnt without pre-treatment, resulting thus to hazardous emissions.
- It leads to the production of biogas that is relatively rich in methane. As the overall process is described as carbon neutral, Anaerobic Digestion offers a potential of energy production than can replace fossil fuel sources and has a direct positive effect on greenhouse gas reduction [63].
- Methane is generally a major greenhouse gas if it is allowed to escape without treatment. Anaerobic Digestion can maximize methane generation without being harmful to the environment [64].
- Unlike other renewable sources, anaerobic digestion produces renewable energy in the form of a gaseous fuel. As a result, energy can be stored and used during peak demand times [64].
- It leads to valuable by-products, being one of the few technologies that are able to convert organic waste into energy without loss of nutrients [20].
- Anaerobic digestion reduces the biological oxygen demand (BOD) and the chemical oxygen demand (COD) of effluents, decreasing the danger of effluents being hazardous for the environment [64].
- In case of wastewater treatment, the volume of effluent sludge is considerably lower to that prior to the treatment. As a result, disposal is easier and of lower cost [20].
- Anaerobic digestion can reduce odor up to 80%.
- During the whole process, there is almost no use of chemicals.

However, there are still remaining issues that are an obstacle to the further development of Anaerobic Digestion.

- Anaerobic Digestion starts at a very low rate and a significant amount of time is needed for the production of biogas.
- It is associated with high initial investment and operational costs. In order that AD is viable, not only biogas but also the by-products must be used.
- The system is very sensitive to changes in operating conditions. Several

parameters such as pH, temperature and alkalinity must be tightly controlled in order to achieve optimum performance [64].

- Microorganisms are sensitive in toxic materials such as heavy metals, ammonia and antibiotics [64]. As heavy metals present in the feedstock cannot be removed during the process, feedstock must be as clean as possible [20].

The produced digestate is acceptable for purposes such as soil amendment or landscaping. Digestate contains nearly 100% of the nutrients from the substrate, being thus a cheaper and environmentally friendly alternative to chemical fertilizers [20]. Digestate can also get composted, in order to obtain a product of higher value [21].

ANAEROBIC DIGESTION MODEL NO.1 (ADM1)

As Anaerobic Digestion is a complex process that is most of the times slow and expensive, the development of suitable models has been a necessity for its better understanding and optimization. Although many different models for Anaerobic Digestion had been introduced prior to the development of ADM1, their use by engineers, technology providers and operators has been very limited. This is due to the wide variety of available models and the fact that many of them were applicable only to a specific type of waste. On the other hand, the presence of a generic model for Anaerobic Digestion is associated with many advantages. ADM1, which was introduced in 2002, is a common basis for the further development of studies, in order to make outcomes more comparable. A general and applicable model may lead to increased application for Anaerobic Digestion in full-scale plants, and will make the process control and optimization easier and more efficient [63].

ADM1 is applicable for a wide range of substrates. For example, it has been used to simulate the anaerobic digestion of organic wastes such as municipal wastewater sludge and two-phase olive pulp [65]. This does not come without cost, as it may not be as accurate as other relevant models. But if the treatment of a specific type of waste is to be applied, the extension of ADM1 in order to include more compounds and processes is possible. For example, Fezzani and Ben Cheikh extended the original model in order to describe the anaerobic degradation of phenol compounds during the digestion of olive mill waste [66].

The reaction system in the anaerobic reactor consists of a number of steps in row or in parallel. Those reactions can be divided into two main types, namely biochemical and physicochemical ones: The reactions that are normally catalyzed by intracellular or extracellular enzymes are defined as **biochemical**, while **physicochemical** reactions are not biologically mediated.

Biochemical reactions act on the pool of biologically available organic material. They are the basic part of ADM1 and any anaerobic system can be described only by the use of those equations. A non biological step, disintegration of complex solids, has been added to the four main steps of the process, namely hydrolysis, acidogenesis, acetogenesis and methanogenesis [67]. Biochemical processes are described below and are graphically illustrated in Fig. (**4**). The numbers shown in the Figure are just indicative and may vary according to the substrate and the operating conditions of the process. The term "substrate" refers to the biologically degradable COD (Chemical Oxygen Demand) and not to the total input of influents inside the digester [64].

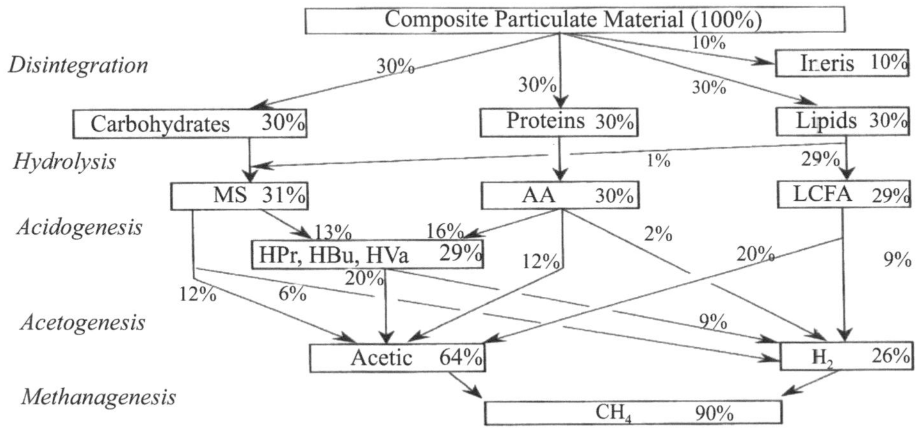

Fig. (4). Schematic illustration of anaerobic digestion steps [63].

- Disintegration: The composite particulates of the substrate are degraded to complex particulate polymers [63].
- During **hydrolysis**, the fermentative bacteria convert the products of disintegration to their soluble monomers. Hydrolysis is the most complex step of Anaerobic Digestion and defines the rate of the whole process [63].
- The further breakdown of the remaining components by fermentative bacteria is called **acidogenesis**. Mixed organic acids are produced, along with carbon dioxide and hydrogen [63].
- During **acetogenesis**, the products of acidogenesis are broken down by acetogenic bacteria to form acetic acid, as well as carbon dioxide and hydrogen [63].
- During the final step (**methanogenesis**), methane is produced by both cleavage of acetate to methane and reduction of carbon dioxide by molecular hydrogen to produce methane [67].

An extra process that is taken into account in the ADM1 is the **decay and growth**

of biomass. Biomass growth is implicit during the substrate uptake. After its decay, dead biomass is maintained in the system as composite particulate material.

Physicochemical reactions include acid-base reactions and gas-liquid transfer. They are of great importance during the modeling of anaerobic systems, while they are used for the expression of a number of factors such as pH, as well as the concentration of free acids and bases and the properties of biogas.

ADM1 is a complex model including 32 dynamic state concentration variables and 19 biochemical rate processes. It can be solved through a set of differential equations. The set of differential equations (DE) of the ADM1 model in liquid phase are as follows: 10 equations are used to model the evolution of soluble matter concentrations, while two DE model inorganic carbon and inorganic nitrogen levels. Twelve DE depict the dynamic behavior of particulate matter and biomass concentrations. From those, two equations are used to model cations and anions levels in and an additional six are involved in acid–base reactions to determine the pH of effluent and to calculate ionized forms of organic acids, free ammonia nitrogen and carbon-dioxide concentrations [66]. As the acid-base processes are rapid compared to the other reactions, they can also be solved by algebraic equations.

The nomenclature used in the model, along with the description of the variables is shown in Table **4**. It is important that there is no unit transformation, as all the equations of the model are based on the suggested units. In Table **5**, the dynamic state variables included in the biochemical processes of the models are listed, along with the carbon and nitrogen contents of all substances. For some cases, those contents are the same for every substrate and are given in the first presentation of ADM1 [63]. These are shown in bold. The others are provided in the model application for wastewater treatment, conducted by Rosen and Jepson [68, 69].

Table 4. Nomenclature and description of parameters and variables [63].

Symbol	Description	Units
C_i	Carbon content of component i	$kmole_C/kg_{COD}$
N_i	Nitrogen content of component i	$Kmole_N/kg_{COD}$
$v_{i,j}$	Rate coefficients for component i on process j	kg_{COD}/m^3
$f_{product,substrate}$	Yield of product on substrate	$kg_{COD,product}/kg_{COD,substrate}$
$K_{a,acid}$	Acid-base equilibrium coefficient	M
K_H	Henry's law coefficient	M/bar
$k_{A/Bi}$	Acid base kinetic parameter	1/(M*d)

(Table 4) cont.....

Symbol	Description	Units
k_{dec}	First order decay rate	1/day
$I_{inhibitor, process}$	Inhibition function	----------
$k_{process}$	First order parameter	1/day
$k_L a$	Gas-liquid transfers coefficient	1/day
$K_{I,inhibit,process}$	50% inhibitory concentration	kg_{COD}/m^3
$k_{m,process}$	Maximum specific uptake rate	$kg_{COD\_S}/(kg_{COD\_X}*day)$
$K_{S,process}$	Half saturation value	kg_{COD_S}/m^3
ñj	Kinetic rate of process j	$kg_{COD}/(m^3*day)$
$Y_{substrate}$	Yield of biomass on substrate	kg_{COD_X}/kg_{COD_S}
$p_{gas,i}$	Pressure of gas i	Bar
P_{gas}	Total gas pressure	Bar
S_i	Soluble component i	kg_{COD}/m^3
T	Temperature	K
V	Volume	m^3
X_i	Particulate component i	kg_{COD}/m^3

Table 5. Dynamic state variable characteristics [63].

Symbol	Number (i)	Name	Unit	C_i	N_i
X_C	13	Composite	kg_{COD}/m^3	0.02786	0.0376/14
X_{ch}	14	Carbohydrates	kg_{COD}/m^3	**0.0313**	**0**
X_{pr}	15	Proteins	kg_{COD}/m^3	0.03	0
X_{li}	16	Lipids	kg_{COD}/m^3	**0.0220**	**0**
X_I	24	Particulate inerts	kg_{COD}/m^3	0.03	0.06/14
S_I	12	Soluble inerts	kg_{COD}/m^3	0.03	0.06/14
S_{su}	1	Monosacharides	kg_{COD}/m^3	**0.0313**	**0**
S_{aa}	2	Amino acids	kg_{COD}/m^3	0.03	0.007
S_{fa}	3	Total LCFA	kg_{COD}/m^3	**0.0217**	**0**
S_{va}	4	Total valerate	kg_{COD}/m^3	**0.0240**	**0**
S_{bu}	5	Total butyrate	kg_{COD}/m^3	**0.0250**	**0**
S_{pro}	6	Total propionate	kg_{COD}/m^3	**0.0268**	**0**
S_{ac}	7	Total acetate	kg_{COD}/m^3	**0.0313**	**0**
S_{h2}	8	Hydrogen	kg_{COD}/m^3	**0**	**0**
S_{ch4}	9	Methane	kg_{COD}/m^3	**0.0156**	**0**
S_{IC}	10	Inorganic carbon	M	**1**	**0**

(Table 5) cont.....

Symbol	Number (i)	Name	Unit	C_i	N_i
S_{IN}	11	Inorganic nitrogen	M	**0**	**1**
$X_{su,aa,fa,C4,pro,ac,h2}$	17-23	Biomass	kg_{COD}/m^3	**0.0313**	**0.00625**

Biochemical Reactions

The substrate to be treated is defined as **composite** (X_C). The particulate polymers formed after the disintegration of substrate are divided into **carbohydrates** (X_{ch}), **proteins** (X_{pr}), **lipids** (X_{li}), and **inerts, both particulate** (X_I) and **soluble** (S_I). During hydrolysis, carbohydrates, proteins and lipids are broken into their soluble monomers, namely **sugars** (S_{su}), **amino acids** (S_{aa}), and **fatty acids** (S_{fa}). The organic ions formed after the completion of acidogenesis are **total valerate** (S_{va}), **total butyrate** (S_{bu}) and **total propionate** (S_{pro}). **Total acetate** (S_{ac}) and **hydrogen** (S_{h2}) are the products of acetogenesis, while methanogenesis leads to the formation of **methane** (S_{ch4}). **Inorganic Carbon** (S_{IC}) and **Inorganic Nitrogen** (S_{IN}) are the two additional soluble components included in the biochemical reactions.

Biomass is grown during the uptake of substrate. It is on particulate form and is divided according to the microorganisms that degrade substrate during each process. In total, seven types of biomass are included. During acidogenesis, three groups of bacteria break down sugars ($\mathbf{X_{su}}$), amino acids ($\mathbf{X_{aa}}$) and fatty acids ($\mathbf{X_{fa}}$). During acetogenesis, one group of microorganisms is used to break down valerate and butyrate ($\mathbf{X_{C4}}$) and another one is used for propionate ($\mathbf{X_{pro}}$). Finally, two additional groups of bacteria are used to degrade acetate ($\mathbf{X_{ac}}$) and hydrogen ($\mathbf{X_{h2}}$) during methanogenesis. After their decay, bacteria are converted into composite particulate ($\mathbf{X_C}$) and may undergo disintegration.

Generally, soluble matter has been assigned with numbers i=1-12 and particulate matter corresponds to i=13-24. Moreover, cations and anions are soluble components that are involved in physicochemical reactions and affect the value of pH. The biochemical processes included in ADM1 are shown in Table **6**. The change of concentration of any component is calculated as shown in equations **(1)** and **(2)**, for soluble and particulate components respectively.

$$\frac{dS_i}{dt} = \frac{q_{in}}{V_{liq}} \cdot (S_{i,in} - S_i) + \Sigma v_{i,j} \cdot \rho_j \qquad \textbf{(1)}$$

$$\frac{dX_i}{dt} = \frac{q_{in}}{V_{liq}} \cdot (X_{i,in} - X_i) + \Sigma v_{i,j} \cdot \rho_j \qquad \textbf{(2)}$$

Table 6. Biochemical rate coefficients for soluble and particulate components [63].

Component i	1	2	3	4	5	6	7	8	9	10	11	Rate
Process j	S_{su}	S_{aa}	S_{fa}	S_{va}	S_{bu}	S_{pro}	S_{ac}	S_{h2}	S_{ch4}	S_{IC}	S_{IN}	
1. Disintegration										$-\Sigma C_i*v_{i,1}$	$-\Sigma N_i*v_{i,1}$	$\rho 1$
2. Hydrolysis of Carbohydrates	1									$C_{ch}-C_{su}$		$\rho 2$
3. Hydrolysis of Proteins		1								$C_{pr}-C_{aa}$		$\rho 3$
4. Hydrolysis of Lipids	$1-f_{fa,li}$		$f_{fa,li}$							$-\Sigma C_i*v_{i,4}$		$\rho 4$
5. Uptake of sugars	-1				$(1-Y_{su})*f_{bu,su}$	$(1-Y_{su})*f_{pro,su}$	$(1-Y_{su})*f_{ac,su}$	$(1-Y_{su})*f_{h2,su}$		$-\Sigma C_i*v_{i,5}$	$-(Y_{su})*N_{bac}$	$\rho 5$
6. Uptake of Amino Acids		-1		$(1-Y_{aa})*f_{va,a}$	$(1-Y_{aa})*f_{bu,aa}$	$(1-Y_{aa})*f_{pro,aa}$	$(1-Y_{aa})*f_{ac,aa}$	$(1-Y_{aa})*f_{h2,aa}$		$-\Sigma C_i*v_{i,6}$	$N_{aa}-(Y_{aa})*N_{bac}$	$\rho 6$
7. Uptake of LCFA			-1				$(1-Y_{fa})*0.7$	$(1-Y_{fa})*0.3$		$-\Sigma C_i*v_{i,7}$	$-(Y_{fa})*N_{bac}$	$\rho 7$
8. Uptake of Valerate				-1		$(1-Y_{c4})*0.54$	$(1-Y_{c4})*0.31$	$(1-Y_{c4})*0.15$		$-\Sigma C_i*v_{i,8}$	$-(Y_{c4})*N_{bac}$	$\rho 8$
9. Uptake of Butyrate					-1		$(1-Y_{c4})*0.8$	$(1-Y_{c4})*0.2$		$-\Sigma C_i*v_{i,9}$	$-(Y_{c4})*N_{bac}$	$\rho 9$
10. Uptake of Propionate						-1	$(1-Y_{pro})*0.57$	$(1-Y_{pro})*0.43$		$-\Sigma C_i*v_{i,10}$	$-(Y_{pro})*N_{bac}$	$\rho 10$
11. Uptake of Acetate							-1		$(1-Y_{ac})$	$-\Sigma C_i*v_{i,11}$	$-(Y_{ac})*N_{bac}$	$\rho 11$
12. Uptake of Hydrogen								-1	$(1-Y_{h2})$	$-\Sigma C_i*v_{i,12}$	$-(Y_{h2})*N_{bac}$	$\rho 12$
13. Decay of X_{su}										$C_{bac}-C_{Xc}$	$N_{bac}-N_{Xc}$	$\rho 13$
14. Decay of X_{aa}										$C_{bac}-C_{Xc}$	$N_{bac}-N_{Xc}$	$\rho 14$
15. Decay of X_{fa}										$C_{bac}-C_{Xc}$	$N_{bac}-N_{Xc}$	$\rho 15$
16. Decay of X_{C4}										$C_{bac}-C_{Xc}$	$N_{bac}-N_{Xc}$	$\rho 16$
17. Decay of X_{pro}										$C_{bac}-C_{Xc}$	$N_{bac}-N_{Xc}$	$\rho 17$
18. Decay of X_{ac}										$C_{bac}-C_{Xc}$	$N_{bac}-N_{Xc}$	$\rho 18$

(Table 6) cont.....

Component i	**1**	**2**	**3**	**4**	**5**	**6**	**7**	**8**	**9**	**10**	**11**	**Rate**
19. Decay of X_{h2}										C_{bac}-C_{Xc}	N_{bac}-N_{Xc}	$\rho19$

Component i	**12**	**13**	**14**	**15**	**16**	**17**	**18**	**19**	**20**	**21**	**22**	**23**	**24**	**Rate**
Process j	S_I	X_C	X_{ch}	X_{pr}	X_{li}	X_{su}	X_{aa}	X_{li}	X_{c4}	X_{pro}	X_{ac}	X_{h2}	X_I	
1. Disintegration	$f_{sI,xc}$	-1	$f_{ch,xc}$	$f_{pr,xc}$	$f_{li,xc}$								$f_{xI,xc}$	$\rho1$
2. Hydrolysis of Carbohydrates			-1											$\rho2$
3. Hydrolysis of Proteins				-1										$\rho3$
4. Hydrolysis of Lipids					-1									$\rho4$
5. Uptake of sugars						Y_{su}								$\rho5$
6. Uptake of Amino Acids							Y_{aa}							$\rho6$
7. Uptake of LCFA								Y_{fa}						$\rho7$
8. Uptake of Valerate									Y_{c4}					$\rho8$
9. Uptake of Butyrate									Y_{c4}					$\rho9$
10. Uptake of Propionate										Y_{pro}				$\rho10$
11. Uptake of Acetate											Y_{ac}			$\rho11$
12. Uptake of Hydrogen												Y_{h2}		$\rho12$
13. Decay of X_{su}		1				-1								$\rho13$
14. Decay of X_{aa}		1					-1							$\rho14$
15. Decay of X_{fa}		1						-1						$\rho15$
16. Decay of X_{C4}		1							-1					$\rho16$
17. Decay of X_{pro}		1								-1				$\rho17$
18. Decay of X_{ac}		1									-1			$\rho18$
19. Decay of X_{h2}		1										-1		$\rho19$

If Table **6** is studied by series, the substances that participate in each process can be found. If a given element $v_{j,i}$ in the row "j" and column "i" of the Table is zero, then the substance "i" does not take part in the process "j". If it is positive, it means that substance "i" is formed during the process "j". If the element is negative, then the substance "i" is broken down during process j. During each process, inorganic carbon and nitrogen must be conserved. The rates of reactions are listed in Table **7**.

Table 7. Biochemical kinetic rate equations for soluble components [63].

Number (i)	Process	Rate ρ ($kg_{COD}/(m^3*d)$)
1	Disintegration	$k_{dis} \cdot X_C$
2	Hydrolysis of Carbohydrates	$k_{hyd,ch} \cdot X_{ch}$
3	Hydrolysis of Proteins	$k_{hyd,pr} \cdot X_{pr}$
4	Hydrolysis of Lipids	$k_{hyd,li} \cdot X_{li}$
5	Uptake of sugars	$k_{m,su} \cdot \frac{S_{su}}{K_{S,su}+S_{su}} \cdot X_{su} \cdot I_{pH\_aa} \cdot I_{IN}$
6	Uptake of Amino Acids	$k_{m,aa} \cdot \frac{S_{aa}}{K_{S,aa}+S_{aa}} \cdot X_{aa} \cdot I_{pH\_aa} \cdot I_{IN}$
7	Uptake of LCFA	$k_{m,fa} \frac{S_{fa}}{K_{S,fa}+S_{aa}} \cdot X_{fa} \cdot I_{pH\_aa} \cdot I_{IN} \cdot I_{h2\_fa}$
8	Uptake of Valerate	$k_{m,c_4} \cdot \frac{S_{va}}{K_{S,c_4}+S_{va}} \cdot X_{c_4} \cdot \frac{S_{va}}{S_{bu}+S_{va}+e^{-6}} \cdot I_{pH\_aa} \cdot I_{IN} \cdot I_{h2\_c4}$
9	Uptake of Butyrate	$k_{m,c_4} \cdot \frac{S_{bu}}{K_S+S_{bu}} \cdot \frac{S_{bu}}{S_{va}+S_{bu}+e^{-6}} \cdot I_{pH\_aa} \cdot I_{IN} \cdot I_{h2\_c4}$
10	Uptake of Propionate	$k_{m,pr} \cdot \frac{S_{pro}}{K_{S,pro}+S_{pro}} \cdot X_{pro} \cdot I_{pH\_aa} \cdot I_{IN} \cdot I_{h2\_pro}$

(Table 7) cont.....

Number (i)	Process	Rate ρ ($kg_{COD}/(m^3*d)$)
11	Uptake of Acetate	$k_{m,ac} \cdot \frac{S_{ac}}{K_{S,ac}+S_{ac}} \cdot X_{ac} \cdot I_{pH\_ac} \cdot I_{IN} \cdot I_{nh3}$
12	Uptake of Hydrogen	$k_{m,h_2} \cdot \frac{S_{h2}}{K_{S,h_2}+S_{h_2}} \cdot X_{h_2} \cdot I_{pH\_h2} \cdot I_{IN}$
13	Decay of X_{su}	$k_{dec,X_{su}} \cdot X_{su}$
14	Decay of X_{aa}	$k_{dec,X_{aa}} \cdot X_{aa}$
15	Decay of X_{fa}	$k_{dec,X_{fa}} \cdot X_{fa}$
16	Decay of X_{C4}	$k_{dec,X_{c4}} \cdot X_{c_4}$
17	Decay of X_{pro}	$k_{dec,X_{pro}} \cdot X_{pro}$
18	Decay of X_{ac}	$k_{dec,X_{ac}} \cdot X_{ac}$
19	Decay of X_{h2}	$k_{dec,X_{h2}} \cdot X_{h_2}$

Inhibition may occur because of low concentration of total nitrogen, high concentration of hydrogen or free ammonia, or due to extreme values of pH. The inhibition equations that are included in ADM1 are listed in Table **8**. It must be noticed that the inhibition equation for pH is applied only when pH is lower than its lower acceptable value. The term "X" in equations for pH and hydrogen inhibition corresponds to the degraded substrate. Inhibition of each process is more intense as the factor "I" is closer to 0. When "I" has a value close to 1, then the process is done almost completely and at a fast rate.

Table 8. Inhibition forms included in ADM1 [63].

Inhibitor	Equation	Process used
Ammonia	$I_{nh3} = \frac{K_{I\_nh3}}{K_{I\_nh3} + S_{nh3}}$	11

(Table 8) cont.....

Inhibitor	Equation	Process used
Hydrogen	$I_{h2\_x} = \frac{K_{I\_h2\_x}}{K_{I\_h2\_x} + S_{h2}}$	7-10
Nitrogen	$I_{IN} = \frac{S_{IN}}{S_{IN} + K_{S\_IN}}$	5-12
pH	$I_{pH\_x} = \exp(-3 * (\frac{pH - pH_{UL\_x}}{pH_{UL\_x} - pH_{LL\_x}})^2)$	5-12

Physicochemical Reactions

In the case of acid-base reactions, the substances included are the acid-base forms of the organic acids produced during acidogenesis, along with total acetate, inorganic carbon and inorganic nitrogen. More specifically, the acid-base pairs are: Valeric acid (S_{hva}) and **Valerate (S_{va-})**, Butyric acid (S_{hbu}) and **Butyrate (S_{bu-})**, Propionic acid (S_{hpro}) and **Propionate (S_{pro-})**, Acetic acid (S_{hac}) and **Acetate (S_{ac-})**, Carbon Dioxide (S_{co2}) and **Bicarbonate (S_{hco3-})**, as well as Ammonium (S_{nh4+}) and **Ammonia (S_{nh3})**. Among those 12 variables, only the 6 are variables of the model and are signed in bold. As the concentration of total form of the ions is already included (i=4-7, 10, 11), the calculation of the base concentration implies that the acid concentration is known. The acid-base reactions and their rates are shown in Table **9**.

Those variables are used for the calculation of the concentration of hydrogen cations (**Eq. 3**). Then, pH can be calculated with the use of an extra variable (Θ) (**Eq. 4, 5**):

$$\Theta = S_{cat^+} + S_{nh_4^+} - S_{hco_3^-} - \frac{S_{ac^-}}{64} - \frac{S_{pro^-}}{112} - \frac{S_{bu^-}}{160} - \frac{S_{va^-}}{208} - S_{an^-} \tag{3}$$

$$S_{H^+} = -\frac{\Theta}{2} + \frac{1}{2} \cdot \sqrt{\Theta^2 + 4 \cdot K_W} \tag{4}$$

$$pH = -\log_{10}(S_{H^+}) \tag{5}$$

Table 9. Acid-base reactions and equation rates [63].

Acid-base on Equilibrium	Differential Equation	Rate ($kg_{COD}/(m^3*day)$)
Valerate	$\frac{dS_{va^-}}{dt} = -\rho_{A,4}$	$\rho_{A,4} = k_{A,Bva} \cdot (S_{va^-} \cdot (K_{a,va} + S_{H^+}) - K_{a,va} \cdot S_{va})$
Butyrate	$\frac{dS_{bu^-}}{dt} = -\rho_{A,5}$	$\rho_{A,5} = k_{A,Bsu} \cdot (S_{bu^-} \cdot (K_{a,bu} + S_{H^+}) - K_{a,bu} \cdot S_{bu})$
Propionate	$\frac{dS_{pro^-}}{dt} = -\rho_{A,6}$	$\rho_{A,6} = k_{A,Bpro} \cdot (S_{pro^-} \cdot (K_{a,pro} + S_{H^+}) - K_{a,pro} \cdot S_{pro})$
Acetate	$\frac{dS_{ac^-}}{dt} = -\rho_{A,7}$	$\rho_{A,7} = k_{A,Bac} \cdot (S_{ac^-} \cdot (K_{a,ac} + S_{H^+}) - K_{a,ac} \cdot S_{ac})$
Inorganic carbon	$\frac{dS_{hco_3^-}}{dt} = -\rho_{A,10}$	$\rho_{A,10} = k_{A,Bco_2} (S_{hco_3^-} \cdot (K_{a,co_2} + S_{H^+}) - K_{a,co2} \cdot S_{IC})$
Inorganic nitrogen	$\frac{dS_{nh_3}}{dt} = -\rho_{A,11}$	$\rho_{A,11} = k_{A,BIN} \cdot (S_{nh_3} \cdot (K_{a,IN} + S_{H^+}) - K_{a,IN} \cdot S_{IN})$

In the case of transfer between the liquid and gaseous phase, the 3 gaseous components included in the ADM1 are hydrogen, methane and carbon dioxide. These are also found as liquids (i=8, 9, 10, Table **5**) and take part in biochemical reactions. For those three substances, the reaction rate is slightly changed. After the methodology described in Table **6** and **7** is followed, the rate of the transfer between the two phases will be subtracted (Table **10**). The equation defining the rate of change for the concentration of the liquid form of hydrogen, methane or carbon dioxide is defined by Eq. (**6**):

$$\frac{dS_i}{dt} = \frac{q_{in}}{V_{liq}} \cdot (S_{i,in} - S_i) + \Sigma v_{i,j} \cdot \rho_j - \rho_{T,i} \quad (6)$$

Table 10. Liquid-gas transfer reaction rate [63].

	Component →	i	8	9	10	Rate (ρTj)
J	Process ↓		S_{h2}	S_{ch4}	S_{IC}	
T8	H_2 Transfer		-1			$k_La \cdot (S_{liq,H_2} - 16 \cdot K_{H,H_2} \cdot p_{gas,H_2})$

(Table 10) cont.....

	Component →	i	8	9	10	Rate (ρTj)
T9	CH_4 Transfer			-1		$k_La \cdot (S_{liq,CH_4} - 64 \cdot K_{H,CH_4} \cdot p_{gas,CH_4})$
T10	CO_2 Transfer				-1	$k_La \cdot (S_{liq,CO_2} - K_{H,CO_2} \cdot p_{gas,CO_2})$

The following set of equations is used in order to calculate the gas flow rate. Concentration of hydrogen, methane and carbon dioxide is defined by equation (7). For the calculation of the partial pressures, along with the total pressure of biogas, equations (**8**) and (**9**) are used. In the case of the generic equation (**8**), the variable N is equal to 16, 64 and 1 for H_2, CH_4 and CO_2 respectively. Finally, equation (**10**) is used for the calculation of gas flow rate.

$$\frac{dS_{gas,i}}{dt} = -\frac{S_{gas,i} \cdot q_{gas}}{V_{gas}} + \rho_{T,i} \cdot \frac{V_{liq}}{V_{gas}} \tag{7}$$

$$p_{gas,i} = S_{gas,i} \cdot R \cdot T / N_i \tag{8}$$

$$P_{gas} = p_{gas,H_2} + p_{gas,CH_4} + p_{gas,CO_2} + p_{gas,H_2O} \tag{9}$$

$$q_{gas} = \frac{R \cdot T}{P_{atm} - p_{gas,H_2}} \cdot V_{liq} \cdot \left(\frac{\rho_{T,8}}{16} + \frac{\rho_{T,9}}{64} + \rho_{T,10}\right) \tag{10}$$

The suggested parameters for the application of ADM1 are listed in the **Appendix A** of this chapter. In Table **A.1**, stoicheiometric parameters for the model are shown. They do not depend on temperature change. Table **A.2** contains the suggested parameters for mesophilic and thermophilic digestion.

Application for Wastewater Treatment

The Anaerobic Digestion Model No 1 was applied in two different wastewater treatment plants, namely Henrikdsal WWTP, Stockholm and Psyttaleia WWTP, Athens. For the simulation, both reactors are assumed to be of CSTR type. The operation of CSTR is characterized by perfect mixing, continuous and steady sludge inflow and outflow, steady volume, constant temperature and gas densities and, finally, little or no deviation in biogas production with time. Before the presentation of the results, a short presentation of each wastewater treatment plant is included.

Henriksdal WWTP

Henriksdal WWTP (Wastewater Treatment Plant) is one of the largest and most modern waste treatment facilities in Europe. It has a wastewater capacity of around 250,000 m^3 per day, a quantity that is equivalent to 780,000 inhabitants and corresponds to two thirds of Stockholm's municipal wastewater. The gas produced is either used for heat and power production or is upgraded to biomethane as fuel vehicles [70].

The facility covers a total area of 300,000 m^2. Approximately 90% of the facilities are located underground, making Henriksdal WWTP one of the biggest underground plants worldwide. The tunnels inside the plant have a length of about 18 km. The whole of the basin block lies inside chambers blasted into the rock, on top of which a residence of about 1600 inhabitants has been built [71].

Henriksdal WWTP initially opened in 1941 and was able to treat about 150,000 m^3 wastewater per day. In 1953, the plant was expanded and its capacity was increased to its current value. A major expansion program that began in 1992 and lasted 5 years, lead to the reduction of nitrogen emissions by 50% and enabled phosphorus treatment [71]. Finally, a reconstruction project was carried out from 2007 to 2011, during which a new hall for rough purification and a new receiving station for grease and waste were constructed. 80,000 m^3 of rock were excavated and the odor producing facilities were placed underground. This enabled the development of 3,000 homes in the area around the plant [70].

Stockholm is among the fastest growing cities in Europe. Despite the expected increase in wastewater treatment demands, the city must meet the requirements set by the Baltic Sea Action Plan (BSAP) and the EU Water Directive. After several studies, the upgrade of the Henriksdal WWTP with membrane bioreactor (MBR) technology was selected as the best potential solution. At 2018, when the project is to be completed, Henriksdal WWTP will be the largest plant worldwide to implement the membrane bioreactor technology [70].

Membrane Bioreactor (MBR) technology is a combination of the conventional activated sludge process, with a micro-filtration and ultra-filtration membrane system. Membrane bioreactors are able to produce water of high quality that can be reused, while their footprint is lower by 30-50% compared to an equivalent conventional activated sludge facility [72]. Moreover, Membrane filtration is a technological solution that allows more capacity within existing volumes [73]. As a result, MBR systems have great potential in wide ranging application including municipal and industrial wastewater treatment and process water recycling. In cases where only medium water quality is required, water treated in a Membrane Bioreactor can be directly reused as process water, for example in commercial laundries and in food and cosmetics industry. Otherwise, if high water quality is required, MBR technology is a very efficient pretreatment stage for a next additional membrane treatment step [72].

The renovation of Henriksdal WWTP will result to the reduction of noise and odor. This will let the creation of a new city of 7,000 residential flats close to the plant [74]. The upgraded plant will be able to treat an average wastewater inflow of 535,680 m^3 every day, while its maximum capacity will be equal to 864,000 m^3 per day. These values correspond to the expected wastewater flows of Stockholm in 2040.

In Henriksdal WWTP, digestion takes place in 7 digesters of a combined volume of about 39,000 m^3 under mesophilic conditions. Although input sludge flow rate is varying with the quantity of treated wastewater, it shows an average of 2040 m^3/day. The necessary data for the simulation are taken from two reports for this particular plant from Lund University [68, 69] (Table **11**).

Table 11. Parameters of ADM1 application in Henriksdal WWTP [68, 69].

Parameter	Value	Unit
Average flow rate	2040	m^3/day
Temperature	35	°C
Total Volume	38900	m^3
Retention time	19.1	Days

In Table **12**, the main simulation results regarding the quantity, composition and the calorific value of the gas produced are presented. It is assumed that in the wastewater treatment plant both heat and power are produced. Typical values of electric and thermal efficiency for a big plant are selected, 30% and 55%, respectively, reaching a combined efficiency of 85%.

Table 12. Results of the ADM1 for Henriksdal WWTP.

Data	Value	Unit
Gas flow rate	46310	Nm^3/day
Methane rate	67.2 %	-
Gas pressure	1.447	bar
Gas Lower Heat value	21.35	MJ/kg
	22.62	MJ/Nm^3
Total energy per day	291	MWh

(Table 12) cont.....

Data	Value	Unit
Electric energy per day	87.3	MWh
Thermal energy per day	160.1	MWh
Electric power	3.64	MW_e
Thermal power	6.67	MW_{th}

It can be concluded that the quality of biogas is quite good, as its methane rate is close to 70% which is the upper limit for biogas produced by anaerobic digestion. The lower heating value of biogas for this particular methane rate is close to the value shown in Table **3** for biogas of methane equal to 70%. As it is shown in Fig. (**5**), the values of both quantity of biogas produced per day and methane rate remain almost after the first 2 days that followed the completion of the process. As the digester is of CSTR type, this indicates its fine operation. In terms of biogas composition, hydrogen is trace. As a result, it can be assumed that carbon dioxide accounts for the rest 32.8% of biogas.

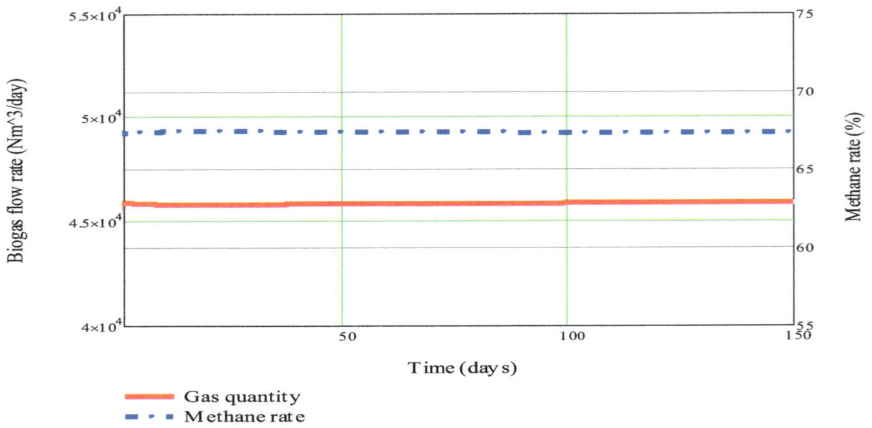

Fig. (5). Change of gas production and methane rate *versus* time.

Psyttaleia WWTP

The main part of bioenergy in Greece is coming through treatment and utilization of waste [75]. Within 2008-2012, a continuous rise in energy production from **biogas** was noticed (Table **13**) [25, 45]. The increase in biogas production in Greece was higher than the average rate in Europe. The total installed capacity of biogas plants is equal to 44 MW_e [76].

Production of energy from **landfill gas** is an established technology in the country. According to European Observatory [25], more than 76% of total biogas production in 2013 came from landfills. Greece had a share of about 2.4% in landfill gas production within the EU, being ranked 6th among the 28 EU members. The power plant in Ano Liossia (Athens) is among the largest power plants from landfill biogas globally, producing 23.5 MW_e.

Table 13. Biogas production in Greece between 2008 and 2012 [25, 45].

Year	Biogas Production (ktoe)	Rate of Increase	EU Share	Rank
2008	33	----	0.5%	14th
2010	49	48.5%	0.57%	14th
2012	88.4	80.8%	0.73%	12th
2013	88.6	0.7%	0.73%	13th

Moreover, a considerable amount of bioenergy is produced by the anaerobic digestion of **wastewater sludge** [75]. In 2013, the share of urban and industrial sewage sludge treatment on the total biogas production in Greece was slightly higher than 18%. During the same year, Greece accounted for about 1.3% of the total biogas production from sewage sludge within the EU, being ranked 13th. More than 70% of the total biogas produced from sewage sludge treatment comes from the wastewater treatment plant in Psyttaleia.

Psyttaleia is a small, inhabited island near Athens where the whole amount of wastewater from 5 million people living in the greater area of Athens is being treated. Psyttaleia WWTP is the larger sewage treatment plant in Europe, with an average volume of the inflow wastewater of 700,000-800,000 m^3 per day.

The treatment of wastewater includes the removal of heavy metals, sand and odor (primary treatment), as well as the removal of more than 90% or organic load and the 80% of nitrogen (biological treatment). During those stages, sludge is collected in the sedimentation tanks. Then, the treated wastewater is sent to the sea and sludge is digested. The produced biogas is burnt in two CHP units of 11.4 MW_e combined power capacity [76]. Those units are used to cover the power and thermal needs of the total facility, together with a 12.9 MW_e natural gas CHP plant. Surplus power is sold to the Greek operator of the transmission system. After Anaerobic Digestion is complete, the effluent sludge is dried and a quantity of 120-150 tons of combustible material per day is additionally produced.

In Table **14**, the average values for wastewater inflow rate in Psyttaleia WWTP, as well as the average values of sludge inflow rate in the anaerobic digesters of the plant are listed, for 10 months of the year. Those data are of 2008, but as the population of Athens has not significantly changed since then, they can also be considered to refer to the current situation.

Table 14. Inflow rate of sludge to digesters of Psyttaleia WWTP during the year.

Month	Wastewater Inflow (m^3/day)	Sludge Inflow (m^3/day)
February	873160	4408
March	860520	4344
April	832040	4672
May	761160	4016
June	707070	3512
July	652970	3352
August	556290	3424
September	699140	4048
October	705320	3864
November	722590	4256

For the process, eight (8) digesters of volume equal to 10,000 m^3 each are used in parallel. As digestion is mesophilic, temperature was set equal to 35 °C. Similar to the case of Henriksdal WWTP, the plant's efficiency for power and heat generation was set equal to 30% and 55% respectively. It was finally assumed that 80% of produced electricity is used to cover the needs of the facility and 20% of this will be sold. According to the Greek operator, electricity from plants of installed power capacity higher to 2 MW_e is sold for 99.45 Euro per Megawatt hour (Table **15**).

Table 15. Assumptions for setup of the simulation for Psyttaleia WWTP.

Parameter	Value	Unit
Number of digesters	8	---------------------------
Total volume	80000	m^3
One digester volume	10000	m^3
Temperature	35	°C
Electricity sale price	99.45	€/MWh

In Table **16**, the results of modeling the Anaerobic Digestion in Psyttaleia WWTP are presented, for all the eight digesters of the plant. Content of methane in biogas and gas calorific value are omitted as they were almost steady ranging between 67.0-67.5% and 21.21-21.54 MJ/kg (22.51-22.72 MJ/m^3) respectively. Electric energy (E_{el}) is calculated on a daily basis.

The thermal power generated during the year is between 11 and 15 MW_{th}. The maximum average electric power does not exceed 8.2 MW_e. Of course, at days that the inflow rate is higher to the month's average, gas production will be higher to the correspondent value shown in Table **16** and electric power may reach the plant's installed capacity (11.4 MW_e). If the 20% of the electricity was sold, the plant would have a monthly income of at least 90,000 Euros.

Table 16. Results of the application of ADM1 to Psyttaleia WWTP.

Month	Q_{gas} (Nm^3/day)	P_{th} (MW)	P_{el} (MW)	E_{el} (MWh)	Income (€)
Feb	99080	14.32	7.81	187.48	104410
Mar	97850	14.14	7.71	185.09	114130
Apr	104120	15.07	8.22	197.25	117700
May	91490	13.20	7.20	172.75	106520
Jun	81560	11.72	6.39	153.41	91540
Jul	78370	11.24	6.13	147.18	90750
Aug	79810	11.46	6.25	149.99	92480
Sep	92120	13.29	7.25	173.96	103800
Oct	88520	12.75	6.96	166.97	102950
Nov	96150	13.89	7.58	181.80	108480

However, ADM1 is a generic model that may take into account some idealized aspects of anaerobic digestion compared to a real situation. In the case of Psyttaleia WWTP, the real data obtained for biogas production were about 35% lower compared to the model results. Moreover, the rate of methane in biogas produced in Psyttaleia WWTP is about 61-65%, while ADM1 resulted in a slightly higher value. This difference may also be caused by abnormal operation of reactors in the treatment plant during some days, or by wrong assumptions about the exact values of the model inputs.

ANAEROBIC DIGESTION SYSTEMS

Anaerobic digestion may be affected by many parameters, prior to and during the process. Three of them that are of significant importance in both the real process will be described here, namely temperature, pH and the number of anaerobic digestion stages. Other factors affecting an anaerobic digestion system include retention time, organic loading rate, total solid content, carbon to nitrogen ratio, mixing *etc*.

The effect of temperature and number of digestion stages were examined by the use of ADM1. The case of Henriksdal Wastewater Treatment plant will be studied. In each case, only one parameter is going to change, with the others remaining steady. The change in total energy production is indicated by the change in power generation. Moreover, the effect of volume and inflow rate has also been examined.

Temperature

The temperature range under which anaerobic digestion is possible is divided into three groups: Psychrofilic (ambient temperature), meshophilic (20-45 °C, usual temperature 35 °C) and thermophilic (50-65 °C, usual temperature equal to 55 °C) [21, 77]. As the process is very slow at ambient temperature, psychrofilic digestion will not be discussed.

As most of digesters are designed to operate at mesophilic range, mesophilic process is a widely used and proven technology [78]. Moreover, it offers a more stable operation and has lower energy demands compared to thermophilic.

Thermophilic digestion is significantly more efficient in terms of biogas production [18]. It offers shorter retention times and needs lower volumes for a given quantity of biogas production. Moreover, the rates of the process are higher under high temperatures [78, 79]. According to experiments, thermophilic degradation can be up to 8 times faster compared to mesophilic one and may lead to 36-38% increase in biogas production [77]. Finally, the destruction of pathogens organisms and weed seeds is more efficient under higher temperatures [80]. On the other hand, thermophilic digestion is more sensitive in a rapid change of temperature. But the main problem of thermophilic digestion is its considerably increased energy demand. The difficulties and high costs associated with the establishment and maintenance of high temperatures within the digester are the main reasons to explain why mesophilic digestion is prevailing despite its lower performance.

In the case of thermophilic digestion, the performance of the digester improves with temperature increase until a temperature around 65 °C. Above that, the reaction rate is going to decrease [63]. Moreover, it is crucial that the temperature inside the digester is maintained constant in order to sustain the gas production rate [81]. It is suggested that daily deviations are lower than 0.6 °C per day in order that operational problems are avoided.

The conditions of initial simulation by the use of ADM1 were mesophilic and its results (Table **17**) will be compared with those coming from Anaerobic Digestion under thermophilic and psychrophilic conditions, respectively. It is clear that gas quantity and energy produced increase with temperature. In the case of thermophilic digestion, gas production was 52% higher compared to mesophilic process. Although the rate of methane decreased, this drop was relatively small and the total energy produced rose up for about 43%. The opposite trends were shown with the drop of temperature. It can be then assumed that if a given quantity of gas were to be produced, less time or a smaller reactor would be needed under thermophilic conditions. Finally, the completion of the process was impossible for temperature of 78 °C and above.

Table 17. Effect of temperature in anaerobic digestion.

	Case 1	**Initial Simulation**	**Case 2**
Temperature (°C)	55	35	15
Gas quantity (Nm^3/day)	70470	46310	38890
Methane rate (%)	63.8	67.2	68.1
Gas calorific value (MJ/Nm^3)	21.30	22.62	23.31
Electric power (MW_e)	5.211	3.638	3.148

pH and Alkalinity

The value of pH is a measure of the acidity of the solution [79]. The optimal values for the acidogenesis and methanogenesis stages are different. Acidogens require a pH higher to 6.4, while methane formers are more active in a narrow range of neutral pH (6.6-7.0) [21]. As a result, pH must be at a value between 6.6 and 7.2. Beyond this range digestion may not proceed, or it will be unstable and of low efficiency. It is important that pH does not drop below 6.2, as high concentration of acids can become toxic to methane bacteria and inhibit the process [79]. The system needs to control the value of pH. If pH gets low, addition of alkalinity is necessary.

During the simulation on ADM1 for both waste treatment plants, pH was stable and its value was between 6.8 and 7.2. For values between 6 and 6.8, the gas produced is of lower quantity and quality. For even lower pH, either low amount of biogas and energy is produced or the process stops completely. This will be discussed in **Appendix B**, where the effect of specific parameters on pH and biogas production will be described.

Volume and Inflow Rate

After that, the effect of reactor volume was examined. In the first case, volume is reduced to 30,000 m^3, while in the second is increased to 48,000 m^3. It is obvious that volume has only a slight effect on quantity and quality of biogas (Table **18**). Generally, larger bioreactors lead to higher gas and energy production. Moreover, retention time increases with volume.

Table 18. Effect of digester volume in anaerobic digestion.

	Case 1	Initial Simulation	Case 2
Volume (m^3)	30000	38900	48000
Retention time (day)	14.7	19.1	23.5
Gas quantity (Nm^3/day)	44560	46310	47720
Methane rate (%)	67.4	67.2	66.9
Gas calorific value (MJ/Nm^3)	22.73	22.62	22.49
Electric power (MW_e)	3.516	3.638	3.727

The effect of sludge inflow in the reactor is shown in Table **19**. In first case, sludge inflow is 25% higher of the average flow rate (2544 m^3/day), while the second case corresponds to a day during which the reactor inflow is equal to 75% of the average flow rate (1530 m^3/day).

Opposite to temperature, a rise in the digester inflow rate increases not only gas quantity but also the methane content of biogas. Gas quantity is almost analogous to the rate of sludge inflow to the digester. An increase of 25% in inflow rate led to the increase of gas and energy production by 20.5% and 21.2% respectively. On the contrary, a 25% inflow drop caused a respective drop of 21.7% and 22.5%. With all the other parameters unchanged, this digester cannot handle sludge inflow rate of 2680 m^3/day or more. This is an inflow increased by 31.4% compared to the average inflow rate of the digester.

Table 19. Effect of wastewater inflow rate in anaerobic digestion.

	Case 1	Initial Simulation	Case 2
Inflow rate (m^3/day)	2544	2040	1530
Retention time (day)	15.3	19.1	25.4
Gas quantity (Nm^3/day)	55870	46310	36250
Methane rate (%)	67.4	67.2	66.7
Gas calorific value (MJ/Nm^3)	22.73	22.62	22.39
Electric power (MW_e)	4.409	3.638	2.818

Single Stage and Multi Stage Anaerobic Digestion

In the single stage digestion, all the reactions take place inside a single reactor. The operating conditions are suitable for all the reactions and no particular phase has been optimized. This scheme offers simple operation, lower costs and maintenance, but has lower gas production and conversion rate [82]. The retention time of one-staged AD is generally between 14 and 28 days [83].

The development of multi-stage Anaerobic Digestion aims at process optimization and the increase in gas production by the use of separate reactors for the different stages [18]. This enables the optimization of each reaction and also enhances organic destruction [83]. Typically, a multi-stage system is consisted by two reactors. Multi-stage processes have significantly better results, as bigger quantity of waste can be processed in a given amount of time compared to single stage AD. Thus, multi stage digestion is faster for a given sludge input and also leads to higher biogas production for a given input and retention time [83]. Other advantages of multi stage digestion include the higher organic rate achieved, the lower volume required for a given sludge input or gas production and the better control of odor [21, 82]. On the other hand, multi stage digestion is associated with higher operation and maintenance cost, as well as complexity [21]. At the moment, multi-staged systems currently represent only about 10% of the current treatment capacity.

There are several types of staged anaerobic digestion. Usually, either the two digesters complete different processes or they operate under different temperature. In the first case, the primary phases of the process (Hydrolysis-Acidogenesi--Acetogenesis) are conducted in the first digester while methane is produced in the second one. In the second, the complete process is done in both digesters, but under different temperature ranges (thermophilic and mesophilic) [82].

In the practical examination conducted by ADM1, the two-staged digestion is conducted in two digesters in a row, of volumes equal to 23,700 m^3 and 15,200 m^3 respectively. Their combined volume is equal to that of the single reactor used in the one-staged digestion. All the inputs and initial outputs of the systems have been defined [67]. Although, two digesters operating in a row seldom conduct the same processes and under the same temperature, equal temperatures were set in both digesters in order to enable direct comparison of the results (Table **20**).

Table 20. Comparison of one-staged and two-staged digestion.

	One-staged Digestion	Two-staged Simulation
Gas quantity, 1st reactor (Nm^3/day)	46310	43000
Gas quantity, 2st reactor (Nm^3/day)	--------	7210
Gas quantity, total (Nm^3/day)	46310	50210
Methane rate, 1st reactor (%)	67.2	67.5
Methane rate, 2nd reactor (%)	--------------	58.1
Gas calorific value, 1st reactor (MJ/Nm^3)	22.62	22.79

(Table 20) cont.....

	One-staged Digestion	Two-staged Simulation
Gas calorific value, 2nd reactor (MJ/Nm^3)	--------------	19.23
Electric power, 1st reactor (MW_e)	3.638	3.403
Electric power, 2nd reactor (MW_e)	-------------	0.481
Electric power, total (MW_e)	3.638	3.884

Two-staged digestion results in higher quantity of gas by about 8.5%. Energy produced also rose up by 6.8%. The main part of gas and energy (85.6% and 87. 6%, respectively) of the two-staged digestion system comes from the first digester. Only a small amount of gas is produced in the second digester, being also of lower methane content and calorific value compared to the gas produced in the first digester.

Gas quantity and power produced in the first digester of the two-staged system are only little lower than those produced in the single digester. This is expected, as the only difference of those two digesters is their volume. As shown in Table **18**, reactor volume does not play a significant role in both gas quantity and calorific value. So, the first digester of the two-staged system accounts for the production of almost the same amount of energy with the single digester of the one-staged system. The considerably less loaded second digester accounts for a small quantity of biogas that finally makes two-staged system advantageous in terms of gas and energy production.

According to the simulation, two-staged digestion was completed much faster compared to the one-staged process. The retention time was equal to 11.6 days for the first digester and 7.5 days for the second, while the retention time for the single digester was 19.1 days. As a result, the two-staged direction needed 7.5 days less to complete, reducing the required time by 40%.

ADM1 MODIFICATION FOR OLIVE-MILL WASTE TREATMENT

In 2012, about 2,900,000 tons of olive oil were produced globally. The largest olive oil producers are Spain, Italy and Greece, followed by Turkey and Tunisia. The countries around the Mediterranean account for about 98% of the global olive oil production. There are two main processes used for olive oil extraction. Both of them generate large amounts of byproducts. The three-phase process yields to about 20% olive oil, 30% to a solid residue known as olive press cake and 50% to an aqueous liquid known as olive-mill wastewater (OMW). Meanwhile, the more contemporary two-phase system results in less volume of OMW but, instead of olive cake, a semi-solid residue containing high organic matter concentration and polluted load is produced [84].

The treatment and disposal of olive mill waste, and especially olive mill wastewater represent one of the main environmental concerns of olive oil producing countries [67]. Around 30 million m^3 of OMW is produced annually in the Mediterranean [84]. These liquid wastes are acid effluents containing low amount of ammonium nitrogen. Moreover, OMW is heavily loaded with pollutants, compared to ordinary types of domestic wastewater. The high polluting activity of OMW is linked with its high content of total solids and dissolved organic compounds such as sugars, carbohydrates (13.79 g_{COD}/L), lipids (18 g_{COD}/L), proteins (3.96 g_{COD}/L), volatile fatty acids (0.3 g_{COD}/L) and phenolic compounds [67]. Except of its low pH, high organic load and high content of total solids, OMW is characterized by dark color and very intense smell.

Mainly due to phenols, which are poisonous caustic crystalline compounds, OMW exhibit highly phytotoxicity and antimicrobial properties. If it is not disposed or treated properly, OMW can result in serious environmental damage [84]. Furthermore, OMW contains high concentration of inorganic components such as potassium, calcium, magnesium, phosphate chloride and other salts. These compounds are undesirable, despite not being toxic [67].

Due to its complex compounds, olive mill waste is not easily biodegradable and needs to be detoxified before it can be properly used in agricultural and industrial processes. OMW cannot be sent to ordinary wastewater treatment systems and is often dumped into rivers and lakes, or used for farming irrigation. This results in contamination of ground water and eutrophication of lakes, rivers and canals. The excessive growth and the consequent decay of algae leads to depletion of the oxygen present in the water and the reduction of aquatic plants and fish populations [84]. Another common way of OMW treatment is their collection and retain in large evaporation basins or ponds. However, this method may result to the accumulation of toxic compounds in the soil, causing land degradation and soil and groundwater contamination. Even a small quantity of OMW in contact with groundwater has the potential to cause significant pollution to drinking water sources [84]. Other methods that have been proposed for OMW treatment are chemical and electrochemical treatment, aerobic biological treatment and anaerobic digestion process [80].

The application of anaerobic digestion for OMW treatment has encountered many difficulties. These difficulties are due to the high concentration of OMW in organic and phenolic compounds (80-140 g_{COD}/L and 8-14 g_{COD}/L respectively). Moreover, the nitrogen deficiency (0.2-0.5 g_N/L) and low alkalinity (1-2 g_{CaCO3}/L) of OMW result in a very low value of pH, between 5 and 5.4 [85]. The anaerobic degradation of phenol compounds is a quite complex process. The antibacterial activity caused by the high levels of phenols, along with the presence of either phenolic compounds or volatile fatty acids (VFA) lead to inhibition during the methane formation [66].

In order to enable the anaerobic treatment of OMW, several approaches have been investigated. These include the pretreatment of OMW by fermentation of aerobic cultures for the removal of phenolic compounds, or the dilution with water (1:10) and the addition of urea as nitrogen supplement. Moreover, the alkalinity of OMW has been often adjusted by $NaHCO_3$, NaOH or $Ca(OH)_2$ [85]. Other solutions include the aerobic biological pre-treatment and the co-digestion of OMW with substrates rich in ammonium nitrogen. The latter is the most interesting alternative, as it does not require any addition of chemical substances and has no negative effects on the potential of biogas production rate from OMW compared to other pretreatment methods [80]. Moreover, co-digestion reduces the total cost of treatment. It has been shown that piggery effluents, swine or cattle manure, dairy wastewater, and olive mill solid waste (OMSW) are substrates suitable for co-digestion with OMW as they are rich in nitrogen and act as a source of ammonium and alkalinity [85].

Among those possible substrates for co-digestion with OMW, Olive Mill Solid Waste seem the most interesting option, as they are produced on the same site with OMW. During the last decade, extensive research was conducted towards the anaerobic co-digestion of OMW with OMSW. Fezzani and Ben Cheikh managed to optimize the amount of OMSW to be used as a co-substrate alongside OMW under mesophilic conditions, without the need of the addition of chemical substances. They concluded that the optimal rate of OMSW was about 56 grams of Total Solids per liter of OMW. They found out that the co-digestion of OMW with OMSW at an optimal ratio increased biogas production and COD removal efficiency from 11.7 to 30.5 $L_{gas}/L_{digester}$ and 44.5% to 83.4% respectively. Finally, the startup time of steady biogas production was reduced from 65 to 28 days [85].

Presentation of the Extended ADM1

In 2009, Fezzani and Ben Cheikh presented a modified version of ADM1 in order to efficiently simulate the anaerobic digestion of olive mill waste. The structure of the original ADM1 was modified to take into account phenol compounds as composite, particulate and soluble substrates. In addition to the original ADM1, the extended model includes disintegration of composite solids like OMSW into particulate phenol compounds and the hydrolysis of particulate phenols to produce soluble phenols. Moreover, it is assumed that soluble phenols are first converted to benzoate which is further converted by acetogens to acetate, hydrogen and carbon dioxide (Fig. **6**) [66].

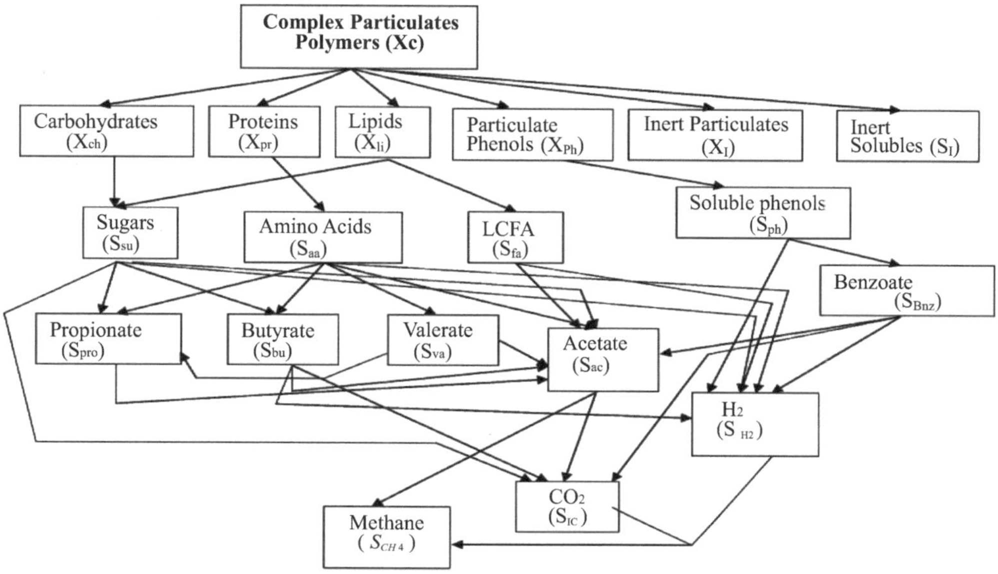

Fig. (6). Biochemical conversion processes according to ADM1 model extended for phenol compounds degradation [66].

The inclusion of phenol degradation requires the addition of five extra variables and six conversion processes. The additional variables correspond to particulate and soluble phenols (**X_{ph}** and **S_{ph}** respectively), total benzoate (**S_{bnz}**), as well as phenol and benzoate degrading microorganisms (**X_{Bph}** and **$X_{B,bnz}$**). The concentration of all those variables is measured in kg_{COD}/m^3, in a similar way to the variables of the original ADM1 shown in Table **5**. The conversion processes correspond to phenol disintegration (**j=1**), particulate phenol hydrolysis (**j=4a**), uptake of soluble phenols and total benzoate (**j=7a** and **7b** respectively) and, finally, the decay of phenol and benzoate biomass (**j=20** and **j=21** respectively). The new kinetic equations, new yield coefficients and new derived stoicheiometry for all steps included in the extended ADM1 model are listed in Table **21** [66].

The mass balance equations used to describe the dynamic behavior of the additional variables of the extended ADM1 follow the general form shown in equations **(1)** and **(2)**. For example, the concentration of soluble benzoate is calculated as shown in Eq. (**11**).

$$\frac{dS_{bnz}}{dt} = \frac{q_{in}}{V_{liq}} \cdot (S_{bnz,in} - S_{bnz}) + f_{bnz,Sph} \cdot (1 - Y_{ph}) \cdot \rho_{7a} - \rho_{7b} \tag{11}$$

For the application of the extended ADM1, Table **21** is used additionally to Table **6**. Except of the five new parameters, the equations defining the concentration of four parameters of the original ADM1 have been modified. These parameters are included in Table **21** and are total acetate (i=7), soluble hydrogen (i=8), inorganic carbon (i=10) and inorganic nitrogen (i=11). In the original ADM1, their concentration is defined by the application of Eq. (**1**) on the data included in Table **6**. For the case of the extended ADM1, the terms occurring from the application of Eq. (**1**) according to the parameters shown in Table **21** will be added. The concentration of acetate in the extended ADM1 is defined by Eq. (**12**). It is clear that Eq. (**12**) occurs from Eq. (**1**) for acetate, augmented by the term representing the uptake of benzoate. Concentrations of soluble hydrogen, inorganic carbon and inorganic nitrogen are calculated in the same way. Moreover, the equation for composite substrate concentration (i=1) was also changed to involve composite substrate issued from phenol biomass decay, as shown in Eq. (**13**) [66].

Table 21. Biochemical rate coefficients for additional variables in extended ADM1 for olive mill waste treatment [66].

Component i ↓→	7	8	7a	7b	10	11	16a	23a	23b	Rate
Process j	$S_{\square}$	S_{h2}	S_{ph}	S_{bnz}	S_{IC}	S_{IN}	X_{ph}	X_{Bph}	$X_{B,bnz}$	
1. Disintegration			$f_{Sph,Xph}$		$f_{Xph,Xc}*C_{Xph}$		$f_{Xph,Xph}$			ρ1
4a. Hydrolysis of particulate phenol					$C_{Xph}-C_{ph}$		-1			ρ4a
7a. Uptake of phenol	$f_{ac,bnz}*(1-Y_{bnz})$	$f_{h2,bnz}*(1-Y_{bnz})$	-1	$f_{bnz,ph}*(1-Y_{ph})$	$-[C_{ph}-(1-Y_{ph})*Cbnz-Y_{ph}*C_{biom}]$	$-Y_{ph}*N_{biom}$		Y_{ph}		ρ7a
7b. Uptake of benzoate				-1	$C_{bnz}-(1-Y_{bnz})*C_{ac}-Y_{bnz}*C_{biom}$	$-Y_{bnz}*N_{biom}$			Y_{bnz}	ρ7b
20. Decay of phenol biomass					$C_{biom}-C_{Xc}$	$N_{biom}-N_{Xc}$		-1		ρ20
21. Decay of benzoate biomass					$C_{biom}-C_{Xc}$	$N_{biom}-N_{Xc}$			-1	ρ21

$$\frac{dS_{ac}}{dt} = \frac{q_{in}}{V_{liq}} \cdot (S_{ac,in} - S_{ac}) + \sum_{j=1-19} v_{ac,j} \cdot \rho_j + (1 - Y_{bnz}) \cdot f_{ac,bnz} \cdot \rho_{7b} \quad (12)$$

$$\frac{dX_c}{dt} = \frac{q_{in}}{V_{liq}} \cdot (X_{c,in} - X_c) + \sum_{j=13-19} \rho_j + \rho_{20} + \rho_{21} - \rho_{21} \quad (13)$$

The kinetic rates of the additional processes are illustrated in Table **22**. Additionally, the kinetic rate total acetate uptake has also been modified due to the inclusion of inhibition of methanogenic steps cause by high levels of phenol and total Volatile Fatty Acids (TVFA). The modified rate is shown in Eq. (**14**), while the expressions for phenols and TVFA inhibition are defined by Eq. (**15**) and Eq. (**16**), respectively.

$$\rho_{11} = I_{pH\_ac} \cdot I_{IN} \cdot I_{nh3} \cdot I_{TVFA} \cdot I_{phenol} \quad \textbf{(14)}$$

$$I_{TVFA} = \frac{1}{1 + \frac{S_{TVFA}}{K_{I\_TVFA}}} \quad \textbf{(15)}$$

$$I_{phenol} = \frac{1}{1 + \frac{S_{phenol}^{2}}{K_{I\_phenol}}} \quad \textbf{(16)}$$

Table 22. Additional biochemical kinetic rate equations for extended ADM1 [66].

Number (i)	Process	Rate ρ ($kg_{COD}/(m^3 * d)$)
4a	Hydrolysis of particulate phenol	$k_{hyd,ph} \cdot X_{ph}$
7a	Uptake of phenol	$k_{m,ph} \cdot \frac{S_{ph}}{K_{S,ph} + \frac{S_{bnz}^{2}}{K_{I\_ph}}} \cdot X_{Bph}$
7b	Uptake of benzoate	$k_{m,bnz} \cdot (\frac{S_{bnz}}{K_{S,bnz}} + S_{bnz}) \cdot X_{Bbnz} \cdot I_{pH\_aa} \cdot I_{IN} \cdot I_{h2\_fa}$
20	Decay of phenol biomass	$k_{dec,Bph} \cdot X_{Bph}$
21	Decay of benzoate biomass	$k_{dec,Bbnz} \cdot X_{B,bnz}$

Finally, the charge balance Eq. (**3**) of the original ADM1 was extended to include the contribution of soluble phenolic compounds and benzoate ion concentrations into acid-base reactions, as shown in Eq. (**17**). Then, hydrogen cations and pH are calculated as in the original model, by Eq. (**4**) and (**5**), respectively. The acid-base reactions and the correspondent rates are shown in Table **23**. They are additional to those shown in Table **9** for the original ADM1 [66].

$$\Theta = S_{H^+} - S_{OH^-} = S_{cat^+} + S_{nh_4^+} - S_{hco_3^-} - \frac{S_{ac^-}}{64} - \frac{S_{pro^-}}{112} - \frac{S_{bu^-}}{160} - \frac{S_{va^-}}{208} - \frac{S_{ph^-}}{220} - \frac{S_{bnz^-}}{210} - S_{an^-} \quad \mathbf{(17)}$$

Table 23. Additional acid-base reactions and equation rates of the extended ADM1 [66].

Acid-base on Equilibrium	Differential Equation	Rate ($kg_{COD}/(m^3*day)$)
Phenol	$\frac{dS_{ph^-}}{dt} = -\rho_{A,ph}$	$\rho_{A,ph} = k_{B,ph} \cdot (S_{ph^-} \cdot (K_{a,ph} + S_{H^+}) - K_{a,ph} \cdot S_{ph})$
Benzoate	$\frac{dS_{bnz^-}}{dt} = -\rho_{A,bnz}$	$\rho_{A,bnz} = k_{B,bnz} \cdot (S_{bnz^-} \cdot (K_{a,bnz} + S_{H^+}) - K_{a,bnz} \cdot S_{bnz})$

All the parameters that are additional to those included in the original ADM1, along with those that their value has to be modified are listed in Table **C.1** and **C.2** of **Appendix C**. For parameters included in the original ADM1 that no new value has been set, the data listed in Tables **A.1** and **A.2** in **Appendix A** must be used.

Model Application in a Small Rural Olive Mill

The extended version of ADM1 was applied in a small rural olive mill. About 1.5-2.5 m^3 of wastewater is produced by a small olive mill on a daily basis. Digestion was set to be mesophilic and the volume of the digester was set equal to 60 m^3. As the reactor is of small size, an electric efficiency of 25% and a thermal efficiency of 50% were assumed. All data are listed in Table **24**.

Table 24. Assumptions for setup of the simulation for olive mill waste treatment.

Parameter	Value	Unit
Flow rate of waste	2	m^3/day
Volume of digester	60	m^3
Temperature	37	°C
Electric efficiency	25	---------------------------
Thermal efficiency	50	---------------------------

The concentration of OMW was set to 80 g_{COD} per liter of OMW, a quantity that corresponds to high organic loading according to Fezzani and Ben Cheikh [67]. The characteristics of OMW, OMSW and sludge that are used to determine the values of input and initial output steady-state variables of the model application are listed in Tables **25** & **26**. The defined optimal rate for OMSW and OMW co-digestion was selected. According to that, 56 grams of OMSW were added to 1 liter of OMW to be digested [66].

Table 25. Characteristics of OMSW used as substrate [66].

Parameter	Units	OMSW	
		Mean Value	Deviation
TCOD	g_{COD}/L	1100	25
Carbohydrates	g_{COD}/L	345	5
Total lignin	g_{COD}/L	440	5
Total proteins	g_{COD}/L	72	5
Total lipids	g_{COD}/L	100	5
Total polyphenols	g_{COD}/L	18	5
TKN	g_N/L	12	1.5

Table 26. Characteristics of the OMW and the sludge used in the extended ADM1 as input main influent and initial conditions, respectively [66].

Parameter	Units	OMW Mean Value	Deviation	Sludge Mean Value	Deviation
TCOD	g_{COD}/L	80	2.5	20.5	1.5
SCOD	g_{COD}/L	60	2.5	0.72	0.1
Total carbohydrates	g_{COD}/L	21	1.5	0.5	0
Total lipids	g_{COD}/L	14	1.5	0.45	0
Total lipids	g_{COD}/L	17	1.5	0.3	0
Total phenols	g_{COD}/L	24	1.2	0	0
Total Inert	g_{COD}/L	4	1.5	19	0
Sugars	g_{COD}/L	13	1.5	0.005	0
Amino Acids	g_{COD}/L	4.5	1.5	0.02	0
LCFA	g_{COD}/L	7	1.5	0.004	0
Soluble phenols	g_{COD}/L	17	0	0	0
Soluble inert	g_{COD}/L	7.5	1.5	0.65	0
Acetic Acid	g_{COD}/L	5.5	0.5	0.01	0

(Table 26) cont.....

Parameter	Units	OMW Mean Value	Deviation	Sludge Mean Value	Deviation
Propionic Acid	g_{COD}/L	1.75	0.5	0.01	0
Butyric Acid	g_{COD}/L	2.85	0.5	0.01	0
Valeric Acid	g_{COD}/L	0.85	0.5	0.01	0
Alcalinity	g_{CaCO3}/L	3.67	0.25	2.5	0.5
NH_4^+-N	mg_N/L	470	50	1300	70
Anions	mol/L	0.053	0.002	0.091	0.002
Cations	mol/L	0.343	0.02	0.027	0.004

During the simulations, a term representing the additional alkalinity was taken into account, so pH was neutral, having a value almost equal to 7. The next graphs (Figs. **7-10**) illustrate the change in biogas production and the methane content of biogas, along with the concentration of soluble and particulate phenols and Total Volatile Fatty Acids.

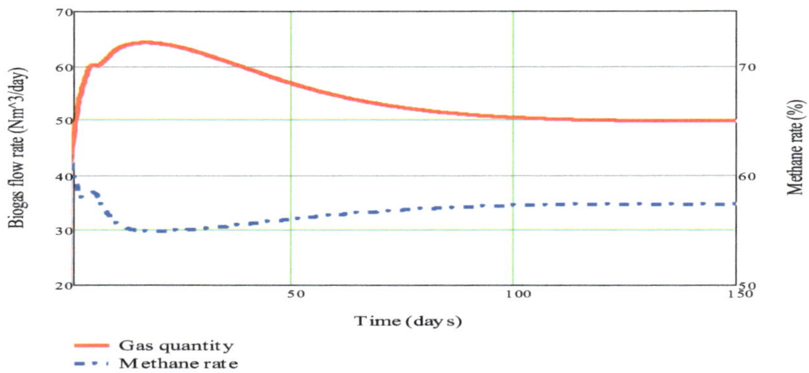

Fig. (7). Change of gas production and methane rate *versus* time.

Biogas quantity is increasing during the first days, reaching 65 Nm^3/day. (Fig **7**) After that, it is steadily dropping and gets a steady value of about 52 Nm^3/day. Opposite to that, the methane rate in biogas drops rapidly during the first few days, but it then increases slightly, getting a final value of about 57%. If those trends are compared with those of biogas produced by sewage sludge treatment (Fig **5**), it is clear that the variation in both parameters is more intense in the case of olive mill waste.

The concentration of effluent soluble phenols is increasing until a value of 10.5 g_{COD}/L (Fig. **8**). If this is compared with the 17 g_{COD}/L of influent soluble phenols (Table **26**), it is clear that the anaerobic treatment of olive mill waste managed to reduce the concentration of soluble form of phenolic compounds.

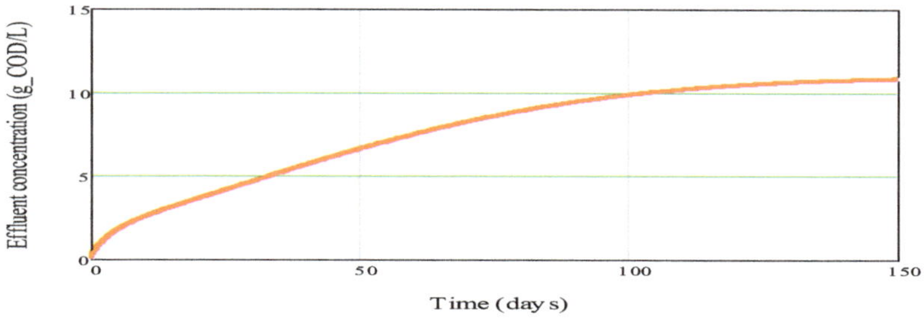

Fig. (8). Concentration of effluent soluble phenols *versus* time.

Furthermore, the concentration of particulate phenols varies in a way similar to that of soluble phenols. (Fig. **9**) Its value increases with time, but is lower to the combined influent particulate phenols and particulate total polyphenols listed in Tables **25** & **26**.

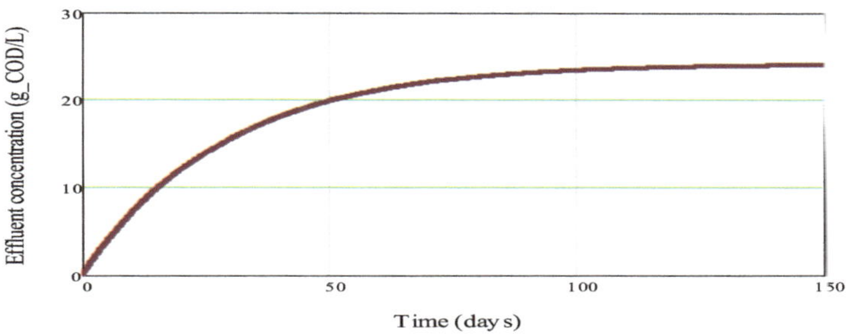

Fig. (9). Concentration of effluent particulate phenols *versus* time.

In Fig. (**10**), the concentration of TVFA is illustrated. Acetic acid accounts for more than 99% of TVFA concentration. This means that acetic acid is accumulated in the reactor, but the process is not inhibited, probably due to the addition of alkalinity.

The results of the model for the selected olive mill are shown in Table **27**. The biogas produced is of medium quality, with its methane content being around 56%. As a result, the energy content of gas is relatively low. Opposite to that, the methane content of biogas produced by the digestion of municipal wastewater sludge is higher to 67% and its LHV is about 21.4 MJ/kg (Table **12**). It can be concluded that olive mill waste is a more difficult substrate compared to wastewater sludge in terms of treatment and, as a result, yields to gas of lower energy content. However, anaerobic digestion of olive mill waste seems to be as effective as that of sludge from municipal wastewater in terms of gas and energy production per quantity of influent waste. Olive mill waste accounted for 26 Nm^3 of biogas and 141.25 kWh of energy for every 1 m^3 of daily inflow in the reactor.

The treatment of municipal wastewater resulted in lower production of biogas, but in slightly higher production of energy (22.7 Nm^3 and 142.6 kWh per m^3 of daily inflow, respectively).

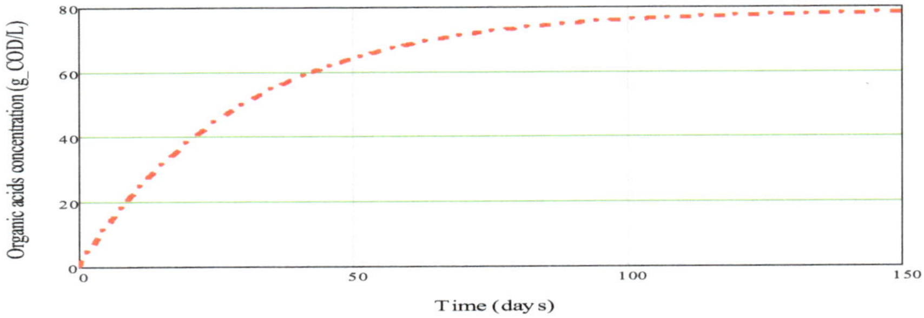

Fig. (10). Concentration of effluent TVFA *versus* time.

Table 27. Results of the extended ADM1 for the olive mill waste treatment.

Data	Value	Unit
Gas flow rate	52.1	Nm^3/day
Methane rate	56.7 %	-
Gas Lower Heat value	16.74	MJ/kg
	19.53	MJ/Nm^3
Total energy per day	282.5	kWh
Electric power	2.94	kW_e
Thermal power	5.89	kW_{th}

In addition to the achieved treatment of a difficult and environmentally dangerous type of waste and the reduction of phenols concentration (Figs. **8 & 9**), almost 3 kW of power and 6 kW of heat are generated by the anaerobic digestion of the waste produced by the modeled small rural olive oil producing facility. The energy produced can be utilized to supply the thermal needs of the digester and the olive-mill, or to heat a rural farming house. Moreover, the generated power may be enough to cover the needs for lighting or even the utilization of home appliances such as a fridge or a microwave.

The model was finally applied to a case of a sudden increase in waste production. The daily inflow rate of OMW was set to increase to 5 m^3 from 2 m^3. The retention time was reduced from 30 to 12 days. In Fig. (**11**), the change of gas production and methane rate *versus* time is illustrated. If compared to Fig. (**7**), an instability in gas production during the first days can be noticed. This may be due to the increased waste inflow and the consequent reduction of retention time lower to 20 days.

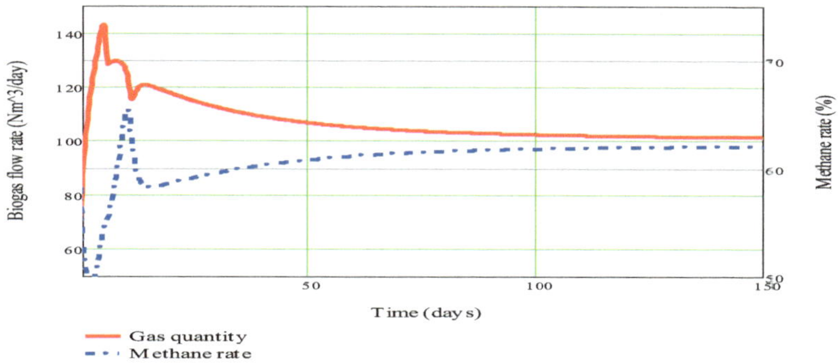

Fig. (11). Change of gas production and methane rate *versus* time for increased waste inflow.

An average of 104.2 Nm^3 of biogas was produced every day. It was shown that the increase in waste inflow rate resulted in gas richer in methane (60.9% methane content). The LHV of the produced biogas was 18.58 MJ/kg. On the other hand, the ratio of gas and energy production per unit of waste inflow rate were reduced by 15-20%, to 20.84 Nm^3 of biogas and 120.56 kWh of energy per m^3 of daily inflow of waste, respectively.

CONCLUSIONS

The need for more extensive use of bioenergy was demonstrated and its potential sources and applications were described and explained. The ways for combining waste treatment with energy production were described, along with the properties and potential uses of solid biomass and biogas. Biogas and solid biomass were compared with other fuels in terms of calorific value. Anaerobic digestion, the most popular method for biogas production in both small and large scale was described and modeled with the use of ADM1, a generic model suitable for many types of organic waste. The model was applied in two wastewater treatment plants for the digestion of sewage sludge that is the main by-product of wastewater treatment. The effect of specific factors on anaerobic digestion process was examined both theoretically and with the use of ADM1. The results of ADM1 were well associated with theory. Furthermore, the extension of ADM1 to include the digestion of olive-mill waste was presented in detail and applied to a rural plant for olive mill wastewater and solid waste. Finally, the effect of selected parameters of ADM1 on the quantity and quality of the produced biogas was examined, presented and discussed.

NOTES

[2**] "CPP" (Central Power Plant) refers to a set of regulations in the United States that was recently finalized.

***3 Member of OECD in order of continents: **Europe:** 22 of the 28 EU members (except Cyprus, Malta, Bulgaria, Lithuania, Romania and Croatia), Switzerland, Iceland, Norway, Turkey. **America:** USA, Canada, Mexico, Chile. **Asia:** Israel, Japan, South Korea. **Oceania:** Australia, New Zealand

CONSENT FOR PUBLICATION

Not applicable.

CONFLICT OF INTERESTS

The authors declare no conflict of interests.

ACKNOWLEDGEMENT

Declared none.

APPENDIX

The suggested parameters of the original ADM1 are illustrated in Appendix A. In some cases, they differ according the temperature of the digester. The effect of selected parameters in the production of biogas, as well as in pH is investigated in Appendix B. In Appendix C, the values of the additional parameters of the modified ADM1 are presented, along with the changes required for the parameters existing in the initial model.

Appendix A: Suggested Parameters for ADM1

The parameters listed in Table **A.1** and **A.2** are necessary for the calculation of stoicheiometric rate coefficients (Table **6**) and the kinetic rates (Table **7**) of the biochemical reactions. Some of the parameters shown in Table **A.2** have different values for mesophilic and for thermophilic digestion.

Table A.1. Sggested stoicheiometric parameters for ADM1 [63].

Parameter	Description	Value
$f_{sI,xc}$	Soluble inerts from composites	0.1
$f_{xI,xc}$	Particulate inerts from composites	0.25
$f_{ch,xc}$	Carbohydrates from composites	0.20
$f_{pr,xc}$	Proteins from composites	0.20
$f_{li,xc}$	Lipids from composites	0.25
$f_{fa,li}$	Fatty acids from lipids	0.95
$f_{h2,su}$	Hydrogen from sugars	0.19

(Table A.1) cont.....

Parameter	Description	Value
$f_{bu,su}$	Butyrate from sugars	0.13
$f_{pro,su}$	Propionate from sugars	0.27
$f_{ac,su}$	Acetate from sugars	0.41
$f_{h2,aa}$	Hydrogen from amino acids	0.06
$f_{va,aa}$	Valerate from amino acids	0.23
$f_{bu,aa}$	Butyrate from amino acids	0.26
$f_{pro,aa}$	Propionate from amino acids	0.05
$f_{ac,aa}$	Acetate from amino acids	0.40

Table A.2. Suggested parameter values for ADM1 [63].

Parameter	Mesophilic (35 °C)	Thermophilic (55 °C)
k_{dis} (d^{-1})	0.5	1.0
k_{hyd_ch} (d^{-1})	10	10
k_{hyd_pr} (d^{-1})	10	10
k_{hyd_li} (d^{-1})	10	10
$t_{res,X}$ (d)	0	0
k_{dec_all} (d^{-1})	0.02	0.04
$K_{S\_NH3\_all}$ (M)	$1*10^{-4}$	$1*10^{-4}$
$pH_{UL\ acet/acid}$	5.5	5.5
$pH_{LL\ acet/acid}$	4	4
k_{m_su} ($COD*COD^{-1}*d^{-1}$)	30	70
K_{S_su} ($kg_{COD}*m^{-3}$)	0.5	1
Y_{su} ($COD*COD^{-1}$)	0.10	0.10
k_{m_aa} ($COD*COD^{-1}*d^{-1}$)	50	70
K_{S_aa} ($kg_{COD}*m^{-3}$)	0.3	0.3
Y_{aa} ($COD*COD^{-1}$)	0.08	0.08
k_{m_fa} ($COD*COD^{-1}*d^{-1}$)	6	10
K_{S_fa} ($kg_{COD}*m^{-3}$)	0.4	0.4
Y_{fa} ($COD*COD^{-1}$)	0.06	0.06
$K_{I,H2\_fa}$ ($kg_{COD}*m^{-3}$)	$5*10^{-6}$	-
k_{m_c4+} ($COD*COD^{-1}*d^{-1}$)	20	30
K_{S_c4+} ($kg_{COD}*m^{-3}$)	0.2	0.4
Y_{c4+} ($COD*COD^{-1}$)	0.06	0.06

(Table A.2) cont.....

Parameter	Mesophilic (35 °C)	Thermophilic (55 °C)
$K_{I,H2\_c4+}$ ($kg_{COD}*m^{-3}$)	$1*10^{-5}$	$3*10^{-5}$
k_{m_pro} ($COD*COD^{-1}*d^{-1}$)	13	20
K_{S_pro} ($kg_{COD}*m^{-3}$)	0.1	0.3
Y_{pro} ($COD*COD^{-1}$)	0.04	0.05
$K_{I,H2\_pro}$ ($kg_{COD}*m^{-3}$)	$3.5*10^{-6}$	$1*10^{-5}$
k_{m_ac} ($COD*COD^{-1}*d^{-1}$)	8	16
K_{S_ac} ($kg_{COD}*m^{-3}$)	0.15	0.3
Y_{ac} ($COD*COD^{-1}$)	0.05	0.05
$pH_{UL\ ac}$	7	7
$pH_{LL\ ac}$	6	6
$K_{I,NH3}$ (M)	0.0018	0.011
k_{m_h2} ($COD*COD^{-1}*d^{-1}$)	35	35
K_{S_h2} ($kg_{COD}*m^{-3}$)	$7*10^{-6}$	$5*10^{-5}$
Y_{h2} ($COD*COD^{-1}$)	0.06	0.06
$pH_{UL\ h2}$	6	6
$pH_{LL\ h2}$	5	5

Appendix B: Effect of Selected Parameters on Produced Biogas

In Appendix **B**, the effect of selected ADM1 parameters on the produced biogas will be examined. The selected parameters were the kinetic rates of disintegration and hydrolysis (k_{dis} and k_{hyd}), the uptake rates (k_m) for the products of hydrolysis and acetic acid, as well as the parameter corresponding to 50% inhibitory concentration of ammonia and hydrogen (K_{I_nh3} and $K_{I,h2}$).

Kinetic Parameters of Disintegration and Hydrolysis

The selected value for the kinetic rate of disintegration is 0.5 day^{-1}. Lower values mean that the process is slower. For rates lower to 0.1 day^{-1}, a significant reduction in biogas production was noticed. Although the methane rate was higher, the trend of energy production followed that of biogas. This means that slow disintegration delays the whole process and may either lead to a significant reduction of gas and energy production, or even cause the process to stop. If disintegration rate is higher to 0.5 day^{-1}, an increase on gas production was shown. However, this increase was relatively small. This is explained by the fact that disintegration is followed by multiple steps. Even if disintegration is completed very quickly, the slow next steps will diminish the effect of the fast initial process. (Fig. **B.1**)

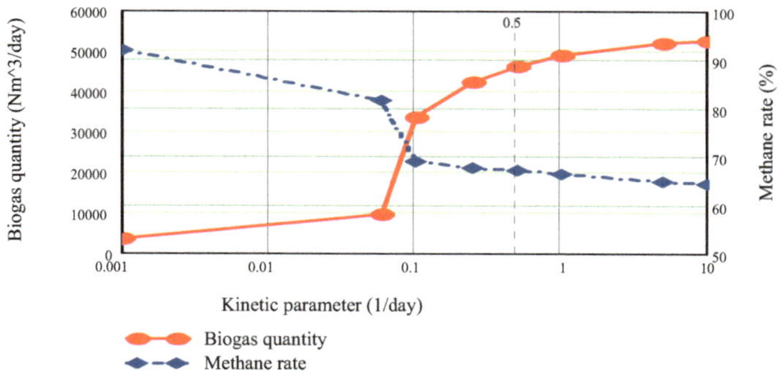

Fig. (B.1). Change of quantity and methane rate of the produced biogas with disintegration kinetic rate k_{dis}.

The hydrolysis kinetic rate is equal to 10 day^{-1} for carbohydrates ($k_{hyd,ch}$), proteins ($k_{hyd,pr}$) and lipids ($k_{hyd,li}$). During every test, the value of the three hydrolysis parameters was changed simultaneously. The effect of the rate of hydrolysis on gas quantity and methane rate seems to be small, compared to that of disintegration rate. However, when hydrolysis got more than 20 times slower (k_{hyd} lower to 0.5 day^{-1}), gas and energy production were reduced by 10%. For faster hydrolysis, the effects were negligible (Fig. **B.2**).

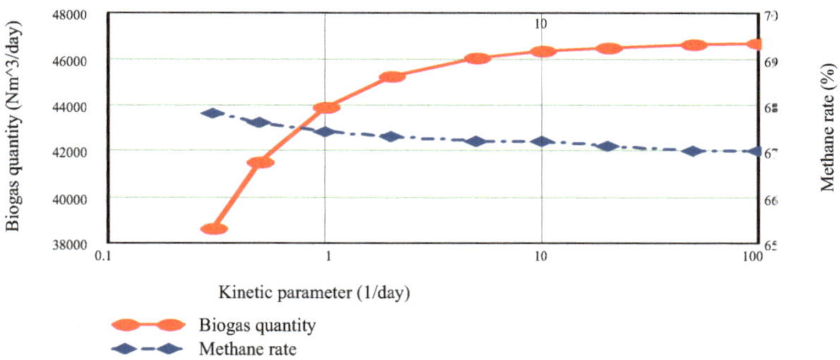

Fig. (B.2). Change of quantity and methane rate of the produced biogas with hydrolysis kinetic rate k_{hyc_ch}, k_{hyc_pr} and k_{hyc_li}.

Specific Uptake Rate of Hydrolysis Products, Organic Acids and Acetic Acid

The initially set values for the specific uptake rate of sugars, amino acids and lipids were equal to 30, 50 and 6 day^{-1} respectively. In order to investigate their effect on the process, these parameters were simultaneously multiplied with a selected number. For ratio of uptake rate higher to 1, the conversion of hydrolysis products to organic acids, carbon dioxide and hydrogen was faster that defined. For ratio lower than 1, acidogenesis was slower.

When the rates of acidogenesis got higher to those that were initially set in the model, no significant change in biogas quality and quantity was observed.

(Fig. **B.3**). However, when the uptake rates of hydrolysis products got lower than the half of their initial values, a decrease in biogas quantity and energy production were noticed. If the rates are reduced by more than 20 times, the process gets very slow and gas production is significantly reduced. The produced energy followed the trend of gas production, despite the observed increase in methane content of gas. For very low acidogenesis rates, corresponding to a ratio lower to 0.02, the whole process may stop.

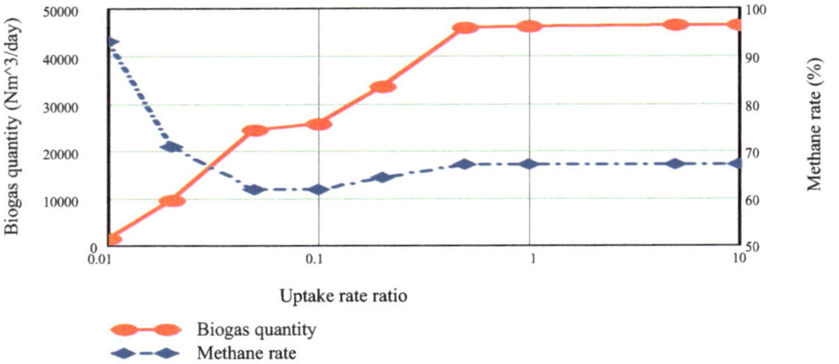

Fig. (B.3). Change of quantity and methane rate of the produced biogas with uptake rate of hydrolysis products k_{m_su}, k_{m_aa} and k_{m_fa}.

The uptake rate of propionate (k_{m_pro}) is equal to 13 day^{-1}. Valerate and butyrate share a common rate, k_{m_c4} equal to 20 day^{-1}. Similarly to the uptake rates of hydrolysis products, k_{m_pro} and k_{m_c4} were simultaneously multiplied with a common ratio. The effect of the uptake rates during acetogenesis is not as important as in the case of acidogenesis. For faster uptake of organic acids, no significant effect was noticed (Fig. **B.4**). For rates lower to 20% of their initially set values, both gas quantity and methane content in biogas started dropping. However, opposite to acidogenesis rates, the maximum drop of gas production did not get higher to 20%.

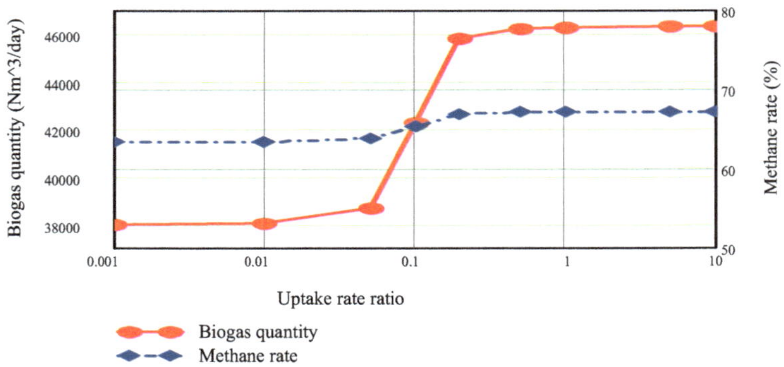

Fig. (B.4). Change of quantity and methane rate of the produced biogas with uptake rate of organic acids k_{m_c4} and k_{m_pro}.

The rate of organic acid uptake during methanogenesis has been set equal to 8 day^{-1}. When it was increased, no change was noticed neither in the quantity of the produced gas nor in the methane content of biogas (Fig. **B.5a**). Opposite to that, a rapid reduction on both gas quantity and methane content of biogas was observed for a value of acetate uptake rate lower to 30% of its initial value. For uptake rate lower than 2 day^{-1}, the daily biogas production dropped by more than 75%, to 10,000 Nm^3. Moreover, the rate of methane in biogas was lower than 10%. As a result, biogas is of very low calorific value and the produced energy is negligible.

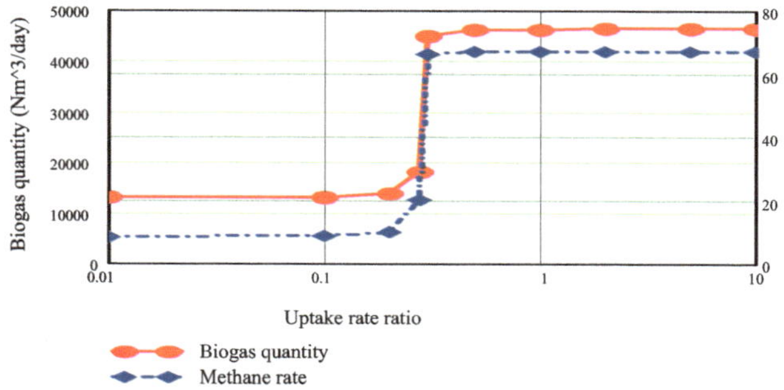

Fig. (B.5a). Change of quantity and methane rate of the produced biogas with uptake rate of acetic acid k_{m_ac}.

The reduced biogas production caused by the slower uptake of acetate, along with the very low methane content is explained by the change on the value of pH (Fig. **B.5b**). When the ratio of acetate uptake rate got higher to 0.3 (k_{m_ac}=2.4 day^{-1}), a neutral pH was observed. For lower ratio, there was a rapid reduction on pH and its value dropped below 5. At that point, the process may still be possible but it is practically not viable as the produced gas is of low quantity and poor energy content.

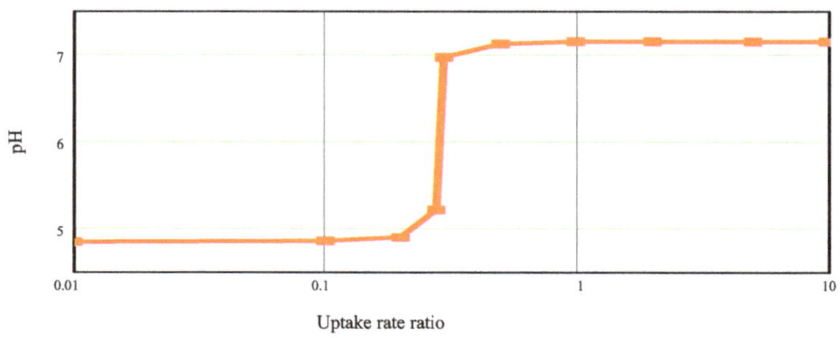

Fig. (B.5b). Change of pH with uptake rate of acetic acid k_{m_ac}.

Opposite to that, in the case of organic acids uptake, where the drop in gas production was considerably smaller, the minimum value of pH for ratio of uptake rates lower to 0.2 was about 6.7. This value of pH enables the completion of the process, even with a reduction in gas production.

Inhibitory Concentration Parameter

The parameters examined were the ammonia inhibition constant during uptake of acetate (K_{I_nh3}) and the constant of hydrogen inhibition during uptake of lipids, valerate/butyrate and propionate ($K_{I\_h2\_fa}$, $K_{I\_h2\_c4}$ and $K_{I\_h2\_pro}$ respectively). A reduction in inhibition constant K_I increases the concentration of hydrogen or ammonia, resulting in more intense inhibition.

The set value of ADM1 for ammonia inhibition constant equals 0.0018 kmole/m^3. An increase in ammonia inhibition constant had no effect on gas production. When the constant was reduced to a value lower than the 10% of the initially selected one, a rapid reduction in both gas production and methane rate was noticed (Fig. **B.6a**). The effect of ammonia inhibition constant is similar to that of the uptake rate of acetate. This is expected, as ammonia inhibition and acetate uptake rate are the two factors that mostly affect the uptake of acetate and the consequent methane formation.

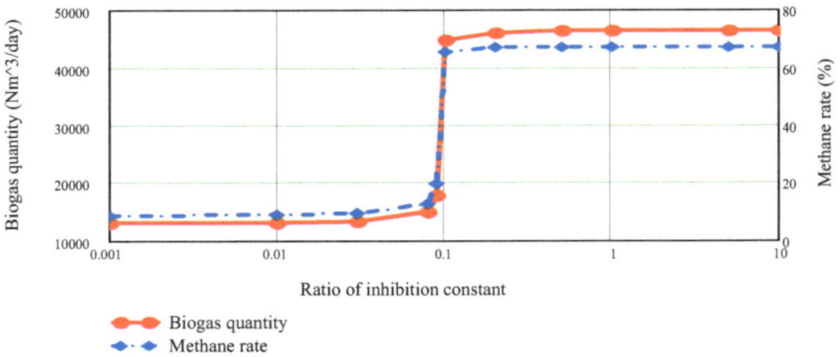

Fig. (B.6a). Change of quantity and methane rate of the produced biogas with inhibition constant of ammonia K_{I_nh3}.

The rapid reduction of gas and energy production for ratio lower to 0.1 (K_{I_nh3}=0.00018 kmole/m^3) is associated with the drop of pH from neutral to a value below 5 at that point (Fig. **B.6b**). High ammonia concentration makes the uptake of acetate slower. As a result and similar to the case of acetate uptake rate, acetic acid is accumulated and the value of pH drops.

According to ADM1 results, it is clear that the uptake of acetate is the key part of anaerobic digestion as it is very sensitive in the accumulation of acids. When the rate of acetic acid uptake is low or the uptake is inhibited by ammonia concentration, the formation of methane is slower than required by the process. Acetic acid is accumulated, as its formation is faster than its conversion to methane. Slow uptake of total acetate results in a drop in the value of pH and will potentially lead to the total inhibition of the process. Even if the process is not inhibited, it is not viable due to the negligible amount of the produced energy.

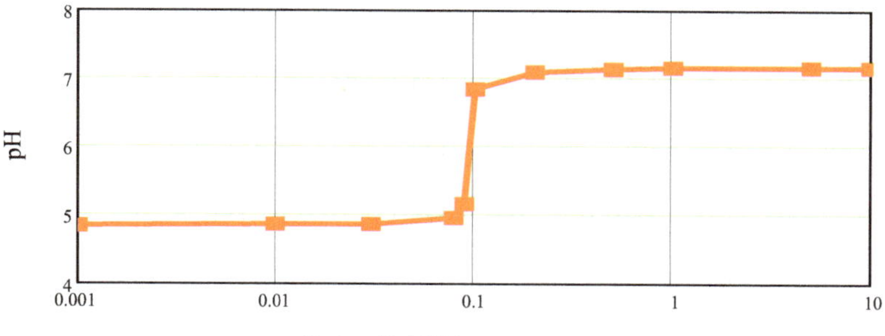

Fig. (B.6b). Change of pH with inhibition constant of ammonia K_{I_nh3}.

Finally, the effect of hydrogen inhibition constants was examined. Gas production and gas methane content were reduced only when the three constants simultaneously dropped lower to 3% of their initially set value. The reduction of biogas production reached 60% at hydrogen inhibition constants reduced by 1000 times. (Fig. **B.7**) The value of pH was reduced at the same way with gas production, but its minimum value did not get lower than 6.8. It is obvious that an extremely low value of hydrogen inhibition constant reduces gas and energy production. However, and opposite to the case of ammonia the reduction is negligible at a ratio of 0.05-0.1. Moreover, the little drop of pH does not fully inhibit the complete process.

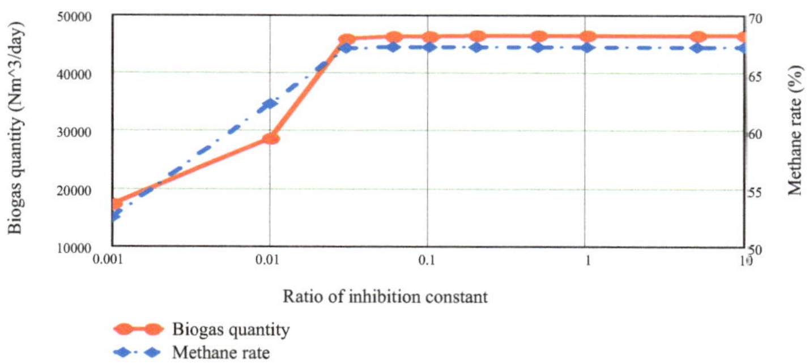

Fig. (B.7). Change of quantity and methane rate of the produced biogas with inhibition constant of hydrogen $K_{I\_h2\_fa}$, $K_{I\_h2\_c4}$ and $K_{I\_h2\_pro}$.

Appendix C: Additional and Modified Parameters for ADM1 Application on Olive-Mill Waste Digestion

In Table **C.1**, the suggested values of the additional stoicheiometric parameters for the extended ADM1 for olive-mill waste treatment are listed. Moreover, the modified values of parameters that are included in the initial ADM1 and are already shown in Table **6** and Table **A.1** are also presented. Additional parameters such as yields, kinetic rates and inhibition constants, as well as the modified values of some of the parameters presented in Table **A.2**, are shown in Table **C.2**.

Table C.1. Modified suggested stoicheiometric parameters for olive-mill waste treatment [66].

Parameter	Description	Value
$f_{sI,Xc}$	Soluble inert fraction in OMSW	0.013
$f_{xI,Xc}$	Particulate inert fraction in OMSW	0.45
$f_{ch,Xc}$	Carbohydrates fraction in OMSW	0.35
$f_{pr,Xc}$	Proteins fraction in OMSW	0.074
$f_{li,Xc}$	Lipids fraction in OMSW	0.1
$f_{Xph,Xc}$	Particulate phenol fraction in OMSW	0.026
$f_{Sph,Xph}$	Soluble phenol fraction issued from particulate phenol degradation	0.1
$f_{bnz,ph}$	Benzoate fraction issued from soluble phenol degradation	0.87
$f_{ac,bnz}$	Acetate fraction issued from soluble phenol degradation	0.51
$f_{h2,Xc}$	Hydrogen fraction issued from soluble phenol degradation	0.51
$f_{h2,bnz}$	Hydrogen fraction issued from benzoate degradation	0.51
N_I	Nitrogen content in inert substrates of OMSW	0
N_{xc}	Nitrogen content in OMSW	0.0107/14
C_{Xph}	Carbon content in particulate phenol compounds	0.033
C_{ph}	Carbon content in soluble phenol compounds	0.0319
C_{bnz}	Carbon content in benzoate	0.034

Table C.2. Modification of suggested parameter values for olive-mill waste treatment [66].

Parameter	Description	Unit	Value
k_{dis}	Disintegration constant	d^{-1}	0.001
k_{hyd_ph}	Phenol hydrolysis constant	d^{-1}	0.0015
k_{m_ph}	Maximum specific rate for phenol uptake	d^{-1}	15
$K_{S,ph}$	Half saturation constant for phenol uptake	kg_{COD}/m^3	30.5
$K_{I,ph}$	Inhibition constant for phenol uptake	kg_{COD}/m^3	50
k_{dec_Bph}	Phenol biomass decay constant	d^{-1}	0.02
Y_{ph}	Phenol yield coefficient	$kg_{COD,X}/kg_{COD,S}$	0.01
k_{m_bnz}	Maximum specific rate for benzoate uptake	d^{-1}	8
K_{S_bnz}	Half saturation constant for benzoate uptake	kg_{COD}/m^3	15.5
$k_{dec,Bbnz}$	Benzoate biomass decay constant	d^{-1}	0.02
Y_{bnz}	Benzoate yield coefficient	$kg_{COD,X}/kg_{COD,S}$	0.0135
$K_{I,TVFA}$	TVFA inhibition constant for acetate utilisers	kg_{COD}/m^3	47.5
K_{I_NH3}	Ammonia inhibition constant for acetate utilisers	$Kmol/m^3$	2.4e-3
$K_{a,ph}$	Phenolic acid equilibrium constant	mol/L	3.16e-10

(Table C.2) cont.....

Parameter	Description	Unit	Value
$k_{b,ph}$	Kinetic rate constant of phenol acid-base reaction	mol/(L*d)	1e+10
$K_{a,bnz}$	Benzoate acid equilibrium constant	mol/L	3.16e-10
$k_{b,bnz}$	Kinetic rate constant of benzoate acid-base reaction	mol/(L*d)	1e+10

REFERENCES

[1] *Energy consumption since 1820 in charts.*, 1820. Accessed 25-09-2016 [Online] Available: https://ourfiniteworld.com/2012/03/12/world-energy-consumption-since-1820-in-charts/

[2] *World Energy consumption..* Accessed 25-09-2016 [Online] Available: https://en.wikipedia.org/wiki/World_energy_consumption

[3] *Global energy Statistical Yearbook,* 2016. Accessed 25-09-2016 Total Energy Consumption. [Online] Available: https://yearbook.enerdata.net/

[4] *Chapter 1. World energy demand and economic outlook,* May 2016. Accessed: 25-09-2016 International Energy Outlook 2016, [Online] Available: https://www.eia.gov/forecasts/ieo/world.cfm

[5] *List of countries by electricity consumption..* Accessed: 25-09-2016 [Online] Available: https://en.wikipedia.org/wiki/List_of_countries_by_electricity_consumption

[6] E. Peduzzi, G. Boissonnet, and F. Marechal, "Biomass modeling: Estimating thermodynamic properties from the elemental composition", *Fuel,* vol. 181, pp. 207-2017, 2016. [http://dx.doi.org/10.1016/j.fuel.2016.04.111]

[7] *Biomass characteristics.* Accessed: 22-09-2016 [Online] Available: http://www.eubia.org/index.php/about-biomass/biomass-characteristics

[8] Accessed 22-09-2016 [Online] Available: http://www.eubia.org/index.php/about-biomass

[9] *Biomass. An energy resource for the European Union,* 2000. Accessed 23-09-2016 [Online] Available: http://www.managenergy.net/download/r31.pdf

[10] *Comission staff working document. State of play on the sustainability of solid and gaseous biomass used for electricity, heating and cooling in the EU,* 2014. Accessed 23-09-2016 [Online] Available: http://ec.europa.eu/energy/sites/ener/files/2014_biomass_state_of_play_.pdf

[11] *Biomass for heat and power - Opportunity and economics,* 2010. Accessed 22-09-2016 [Online] Available: https://europeanclimate.org/documents/Biomass_report_-_Final.pdf

[12] *Global biomass potential under sustainable constraints,* 2010. Accessed 23-09-2016 [Online] Available: http://ec.europa.eu/environment/integration/research/newsalert/pdf/197na5_en.pdf

[13] *Statistical report. 2015 Key Findings.* Accessed 23-09-2016 [Online] Available: http://www.aebiom.org/library/statistical-reports/statistical-report-2015/

[14] H. Li, Q. Chen, X. Zhan, K.N. Finney, V.N. Sharifi, and J. Swithenbank, "Evaluation of a biomass drying process using waste heat from process industries: A case study", *Appl. Therm. Eng.,* vol. 35, pp. 71-80, 2012. [http://dx.doi.org/10.1016/j.applthermaleng.2011.10.009]

[15] Accessed 23-09-2016 [Online] Available: http://www.eubia.org/index.php/ about-biomass/biomas--pelleting

[16] N.S. Bentsen, and C. Felb, *Biomass for energy in the European Union-a review of bioenergy resource assessments,* 2012. Accessed 09-10-2016 [Online] Available: https://biotechnologyforbiofuels.biomedcentral.com/articles/10.1186/1754-6834-5-25 [http://dx.doi.org/10.1186/1754-6834-5-25]

[17] J. Manders, "Integrated waste management of MSW across Europe. Waste to Energy as a professional route to treat residual waste", Accessed 26-09-2016 [Online] Available: http://www.elfm.eu/Uploads/ELFM/FILE_cda3ca83-aea6-432b-bc5a-f2b3782c982d.pdf

[18] *Landfill gas utilization.*. Accessed 27-09-2016 [Online] Available: https://en.wikipedia.org/wiki/Landfill_gas_utilization

[19] Y. Huang, Y.D. Wang, S. Rezvani, D.R. McIlveen-Wright, M. Anderson, and N.J. Hewitt, "Biomass fuelled trigeneration system in selected buildings", *Energy Convers. Manage.*, vol. 52, no. 6, pp. 2448-2454, 2011.
[http://dx.doi.org/10.1016/j.enconman.2010.12.053]

[20] *Biogas, December,* 2011. Accessed 26-09-2016 [Online] Available: http://european- biogas.eu/wp-content/uploads/files/2013/10/EBA-brochure-2011.pdf

[21] F. Monnet, "An introduction to Anaerobic Digestion of organic wastes", *Final report, November,* 2003. Accessed 26-09-2016 [Online] Available: http://www.remade.org.uk/media/9102/an%20introduction%20to%20anaerobic%20digestion%20nov%202003.pdf

[22] S.N.A. Rahman, M.S. Masdar, M.I. Rosli, E.H. Majlan, T. Husaini, S.K. Kamarudin, and W.R.W. Daud, "Overview biohydrogen technologies and application in fuel cell technology", *Renew. Sustain. Energy Rev.*, vol. 66, pp. 137-162, 2016.
[http://dx.doi.org/10.1016/j.rser.2016.07.047]

[23] D. Ludington, *Calculating the Heating Value of Biogas*. Accessed 26-09-2016 [Online] Available: http://masters.donntu.org/2013/fkita/alexandrova/library/Heating_Value_of_Biogas_copy.pdf

[24] *Solid biomass barometer 2015,* 2015. Accessed 27-09-2016 [Online] Available: http://www.eurobserv- er.org/solid-biomass-barometer-2015/

[25] *Biogas barometer,* 2014. Accessed 27-09-2016 [Online] Available: http://www.eurobserv-er.org/biogas-barometer-2014/

[26] *Biofuels barometer,* 2015. Accessed 28-09-2016 [Online] Available: http://www.eurobserv-er.org/biofuels-barometer-2015/

[27] K. Ericsson, and S. Werner, "The introduction and expansion of biomass use in Swedish district heating systems", *Biomass and Bioenergy,* vol. 94, pp. 57-65, 2016. Accessed 09-10-2016 [Online] Available: http://www.sciencedirect.com/science/article/pii/S0961953416302793
[http://dx.doi.org/10.1016/j.biombioe.2016.08.011]

[28] B. Rezaie, and M.A. Rosen, "District heating and cooling: Review of technology and potential enhancements", *Appl. Energy,* vol. 93, pp. 2-10, 2012.
[http://dx.doi.org/10.1016/j.apenergy.2011.04.020]

[29] D. Champier, J.P. Bedecarrats, M. Rivaletto, and F. Strub, "Thermoelectric power generation for biomass cook stoves", *Energy,* vol. 35, no. 2, pp. 935-942, 2010.
[http://dx.doi.org/10.1016/j.energy.2009.07.015]

[30] M. Kimming, C. Sundberg, A. Nordberg, A. Baky, S. Bernesson, O. Noren, and P.A. Hansson, "Biomass from agriculture in small-scale combined heat and power plants – A comparative life cycle assessment", *Biomass Bioenergy,* vol. 35, no. 4, pp. 1572-1581, 2011.
[http://dx.doi.org/10.1016/j.biombioe.2010.12.027]

[31] A.M. Panataleo, S.M. Camporeale, and N. Shah, "Thermo-economic assessment of externally fired micro-gas turbine fired by natural gas and biomass: Applications in Italy", *Energy Convers. Manage.,* vol. 75, pp. 202-213, 2013.
[http://dx.doi.org/10.1016/j.enconman.2013.06.017]

[32] T. Stringfellow, *An independent engineering evaluation of waste-to-energy technologies,* 2014. [Online] Available: http://www.renewableenergyworld.com/articles/2014/01/ an-independen--engineering-evaluation-of-waste-to-energy-technologies.html

[33] J.A. Ruiz, M.C. Juarez, M.P. Morales, P. Munoz, and P. Mendivil, "Biomass gasification for electricity generation: Review of current technology barriers", *Renew. Sustain. Energy Rev.*, vol. 18, pp. 174-183, 2013.
[http://dx.doi.org/10.1016/j.rser.2012.10.021]

[34] S. Koppatz, C. Pfeifer, R. Rauch, H. Hofbauer, T. Marquard-Moellenstedt, and M. Specht, "H2 rich product gas by steam gasification of biomass with in situ CO2 absorption in a dual fluidized bed system of 8 MW fuel input", *Fuel Process. Technol.*, vol. 90, no. 7-8, pp. 914-921, 2009.
[http://dx.doi.org/10.1016/j.fuproc.2009.03.016]

[35] K. Goeransson, U. Soederlind, J. He, and W. Zhang, "Review of syngas production via biomass DFBGs", *Renew. Sustain. Energy Rev.*, vol. 15, no. 1, pp. 482-492, 2011.
[http://dx.doi.org/10.1016/j.rser.2010.09.032]

[36] U. Lee, E. Balu, and J.N. Chung, "An experimental evaluation of an integrated biomass gasification and power generation system for distributed power applications", *Appl. Energy*, vol. 1010, pp. 699--, 2013.
[http://dx.doi.org/10.1016/j.apenergy.2012.07.036]

[37] M. Gassner, and F. Marechal, "Thermo-economic optimization of the polygeneration of synthetic natural gas (SNG), power and heat from lignocellulosic biomass by gasification and methanisation", *Energy Environ. Sci.*, vol. 5, no. 2, pp. 5768-5789, 2012.
[http://dx.doi.org/10.1039/c1ee02867g]

[38] H. Balat, and E. Kirtay, "Hydrogen from biomass – Present scenario and future prospects", *Int. J. Hydrogen Energy*, vol. 35, no. 1, pp. 7416-7426, 2010.
[http://dx.doi.org/10.1016/j.ijhydene.2010.04.137]

[39] S. Rapagnà, K. Gallucci, M. Di Marcello, M. Matt, M. Nacken, S. Heidenreich, and P.U. Foscolo, "Gas cleaning, gas conditioning and tar abatement by means of a catalytic filter candle in a biomass fluidized-bed gasifier", *Bioresour. Technol.*, vol. 101, no. 18, pp. 7134-7141, 2010.
[http://dx.doi.org/10.1016/j.biortech.2010.03.139] [PMID: 20413303]

[40] "A biogas roadmap of Europe", Accessed 22-09-2016 [Online] Available: http://www.seai.ie/Renewables/AD_In_Ireland_22nd_October/A_Biogas_Roadmap_for_Europe.pdf

[41] R.A. Voloshin, M.V. Rodionova, S.K. Zharmukhamedov, T.N. Veziroglou, and S.I. Allakverdiev, "Review: Biofuel production from plant and algal biomass", *Int. J. Hydrogen Energy*, vol. 41, pp. 17257-17273, 2016.
[http://dx.doi.org/10.1016/j.ijhydene.2016.07.084]

[42] *Biofuels for Transport.*. Accessed 23-09-2016 [Online] Available: http://www.eubia.org/index.php/about-biomass/biofuels-for-transport

[43] C.M. Liu, and S.Y. Wu, "From biomass waste to biofuels and biomaterial building blocks", *Renew. Energy*, vol. 96, pp. 1056-1062, 2016.
[http://dx.doi.org/10.1016/j.renene.2015.12.059]

[44] M. Phanphanich, and S. Mani, "Impact of torrefaction on the grindability and fuel characteristics of forest biomass", *Bioresour. Technol.*, vol. 102, no. 2, pp. 1246-1253, 2011.
[http://dx.doi.org/10.1016/j.biortech.2010.08.028] [PMID: 20801023]

[45] *Anaerobic Digestion.*. Accessed: 27-09-2016 [Online] Available: http://www.eubia.org/index.php/about-biomass/anaerobic-digestion

[46] F.A. Al-Sulaiman, F. Hamdullahpur, and I. Dincer, "Greenhouse gas emission and exergy assessments of an integrated Organic Rankine Cycle with a biomass combustor for combined cooling, heating and power production", *Appl. Therm. Eng.*, vol. 31, no. 4, pp. 439-446, 2011.
[http://dx.doi.org/10.1016/j.applthermaleng.2010.09.019]

[47] A. Schuster, S. Karellas, and E. Kakaras, "amd H. Spliethoff, "Energetic and economic investigation of Organic Rankine Cycle applications", *Appl. Therm. Eng.*, vol. 29, no. 8-9, pp. 1809-1817, 2009.

[http://dx.doi.org/10.1016/j.applthermaleng.2008.08.016]

[48] G. Qiu, Y. Shao, J. Li, H. Liu, and B.S. Riffat, "Experimental investigation of a biomass-fired ORC-based micro-CHP for domestic applications", *Fuel,* vol. 96, pp. 374-382, 2012. [http://dx.doi.org/10.1016/j.fuel.2012.01.028]

[49] F. Cotana, A. Messineo, A. Petrozzi, V. Coccia, G. Cavalaglio, and A. Aquino, "Comparison of ORC Turbine and Stirling Engine to Produce Electricity from Gasified Poultry waste", *Sustainability,* vol. 6, no. 9, pp. 5714-5729, 2014.

[50] G. Taljan, G. Verbic, M. Pantos, M. Sakulin, and L. Fickert, "Optimal size of biomass-fired Organic Rankine Cycle CHP system with heat storage", *Renew. Energy,* vol. 41, pp. 29-38, 2012. [http://dx.doi.org/10.1016/j.renene.2011.09.034]

[51] Y. Huang, D.R. McIlveen-Wright, S. Rezvani, M.J. Huang, Y.D. Wand, A.P. Roskilly, and N.J. Hewitt, "Comparative techno-economic analysis of biomass fuelled combined heat and power for commercial buildings", *Appl. Energy,* vol. 112, pp. 518-525, 2013. [http://dx.doi.org/10.1016/j.apenergy.2013.03.078]

[52] F. Biedermann, H. Carlsen, I. Obernberger, and M. Schoech, "Small-scale CHP Plant based on a 75 kWel Hermetic Eight Cylinder Stirling Engine for Biomass Fuels – Development, Technology and Operating experiences", *2nd World Conference and Exhibition on Biomass for Energy, Industry and Climate Protection,* 2004 Rome, Italy

[53] M. Szewczyk, and T. Trzepiecinski, "Application of biomass-powered Stirling engines in cogenerative systems", *Econtechmod,* vol. 01, no. 2, pp. 53-56, 2012.

[54] F. Biedermann, H. Carlsen, M. Schoech, and I. Obernberger, *Operating experiences with a small-scale CHP pilot plant based on a 35 kWel hermetic four cylinder Stirling engine for biomass fuels,* 2013. Accessed 29-09-2016 [Online] Available: http://orbit.dtu.dk/en/ publications/operating-experience--with-a-smallscale-chp-pilot-plant-based-on-a-35-kwel-hermetic-four-cylinder-st-rling-engine-for-biomass-fuels(78981c77-a49d-4460-b042-d2608a539eba).html

[55] G. Marinitsch, F. Biedermann, H. Carlsen, J.K. Bovin, M. Schoech, and I. Obernberger, *Development of a hot gas heat exchanger and a cleaning system for a 35 kWel hermetic four cylinder Stirling engine for solid biomass fuels,* 2005. Accessed 29-09-2016 [Online] Available: https://www.researchgate.net/publication/242408729_Development_of_a_hot_gas_heat_exchanger_and_a_cleaning_system_for_a_35kWel_hermetic_four_cylinder_Stirling_engine_for_solid_biomass_fuels

[56] A. Koellin, W. Siemers, U. Hellwig, N. Sachno, S. Schroeder, and N. Senkel, "High Temperature Biomass Fired Stirling Engine (HTBS)", *International Conference on Renewable Energies and Power Quality (ICREPQ '14),* 2014.

[57] A. Nishiyama, H. Shimojima, A. Ishikawa, Y. Itaya, S. Kambara, H. Moritomi, and S. Mori, "Fuel and emissions properties of Stirling engine operated with wood power", *Fuel,* vol. 86, no. 15, pp. 2333-2342, 2007. [http://dx.doi.org/10.1016/j.fuel.2007.01.040]

[58] D. Maraver, A. Sin, J. Royo, and F. Sebastian, "“Special Issue on Advances in sustainable biofuel production and use”, 19th International Symposium on Alcohol Fuels – ISA", *Appl. Energy,* vol. 102, pp. 1303-1313, 2013. [http://dx.doi.org/10.1016/j.apenergy.2012.07.012]

[59] M. Rokni, "Thermodynamic and thermoeconomic analysis of a system with biomass gasification, solid oxide fuel call (SOFC) and Stirling engine", *Energy,* vol. 76, pp. 19-31, 2014. [http://dx.doi.org/10.1016/j.energy.2014.01.106]

[60] M. Rokni, "Biomass gasification integrated with a solid oxide fuel cell and Stirling engine", *Energy,* vol. 77, pp. 6-18, 2014. [http://dx.doi.org/10.1016/j.energy.2014.01.078]

[61] E. Dellali, S. Begot, F. Lanzetta, E. Gavignet, R. Chutani, and J.Y. Rauch, "Design, fabrication and

CFD modeling of a Stirling engine microgenerator", *17th International Stirling Engine Conference,* 2016

[62] F. Toda, and M. Matsubara, "Development of low-temperature difference Stirling engine", *17th International Stirling Engine Conference,* 2016

[63] D.J. Batstone, J. Keller, I. Angelidaki, S.V. Kalyuzhnyi, S.G. Pavlostathis, A. Rozzi, W.T.M. Sanders, H. Siegrists, and V.A. Vavilin, *IWA Task Group for Mathematical Modelling of Anaerobic Digestion Processes, Anearobic Digestion Model No.1,* 2002.

[64] P. Stuart, "The advantages and disadvantages of anaerobic digestion as renewable energy source", *Loughborough University,* 2002.

[65] F. Boubaker, and B.C. Ridha, "Modelling of the mesophilic anaerobic co-digestion of olive mill wastewater with olive mill solid waste using anaerobic digestion model No. 1 (ADM1)", *Bioresour. Technol.,* vol. 99, no. 14, pp. 6565-6577, 2008. [http://dx.doi.org/10.1016/j.biortech.2007.11.035] [PMID: 18187320]

[66] B. Fezzani, and R. Ben Cheikh, "Extension of the anaerobic digestion model No. 1 (ADM1) to include phenol compounds biodegradation processes for simulating the anaerobic co-digestion of olive mill wastes at mesophilic temperature", *J. Hazard. Mater.,* vol. 172, no. 2-3, pp. 1430-1438, 2009. [http://dx.doi.org/10.1016/j.jhazmat.2009.08.017] [PMID: 19783366]

[67] B. Fezzani, and R. Ben Cheikh, "Implementation of IWA anaerobic digestion model No.1 (ADM1) for simulating the thermophilic anaerobic co-digestion of olive mill wastewater with olive mill solid waste in semi-continuous tubular digester", *Chem. Eng. J.,* vol. 141, pp. 75-78, 2008. [http://dx.doi.org/10.1016/j.cej.2007.10.024]

[68] C. Rosen, and U. Jeppson, *Aspects on ADM1 implementation within the BSM2 framework.* Lund University: Lund, Sweden, 2007.

[69] U. Jeppson, *Investigation of Anaerobic Digestion Alternatives for Henriksdal's WWTP.* Lund University: Lund, Sweden, 2007.

[70] *Henriksdal wastewater treatment plant, Stockholm, Sweden..* Accessed 22-09-2016 [Online] Available: http://www.water-technology.net/projects/henriksdal-wastewater-treatment-plant-stockholm

[71] *The Henriksdal Wastewater Treatment Plant.* Accessed 22-09-2016 [Online] Available: http://programme.worldwaterweek.org/sites/default/files/henriksdal_eng_04.pdf

[72] J. Hoinkis, S.A. Deowan, V. Panten, A. Figoli, R.R. Huang, and E. Drioli, *Membrane Bioreactor (MBR) Technology – a Promising Approach for Industrial Water Reuse.* vol. Vol. 33. Procedia Engineering, 2012, pp. 234-241. Accessed 29-09-2016 [Online] Available: http://www.sciencedirect.com/science/article/pii/S1877705812012301

[73] *World water, Rebuild for world's largest underground WWTP,* 2006. Accessed 22-09-2016 [Online] Available: http://www.edie.net/library/Rebuild-for-worlds-largest-underground-WWTP/3541

[74] G. Moen, D. Stensel, R. Lepistoe, and J.F. Ferguson, "Effect of solids retention time on the performance of thermophilic and mesophilic digestion of combined municipal wastewater sludges", *Water Environmental Research,* vol. 75, 2003pp. 539-548. Accessed 22-09-2016 [Online] Available: https://www.jstor.org/stable/25045732?seq=1#page_scan_tab_contents [http://dx.doi.org/10.2175/106143003X141330]

[75] K. Sioulas, *Centre for Renewable Energy Sources (CRES), Biogas Potential in Greece – Summary report,* 2009. Accessed 26-09-2016 [Online] Available: http://www.big-east.eu/ downloads/IR-reports/ANNEX%202-11_WP2_D2.2_Summary-Greece_EN.pdf

[76] "Biogas Projects Development in Greece", Accessed 26-09-2016 [Online] Available: http://www.lignite.gr/events/5.pdf

[77] P. Vindis, B. Mursec, M. Janzekovic, and F. Cus, "The impact of mesophilic and thermophilic anaerobic digestion on biogas production", *Journal of Achievements in Materials and Manufacturing*

Engineering, vol. 36, 2009no. 2, . Accessed 29-09-2016 [Online] Available: http://citeseerx.ist.psu.edu/viewdoc/download?doi=10.1.1.466.555&rep=rep1&type=pdf

[78] *Anaerobic Digestion – Mesophilic vs. Thermophilic*. Accessed 29-09-2016 [Online] Available: http://www.theecoambassador.com/Anaerobic-Digestion-Temperature.html

[79] A. Abdelgadir, X. Chen, and J. Liu, "Characteristics, Process Parameters and Inner Components of Anaerobic Bioreactors", *BioMed Research International,* vol. 2014, 2014.841573 Accessed 30-0--2016 [Online] Available: https://www.hindawi.com/journals/bmri/2014/841573/

[80] B. Fezzani, and R. Ben Cheikh, "Thermophilic anaerobic co-digestion of olive mill wastewater with olive mill solid wastes in a tubular digester", *Chem. Eng. J.,* vol. 132, pp. 195-203, 2007. [http://dx.doi.org/10.1016/j.cej.2006.12.017]

[81] L. Kardos, A. Juhasz, G. Palko, J. Olah, K. Barkacs, and G. Zaray, "Comparing of mesophilic and thermophilic anaerobic fermented sewage sludge based on chemical and biochemical tests", *Applied ecology and environmental research,* vol. 9, 2011pp. 293-302. Accessed 29-09-2016 [Online] Available: http://www.aloki.hu/pdf/0903_293302.pdf [http://dx.doi.org/10.15666/aeer/0903_293302]

[82] "The Eco Ambassador, Staged Anaerobic Digestion", Accessed 30-09-2016 [Online] Available: http://www.theecoambassador.com/StagedAnaerobicDigestion.html

[83] Accessed 29-09-2016 [Online] Available: http://www.wtert.eu/default.asp?Menue=13&ShowDok=17

[84] C. Hansen, *Environmental Impact of Olive Oil Processing Wastes.*. Accessed 29-09-2016 [Online] Available: http://www.ecomena.org/olive-oil-wastes/

[85] B. Fezzani, and R. Ben Cheikh, "Optimisation of the mesophilic anaerobic co-digestion of olive mill wastewater with olive mill solid waste in a batch digester", *Deslination,* vol. 228, pp. 159-167, 2008. [http://dx.doi.org/10.1016/j.desal.2007.09.007]

CHAPTER 5

Wind Power Contribution in Achieving Global Renewable Energy Targets: Current Status and Future Prospects

John K. Kaldellis[1,*] and **Dimitrios Apostolou**[1,2]

[1] *Soft Energy Applications and Environmental Protection Laboratory, University of West Attica, P.O. Box 41046, Athens, 12201, Greece*

[2] *Department of Business Development and Technology, Centre for Energy Technologies, Aarhus University, Birk Centerpark 15, 7400 Herning, Denmark*

Abstract: The exploitation of wind energy dates thousands of years of history, where societies used wind as solution for many of their daily applications. During the last century, globally electricity needs were covered almost exclusively from fossil-fuel based technologies resulting in a negative impact on the environment. Since 1980, wind energy applications acquired great interest from the global community in order to cope with security of energy supply and environmental issues. This study presents the evolution of wind energy throughout the last 30 years, along with its prospects for covering a considerable percentage of the future global electrical demand. Furthermore, available information concerning the major wind energy markets has been analysed and revealed in general the existing trends of wind energy for the next years to come. In this context, technology and financial aspects along with environmental issues arising from wind power projects' implementation are investigated in order to provide all the necessary data for acquiring an integrated view of the wind energy future.

Keywords: Economics of Wind Energy, Environmental Impacts, Future Prospects, Historical Review, Life Cycle Energy Intensity, Onshore-Offshore Installations, Rotor Diameter, Wind Energy, Wind Energy Markets, Wind Power Capacity, Wind Technology Improvements.

INTRODUCTION

The wind energy made its first steps centuries ago, although simpler devices using wind date back thousands of years ago. Vertical axis windmills were discovered at Persian-Afghan borders dated around 200BC, while between 1300 and 1875 AD

* **Corresponding Author John K. Kaldellis:** Soft Energy Applications and Environmental Protection Laboratory, University of West Attica, P.O. Box 41046, Athens, 12201, Greece; Tel: 0030-2105381348; Fax: 0030-2105381467; Email: jkald@teipir.gr

Emmanuel D. Rogdakis & Irene P. Koronaki (Eds.)

horizontal-axis were found in the region of the Netherlands and the Mediterranean [1 - 3]. In Europe in 1800s, around 20,000 windmills were under operation in France alone, while in Netherlands wind energy was covering the 90% of the power used for industrial purposes [4]. Similarly, wind-driven systems found in USA during the 19th century, were used mostly for water pumping purposes and specifically, between 1850 and 1970, 6 million of these small machines were under operation at the country (see Fig. **1**).

The first large wind machine used to produce electricity was a low speed and high-solidity 12kW wind turbine installed in Cleveland, Ohio in 1888. During the World War I, wind energy was widespread mostly in Denmark with 25kW turbines in order to overcome power shortages. Small scale machines (1-3kW) driving direct current generators being very popular in the rural areas of the US (provided lighting for farms and charged batteries used to power crystal radio sets). Innovations introduced from the airplane industry along with research in Denmark, France, Germany and the UK concerning large-scale wind turbines (WTs) resulted to a rapid development until the 2nd World War. However, it was after the war when a large three-blade upwind rotor WT operated successfully in Denmark, the Gedser mill 200kW at the early 60s [5, 6].

Fig. (1). Early stages of wind energy and the Californian outbreak [10].

Following the oil crisis of 1973, the USA government decided to reduce its oil imports and to involve seriously in the research and development (R&D) of wind energy technologies resulting in one of the most important milestones of the wind energy history [7, 8]. This was the penetration outbreak encountered in California, where over 16,000 machines, ranging from 20 to 350kW (a total of 1.7GW), were installed between 1981 and 1985, as a result of the incentives (such as the federal investment and energy credits) given by the USA government [6, 9].

In northern Europe on the other hand, wind farm installations increased steadily through the 80s and the 90s, with the higher cost of electricity and the excellent wind resources leading to the creation of a small but stable market. After 1990, Europe was the major wind energy market, where between 1995 and 1999 75% of all new installed wind turbines have been implemented in the European continent [4].

TWENTY-YEARS OF WIND POWER EVOLUTION

Global Wind Energy Facts

Since the middle 90s, wind energy presented remarkable growth particularly in the European continent. European Union's (EU) policy for supporting Renewable Sources of Energy (RES) along with the high wind potential found at many regions in Europe were two of the most significant factors for the development occurred in the wind energy sector. On the other hand, the rest of the world's wind capacity began to present significant growth after the beginning of the new millennium where markets mainly from Asia have been emerged. Based to the latest official data [11, 12], the world's annual wind capacity rate was increasing since 2000 with a considerable growth after 2007. The 2015 new installations were around 63 MW (see Fig. **2a**), resulting to a cumulative capacity of 433GW, more than two times higher compared to 2010 (see Fig. **2b**).

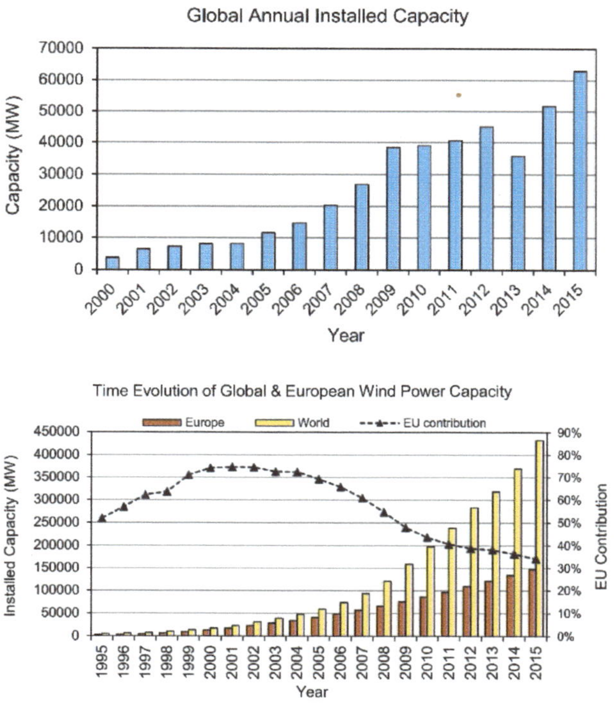

Fig. (2). (a) Global annual installed capacity, (b) Evolution of Global and Europe's wind capacity [10, 11].

Although Europe was until mid-2000s the continent with the highest share of installations (above 70%), a rapid growth of wind industry in other countries since 2005 resulted in a reduction of its share to around 30%. In 2015, Europe's installed capacity reached 148GW and its annual installation were 13.8GW, 22% of global new capacity. In this context, the significant growth of Europe's wind installations capacity which occurred mainly between 1995 and 2005, contributed to a 70% share of the global produced wind energy. In fact, the global wind energy production was almost the same as the one in Europe until the early 2000s. After 2005, this share began to decrease reaching 38% in 2013. In the meantime, growth of the N. America (mainly USA) and Asian (mainly China) markets, resulted in increase of the annually generated wind energy above 630TWh (see Fig. **3**) [10, 12, 13].

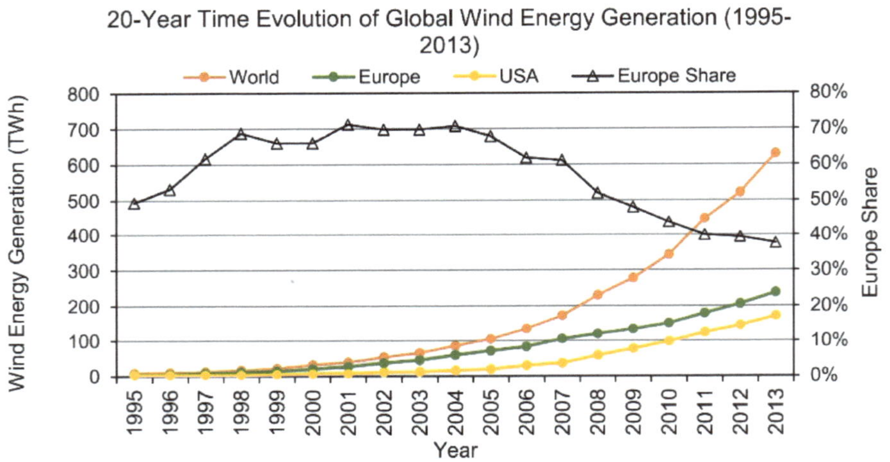

Fig. (3). Wind energy generation during the last 20 years [10, 12].

The evolution of wind energy generation at different markets during the last 15 years was significant, and is depicted at the distribution of global wind energy production by region (Fig. **4**). This distribution shows clearly the reduction of the European share compared to others, mainly after 2005, and the obvious increase concerning the N. America and Asia share. At 2012, electricity generation from wind installations is actually shared between three regions, Europe, Asia-Oceania, and North America [13].

Nowadays, the wind energy markets present differences, compared to the previous years. In fact, Fig. (**5**) indicates the new installations during 2015, along with the total wind capacity share of main markets in 2015. According to the official data [12], China is by far the largest market by adding in 2015 30.7GW approximately

(25% and 40% of EU and US cumulative capacity, respectively), reaching 145.3GW (almost the same cumulative installations of EU). The second largest market was the USA with 2015 added capacity of 8.6GW, while the third was Germany with around 6GW. France, UK and Turkey 2015 installations were around 1GW for each one, and were amongst the 10 biggest markets. It is worthy to mention that large investments in the wind energy sector during 2015 occurred in Brazil and India with 2.7GW and 2.6GW, respectively. In the European continent, Poland is an emerging market with 1.3GW added in 2015, 35% of its cumulative capacity. In fact, although its cumulative installations are at this time around 5GW, its share in the wind energy market reached 2% in 2015.

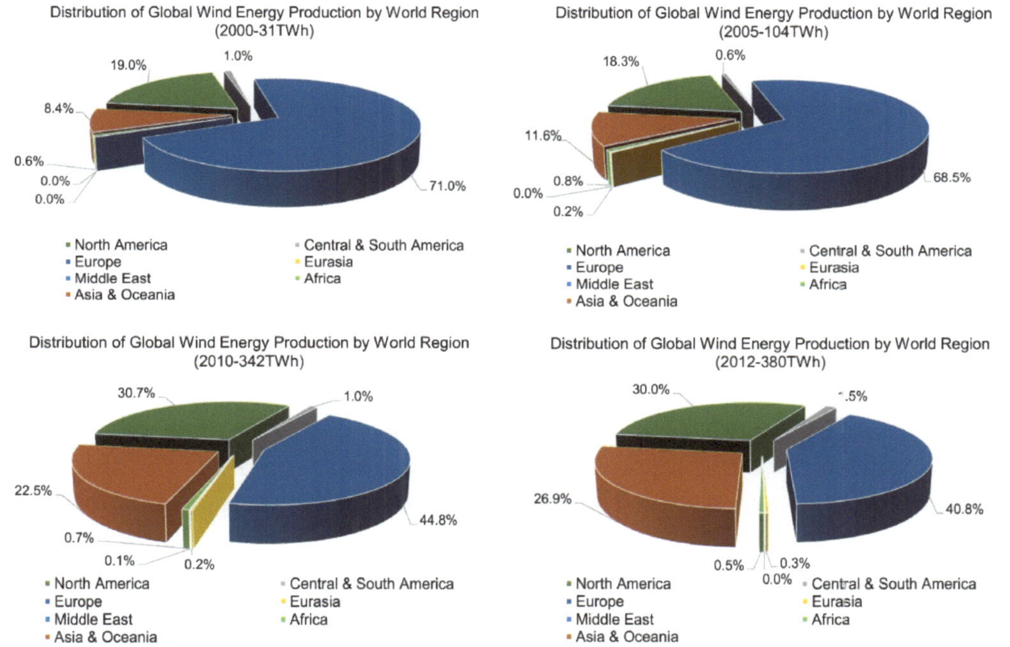

Fig. (4). Evolution of wind energy generation between world regions [12].

India following a steady growth rate, is the China's most important ally having the second higher installed capacity in Asia reaching in 2015, 25GW. In the region of Oceania, Australia continues the wind power investments and in 2015 managed to install almost 0.4GW, thus increasing the Pacific capacity to more than 4.8GW. In Latin America, as it was aforementioned, Brazil entered the world market dynamically during the last years with 2015 installations of 2.7GW, reaching totally 8.7GW, representing more than 70% of the region.

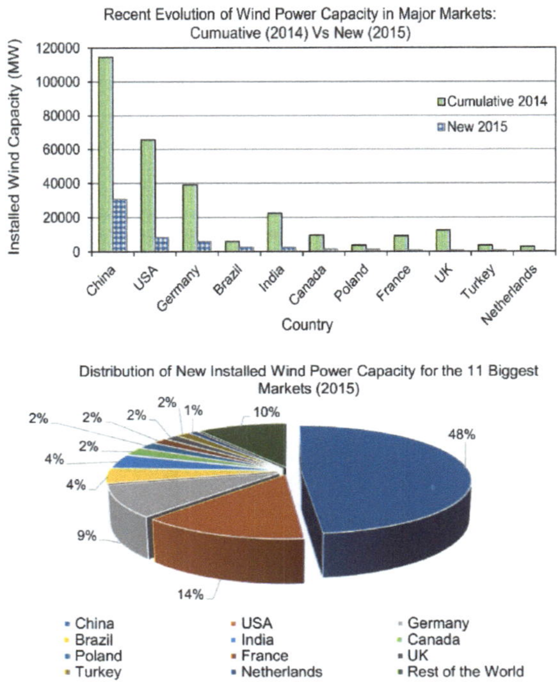

Fig. (5). (a) Evolution of wind capacity in 2014-2015, (b) Market share in 2015 [11].

The second largest market in South America in 2015 was Uruguay with 316MW of new installations (8.6% market share in the region). The area with the lowest wind investments consists of African and Middle East countries with cumulative installations in 2015 of almost 3.5GW. S. Africa, Egypt, Morocco and Tunisia are the main active African countries (a total of more than 3.4GW) with Iran & Jordan being the major Middle East countries found to considerably exploit their local wind potential (~100MW, ~119MW, respectively) [12].

EU Wind Energy Facts

During the last 15 years, wind energy investments in European Union presented an increasing trend. Total capacity in 2015 reached 142GW with many countries being among the 10 biggest markets. Germany has by far the highest capacity with around 45GW of cumulative power, followed by Spain (23GW), where new installations since 2012 are insignificant. The third country in terms of wind capacity for 2015 was the UK with 13.6GW, and 2015 annual installations of around 1GW. France ranked 4th at cumulative capacity in the EU with 10.3GW and 1GW new capacity (in 2015), while Italy had the fifth higher total capacity reaching 8.9GW (Fig. **6**) [12].

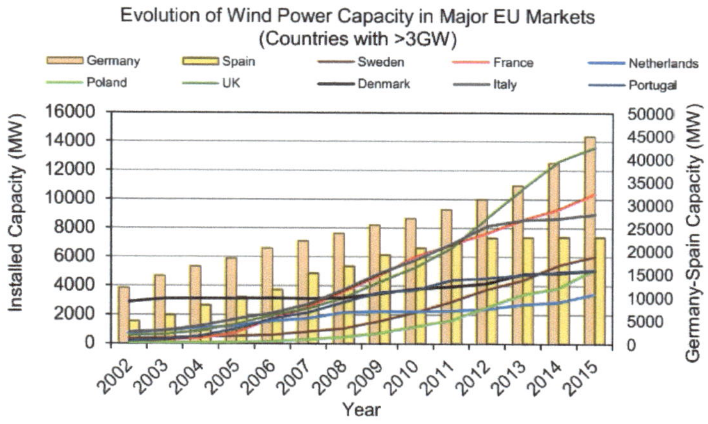

Fig. (6). Wind power capacity evolution in major EU markets [12].

In Denmark, on the other hand, although at the beginning of 2000 it was the third market in EU, wind energy investments were much lower compared to other EU countries during the last ten years. In 2015 annual installations were only 217MW compared to previous years (*e.g.* 645MW in 2000) totalling to a cumulative capacity of 5GW [11, 14]. Considerable growth of wind energy during the last 3 years has been observed in Poland, where in 2013, 2014 and 2015, 894MW, 444MW and 1.3GW have been installed, respectively [12].

The EU member state market shares in 2015 were dominated by Germany, followed by Poland, France and UK. From the 28 member states, only 19 countries presented new installations, totalling 12.8GW (see Fig. **7**). It is necessary to mention at this point that Spain although ranks in the second place concerning cumulative capacity, there were no new investments during 2015.

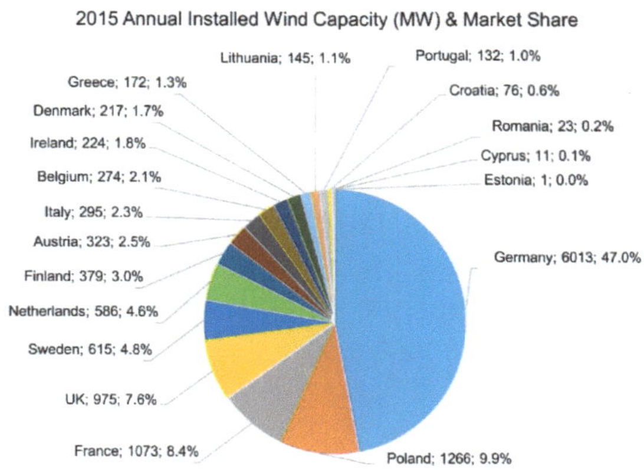

Fig. (7). EU-28 2015 wind energy market [11].

In order to have a wider opinion for the evolution of power capacity in EU, it is interesting to compare the trend of both fossil and renewable power technologies installations. Hence, according to Fig. (**8**), conventional plant's capacity presented an increase between 2003 and 2012 because mainly of the Natural Gas (N.G) fired installations that have been implemented (see also Fig. **9**). Nuclear capacity reduced from 137GW in 2000 to around 121GW in 2014, while hydroelectric cumulative power showed a small increase of 4GW (*i.e.* from 97GW to 101GW).

On the other hand, the most significant contribution to the EU's power capacity was the installations of wind and solar energy based power stations. Wind energy in EU reached 128GW in 2014 compared to 12GW in 2000 ranking at the second place after the fossil fuel based plants. Solar photovoltaics (PV) reached 87GW in 2014 from 0.2GW at the beginning of the millennium. In fact, the growth of wind energy during the last 15 years was significantly higher compared to other power generation sources, with new installations above 115GW, whereas the installations of solar PV were around 86.5GW [12, 15, 16].

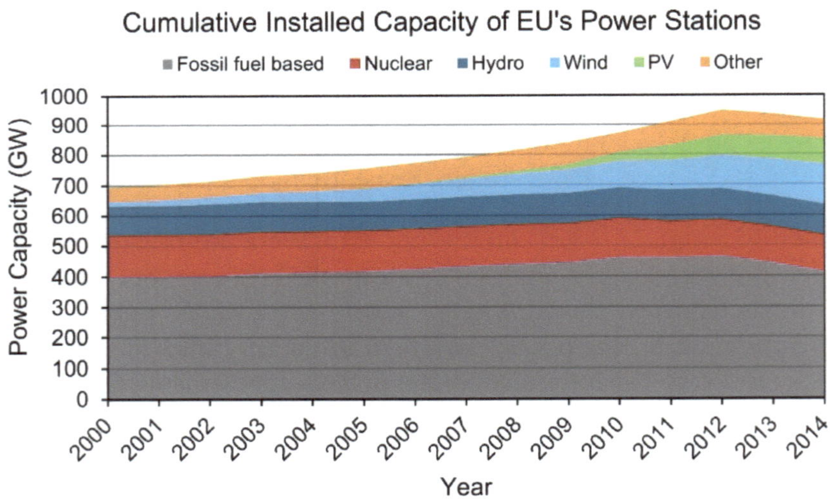

Fig. (8). Evolution of cumulative power capacity in EU by technology type [12, 15, 16].

Of great importance is the annual installed capacity in EU since 2000, where it is obvious that during the last 2 years, new installations concern mainly RES, compared to previous years where apart from PV installations, N.G. power plants were one of the main investments in the sector of energy. On the other hand, wind energy market in EU presented during the last 15 years an increasing trend, compared to other technologies (*e.g.* PV, N.G.). Specifically, new PV installations demonstrated a peak at 2011 that was followed by a gradual decline after 2012 (Fig. **9**) [11].

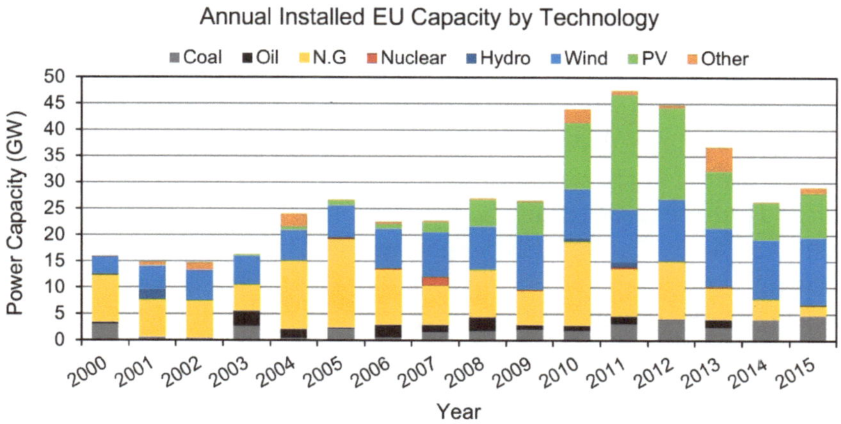

Fig. (9). Annual EU installed electric power capacity [11].

Policies promoted from the EU contributed to achieve a 5.9% coverage of gross electricity generation coming from wind energy. Specifically, nowadays, Denmark is covering above 35% of its national generation from wind energy, while other European countries exceed 15% (*e.g.* Spain, Ireland) or even 20% (*i.e.* Portugal). On the other hand, there are still countries (referred to 2012) where wind energy contribution to electricity generation is equal to zero such as Malta, Slovenia, and Iceland due to lack of wind power investments [13] (see Fig. **10**).

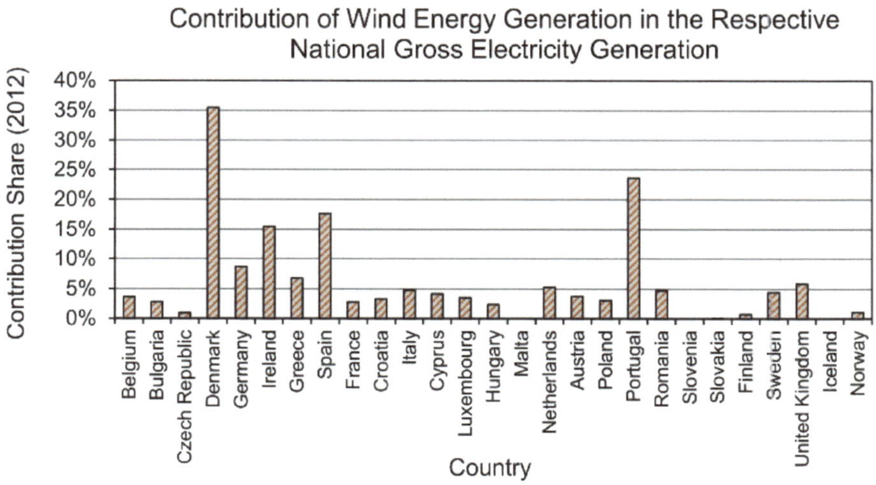

Fig. (10). Wind energy contribution to national electricity generation [13].

Wind Energy Market in Greece

The power generation system in Greece is divided to two separate sectors including the mainland grid and the islands' non-interconnected system. The mainland system is highly dependent on centralised thermal power stations, while the islands' remote grids are basically based on oil-fired plants [17]. Contribution of RES (Large Hydro excluded) to the national interconnected power mix was for 2015 around 25.4%. Total RES capacity in 2015 (interconnected system excluding large Hydro) reached 4.6GW from which 1.8GW concern wind power installations [18]. On the other hand, RES capacity at the non-interconnected system reached in 2015 317MW (Wind) and 136MW (PV), totalling 453MW with a share of 20.5% of total installations (Thermal and RES) [19]. Based on the above, it is obvious that the wind energy in Greece comprise the main RES contributor to the national power mix with a total capacity of 2.15GW in 2015, and a share of 10.6%.

Wind energy production in Greece dates back to the early 80s (first Mediterranean 5x20kW wind park in Kithnos island, 1982), but it was not until the end of 1990s when wind power investments have been triggered. In Fig. (**11**), the evolution of wind power capacity in Greece during the last 25 years is depicted.

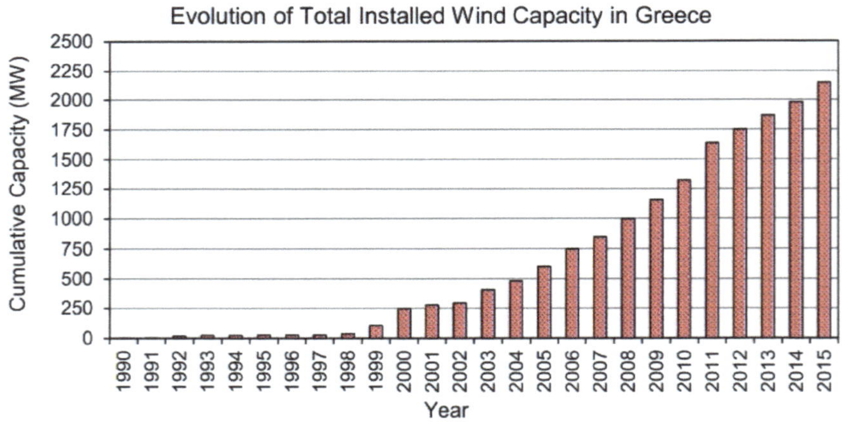

Fig. (11). Wind energy cumulative capacity evolution in Greece [20].

According to Fig. (**11**), wind energy in Greece presents an increasing trend, particularly after 1998, while the year with the higher new installations was 2011, reaching 311MW and ranked between the 10 largest markets in the EU [11].

In this context, the evolution of the wind market in Greece during the last 15 years indicates that apart from the beginning of the millennium and the end of 2011, where installations were confined and increased respectively, wind energy market presented an almost steady profile. Fig. (**12**) shows the wind annual investments in Greece along with the share of manufacturers.

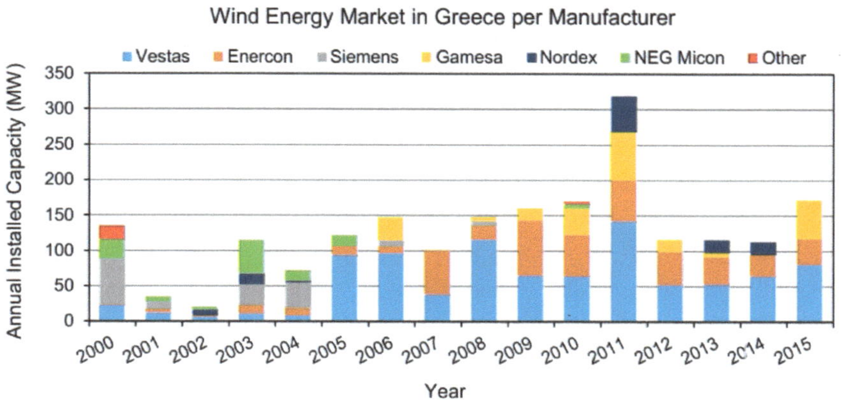

Fig. (12). Annual wind power installations in Greece per manufacturer [20].

New installations at this period ranged between 100MW and 170MW, where the major share of the market was held by Vestas S/A, followed by Enercon GmbH and Gamesa S.A [20]. According to the EU's energy plan, known as "20-20-20" targets, 40% of the electricity generation in Greece should be supplied from RES. In 2015, renewables' share in the total generation mix equals to 19.7% approximately. Hence, it is obvious that it is essential for the Greek State to promote supporting schemes that would encourage investors to exploit the excellent RES potential found in many mainland areas and especially in the islands' region [21]. For example, by comparing Fig. (**13a**) and Fig. (**13b**), it is obvious that there are areas in the Aegean Sea, where new wind power investments would enhance the contribution of renewables to the total electricity production mix.

Offshore Evolution

Apparently, wind energy all over the world and especially in Europe has been established as one of the mainstream power generation technologies, able to meet a large share of the electricity demand in the years to come. In addition, the wind power sector is considered as a competitive and substantial way for achieving future carbon reductions caused mainly by the conventional power generation technologies [22]. Up to now, the major investments of wind energy occur onshore with capacity reaching 420.8GW, compared to offshore farms totalling in 2015 12.1GW [12].

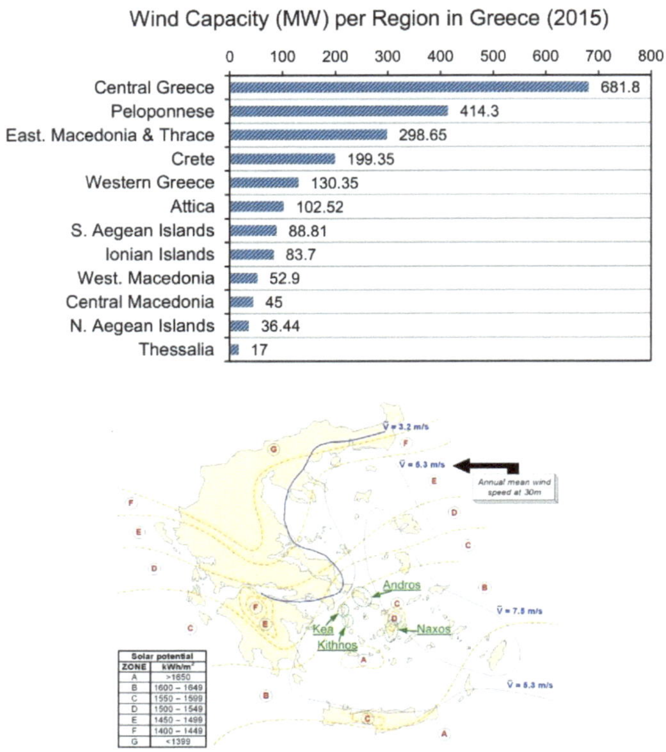

Fig. (13). (a) Cumulative capacity in Greece per region, (b) Wind & Solar energy potential in Greece [20, 21].

However, by taking into consideration that in main markets future development of new installations are subject to environmental constraints which make difficult for the energy planners to localize suitable sites, along with the fact that resource potential is greater at open seas, interest has been turned during the last decade to offshore installations [22, 23]. Fig. (**14**) shows that only in 2015 offshore capacity in Europe increased by 3.02GW, reaching totally 11GW in 2015, which represents above 90% of the global installations [24]. The increasing trend in the Europe's offshore wind markets during the last 5 years is depicted in Fig. (**15**), where the market share indicates that the UK presents a steady rate of annual installations of at least 500MW, while Germany although it was the biggest market for 2015, at the beginning of the decade it was the fifth market globally. Of great interest is that apart Germany and the UK, significant investments have been occurred in other European countries too, such as Belgium, Denmark and Netherlands with total installations at this period of 682MW, 606MW and 180MW, respectively.

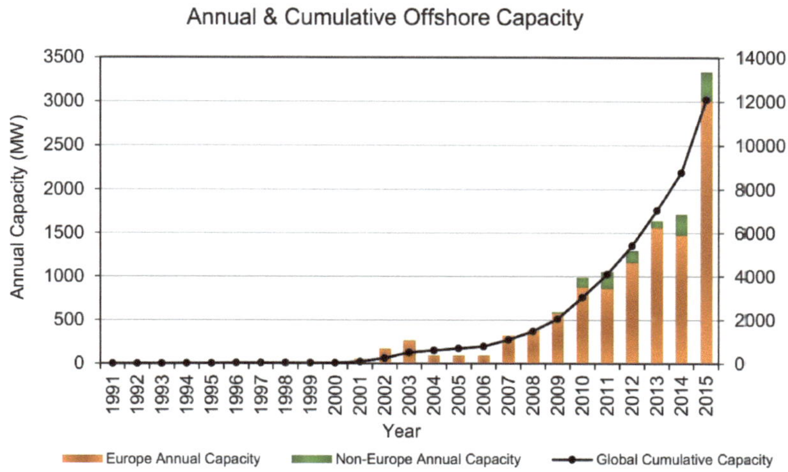

Fig. (14). Annual and cumulative offshore installations since 1991 [11, 12, 25, 26].

Additionally, the market share between 2010 and 2015 indicates that except China which presents an increasing trend, ranking third between wind markets in 2015, no other non-European country has invested significantly to offshore installations (see Fig. **15**) [12].

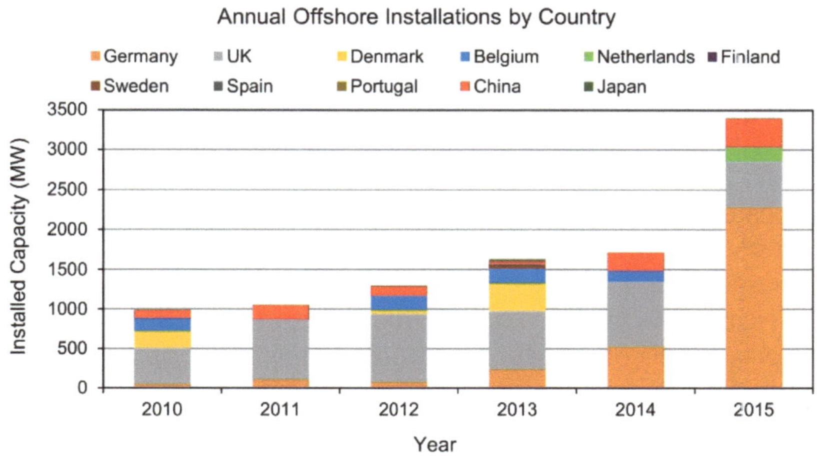

Fig. (15). Evolution of annual offshore installations by country [12].

This shift to offshore wind investments in Europe can be also seen from Fig. (**16**), where the short-term plans indicate that the central scenario from European Wind Energy Association (EWEA) for 2020, includes additional 23.5GW of offshore capacity compared to 2014 installations (*i.e.* 8GW) exceeding totally 30GW. In this scenario, the highest amount of wind capacity is estimated to be installed from the UK, Germany, Denmark, and Sweden with approximately 21.8GW [27]. At the end of 2015, 6 large projects were under construction (at stage where foundations have been installed) in the Netherlands (744MW) and Germany (1194MW), which once completed, total offshore capacity in Europe is going to increase by 1.9GW, reaching totally 12.9GW [24].

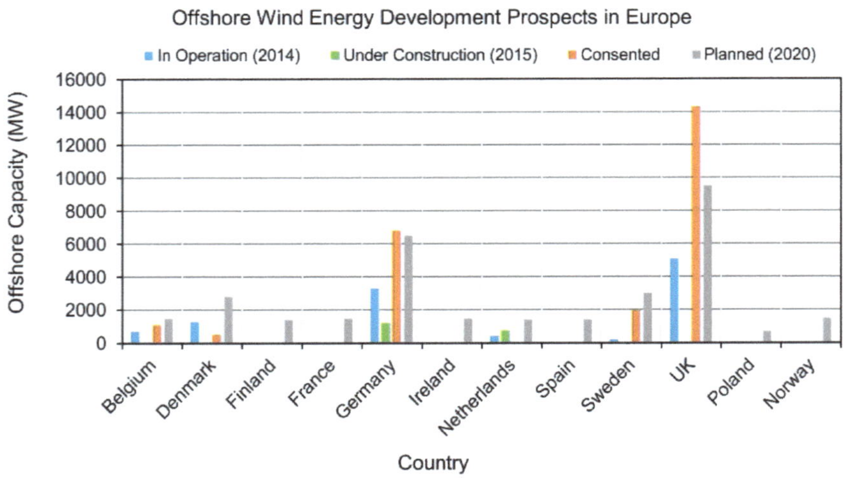

Fig. (16). Current status and short-term plans for offshore installations in EU [24].

TECHNOLOGY ISSUES

During the last 35 years, the development of wind turbines can be seen from their gradual upscale for better land exploitation, the reduced maintenance and operation (O&M) requirements, and the numerous funding development programs concerning large scale machines. Fig. (**17**) indicates that the almost exponential rotor diameter increase recorded up to 2000 [28], is then followed by a stabilizing trend in the power area of 2-3MW. Since 2012, new commercial turbines up to 7-8MW, mainly for offshore applications for limiting the cost requirements of sea installations, have been developed, resulting to further increase of the rotor above 160m (see Table **1**). In fact, turbines of 10MW are already developed and are ready for commercialization (*e.g.* AMSC SeaTitan with rotor diameter 190m, Sway Turbine ST10).

Nowadays, the most widely common solution concerning the towers used for the wind turbines consist of tubular steel design. However, due to the growing trend of larger machines and higher hub heights, development of new alternative concepts is under research. Many manufacturers including Enercon, Nordex and

Senvion deploy hybrid installations using concrete and steel, where concrete is used for the base construction, while steel tubular section is used for the upper parts of the tower [29].

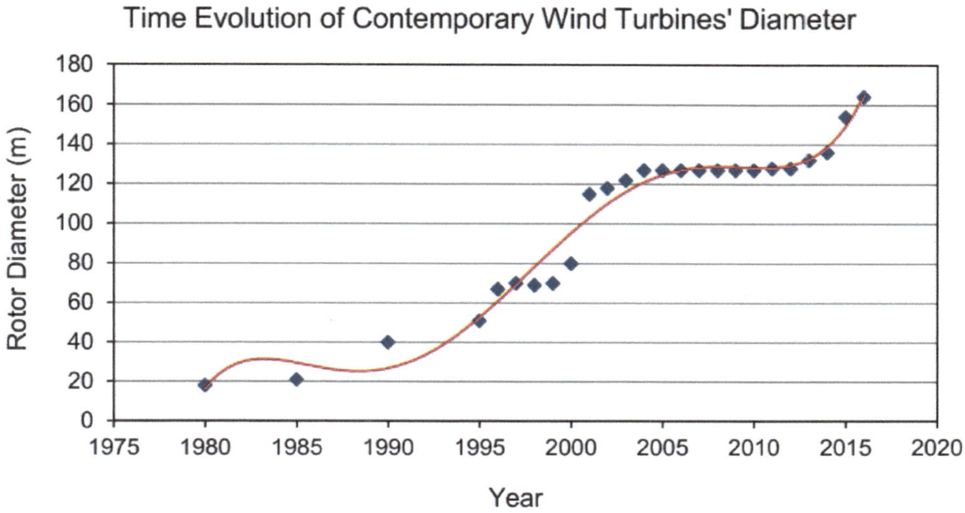

Fig. (17). Time-evolution of WTs' rotor diameter.

Table 1. Contemporary commercialized wind turbines.

	WT Model	Rated Power (MW)	Rotor (m)
Vestas	V164	8	164
GE	3.2 103	3.2	103
Gamesa	G132	5	132
Enercon	E-141	4	141
Suzlon	S97	2.1	97
Siemens	SWT-7.0-154	7	154
Sinovel	SL6000	6	128
Acciona	AW-132/3000	3	132
Nordex	N131/3000	3	131
Senvion	6.2M152	6.2	152

On the other hand, technologies used for offshore installations, are highly associated with the water depth and the location of the investment. By the end of 2015, 3313 substructures have been installed in Europe, from which the most common were the monopiles with a share above 80%. Gravity- based foundations used similarly with monopiles at swallow water depths below 30m comprise 9.1%

of total installations, while jackets and tripods presents an increase compared to 2014 reaching 5.4% and 3.6% share respectively (see Fig. **18**). At this time, new technologies are at the phase of development in order to enable installations to deeper waters. These include floating concepts such as the "Hywind" unit in Norway at 220m depth and the "Blue H" wind turbine concept in Italy at 110m depth [23]. Outside Europe, in 2015, the largest floating wind turbine is in operation with rated power 7MW in Japan [30].

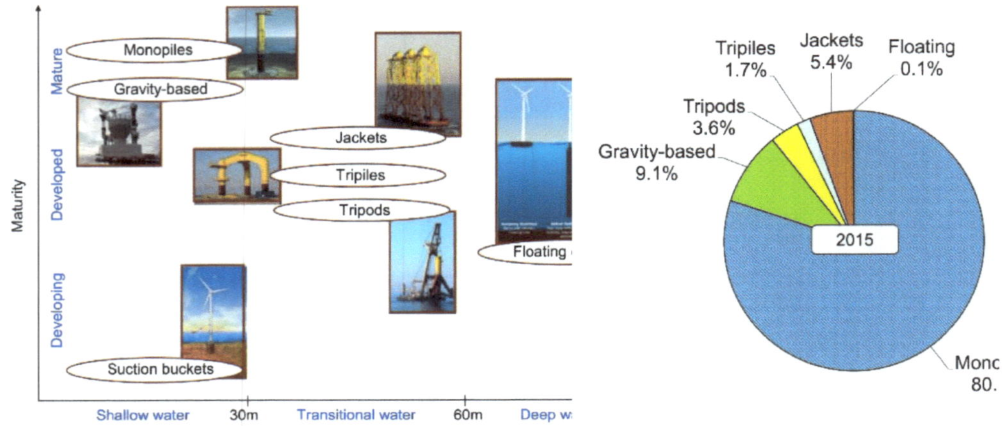

Fig. (18). Offshore wind farm's support structures and 2015 European cumulative share [23, 24].

Concerning the power regulation there are actually four main technologies used for control. Passive Stall Control (PSC), Active Stall Control (ASC), Pitch Control (PC), and Individual Pitch Control (IPC).

- PSC is the simpler form of power control where the design of the blades induces airflow to stall at high wind speeds. However, although it is the less costly method, production is not maximised and at extreme weather conditions blades are subject to high stress.
- In ASC, blades are pitched at wind speeds above the rated value in order to achieve stall conditions. Its main advantage comprises a smoother power output.
- The method of pitch control is more sophisticated and is used for increasing energy capture below rated power, along with maintaining power output at high wind speeds.
- One of the most promising technologies for power regulation is the IPC, where each blade pitches during the movement of the rotor. In this way, loads and specifically asymmetric ones can be reduced to minimum resulting to an increased lifetime of wind turbines. Nevertheless, this technology is more complex and presents higher cost [28, 29].

During the last decade newest large-scale wind turbines are pitch controlled, resulting to better energy capture below the rated power in favour of reducing loads. In Fig. (**19**), the evolution of power control technologies since 2005 in Europe indicates that from 2008, IPC share increased to around 20% of annual installations while the rest 70%-80% belongs to regular pitch control. On the other hand, PSC and ASC are actually almost absent at new installations suggesting that this technology for large machines is practically non-viable option [29].

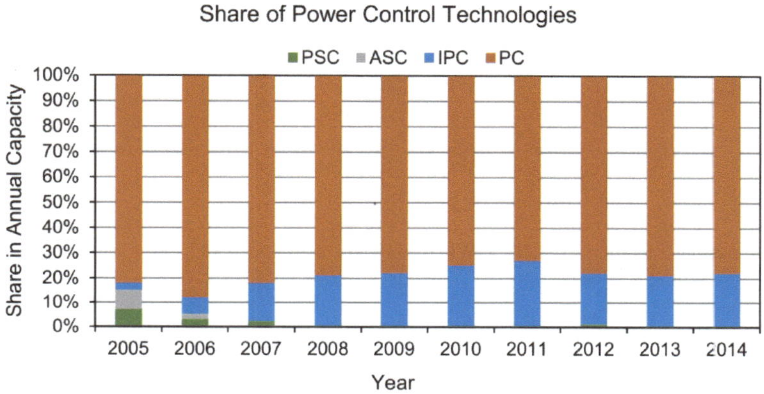

Fig. (19). Share of wind power control technologies for annual installations in Europe [29].

Nowadays, R&D efforts are focused mostly to improve performance of the turbine by structural load reduction without increasing the cost significantly and by energizing the flow around the blades' surface. A good example comprise the " Vortex Generators" made from highly durable thermoplastic which are installed on the turbine's blades [31].

In this context, the effects of technological improvements occurred during the last 20 years are shown at the mean annual capacity factor (CF) of wind installations, both in EU and globally. At this time CFs exceed 22%, whereas in 1995 this value was below 18% (see Fig. **20**) [13, 32]. This long term increase of the mean annual CF was due to:

- Improved efficiency of contemporary wind turbines.
- Sophisticated assessment of the local wind potential.
- Considerable reduction of the downtime periods.
- Offshore applications (higher CFs).

Apart from the large scale wind turbines, a wide range of applications are covered from small scale machines. Off grid applications include remote houses-farms-villages, water pumping, desalination, street lighting and signalling, sailboats, while on grid applications include building integration (rooftops), industrial, farms, dwellings, street lighting, *etc.*

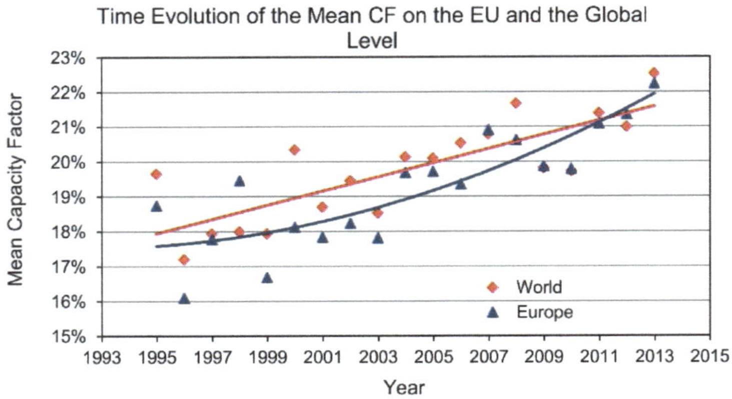

Fig. (20). Mean annual CF evolution in EU - World's wind installations [13, 32].

Considerable interest exhibits the market of very small off-grid systems (1000s/year for powering boats/yachts). On the other hand, example of off-grid applications is the use of 250,000 wind home systems in the Inner Mongolia, the wide spread diffusion of wind hybrid systems all around the world (*e.g.* the Rural Electrification Program of Chile), and the wind diesel systems serve many remote consumers of Alaska, Canada and Australia.

ECONOMIC ASPECTS

The economics of wind energy can be divided into four main categories. The investment cost, the lifecycle cost of energy (LCOE), the support mechanisms, and the employment opportunities. The major trends dominating the market lately encompass the size increase of contemporary WTs, the improvement of efficiency, and the long term reduction of the specific investment cost per kW during the years. During the last two years R&D spending for wind power ranked second among the other RES technologies by remaining the same at $1.8 billion, 2.5 times lower than solar power [30].

At the same time improved wind potential forecasting techniques along with geographical distribution of the wind parks in suitable regions definitely help to minimize grid related effects such as dynamic stability and power quality issues due to the stochastic character of the wind, while minimizing at the same time the hidden cost of wind energy. Additionally, the interconnection of different electrical grids and application of different power technologies, including hybrid stations and energy storage is also quite supportive especially for autonomous electrical systems (*e.g.* remote islands) [33].

Investment Cost

Concerning the investment cost of onshore installations, there was a decrease from around 3700€/kW in early 80s to below 1300€/kW at the beginning of 2000, where the cost has been stabilized. However, since 2010, a rise has been observed due to increased sophistication and upscaling of new machines (see Fig. **21**). The 2014 cost was at the area of 1450€/kW, with ±20% deviation between countries. Countries with higher cost include Japan, Austria, Ireland, Switzerland, and Italy with installation costs above 1700 €/kW, while countries with relatively lower cost include China, Portugal, Denmark, Greece, and India with installations costs below 1300€/kW (see Fig. **22**) [28, 34, 35].

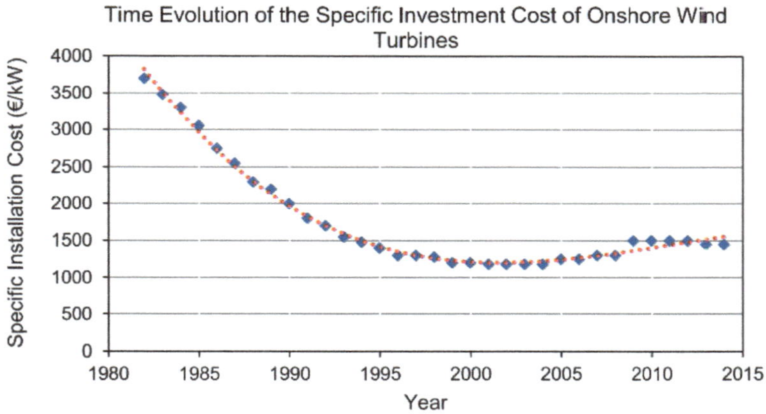

Fig. (21). Investment cost of onshore wind farms [28, 34].

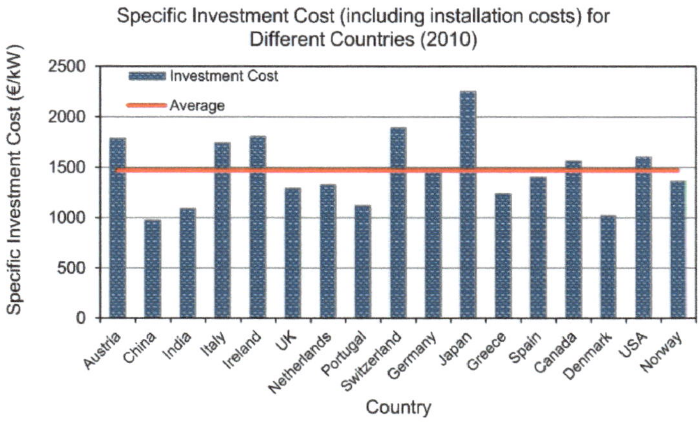

Fig. (22). Specific and average investment cost of onshore wind farms by country [36].

On the other hand, the offshore wind farm investment cost is much higher than the onshore one, because of the huge installation costs, including transportation of the components, construction activities, and connection with the grid. According to Fig. (**23**) the cost at the beginning of 2000 was decreased to around 1500€/kW. However, since 2008 Offshore Capital Cost increased due to larger machines, higher depths and distance from shore reaching at 2014 4000€/kW [36]. This is the most significant challenge that offshore wind industry is facing due to the price difference with fossil fuel power generation technologies. In order to achieve reduction of the cost to competitive level, it would be necessary to establish policy stability in the major markets in order to accomplish scale and low-cost financing [30].

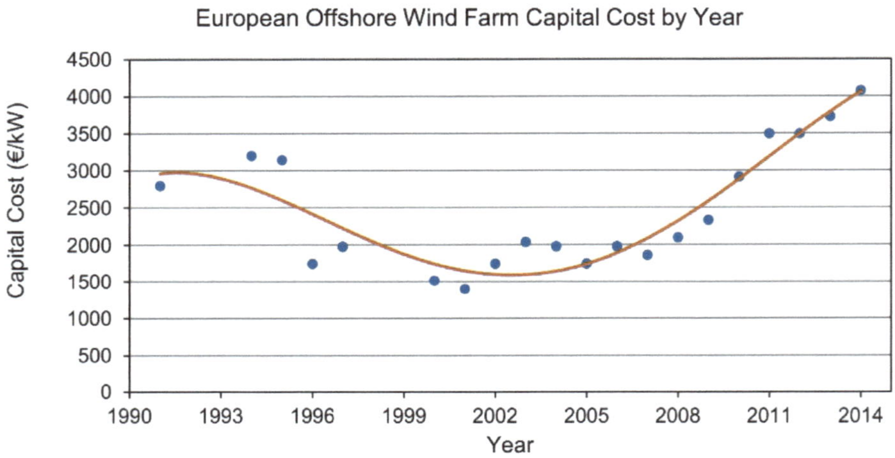

Fig. (23). Capital cost evolution of offshore projects [36].

Lifecycle Cost of Energy (LCOE)

The LCOE includes the M&O costs along with the investment and the decommissioning costs. Fig. (**24**) indicates the 2020 costs for different generation resources. The onshore LCOE ranges between 50 and 70€/MWh, *i.e.* similar to conventional technologies. Hence, it is obvious that LCOE does not comprise a drawback for onshore wind energy investments while for some cases of advanced conventional technologies where LCOE exceeds 110€/MWh, onshore wind investments are more beneficial (see Fig. **24**). Additionally, R&D on materials and manufacturing techniques aims to lower further the cost and to optimise power generation at low wind speeds [37].

On the other hand, offshore installations are costlier in a lifecycle basis ranging between 125 and 200€/MWh (for plants entering service in 2020) due to difficulties occurred at the sea during all the stages of an investment's lifecycle. Furthermore, the trend of moving to higher distances from shore and deeper waters tends to add to the cost from the construction to the decommissioning phases of new offshore installations because of the difficult conditions found at open seas [37].

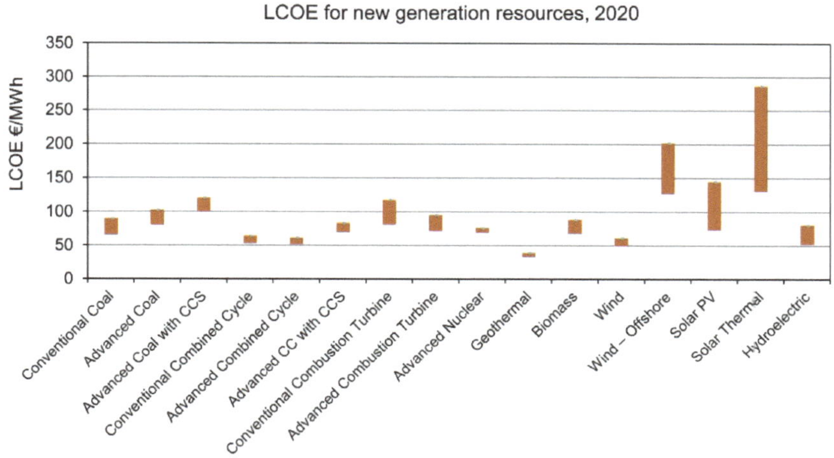

Fig. (24). 2020 LCOE of generation technologies [37].

In comparison with other RES in Europe, LCOE of onshore wind energy is now the most cost competitive power source with a weighted average of around 60€/MWh.

On the other hand, although offshore installations are considered more costly than other conventional technologies, compared to other RES, its weighted average is at the same range as geothermal power (*i.e.* ≈120€/MWh), and much lower than the solar thermal (CSP) which equals to 340€/MWh approximately (see Fig. **25**). In this context, a new declaration signed at 2015 by wind turbine manufacturers and the developer company DONG, sets a goal of reducing the LCOE below 85€/MWh by 2020 through the implementation of new methods concerning grid connection and components' transportation [30].

Furthermore, as it is obvious from Fig. (**26**), wind production cost (both onshore and offshore) concerns mostly the investment cost compared to other technologies where for N.G. based plants for example the generation cost rises from the fuel requirements. Carbon-based technologies also include the carbon cost compared to nuclear and wind power systems, which represents the CO_2 emitted prices. M&O cost is higher in the case of offshore installations due to the harsh conditions existing at the marine environment [38].

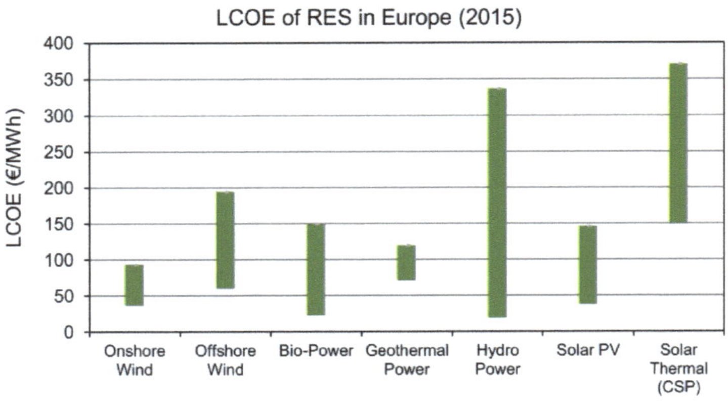

Fig. (25). LCOE of RES power sources in Europe in 2015 [30].

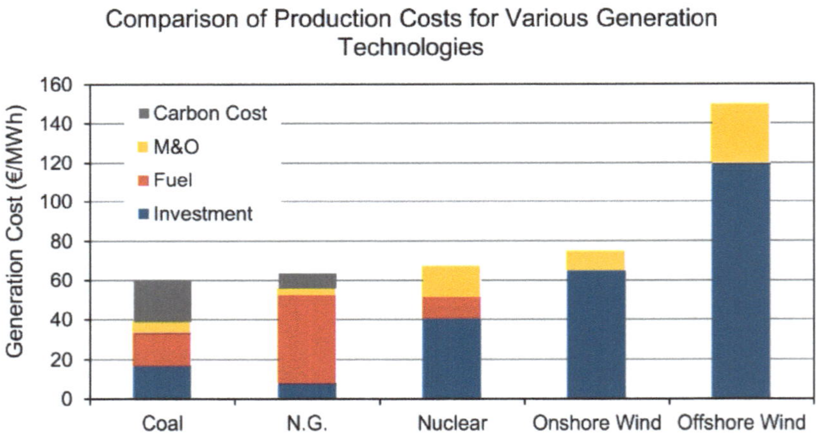

Fig. (26). Breakdown of LCOE for different technologies [38].

Support Mechanisms

Since the 1970s, support from State has been promoted to RES investments in order to meet environmental, financial and social targets. An early example comprises the Californian outbreak where USA encouraged renewable investments to cope with a possible oil crisis similar to the one of 1973. Governmental support was one of the key elements for the remarkable growth of the wind energy. Policy measures on the basis of support mechanisms include price- and quantity-driven instruments which are either directly or indirectly applied for wind energy exploitation [28].

The main incentive programs (direct and indirect) applied to wind energy promotion during 2014- 2015 of International Energy Association (IEA) countries include [34]:

- Carbon Tax: This tax encourages investments towards RES indirectly because renewables do not have the obligation to participate due to zero CO_2 emissions. It is implemented in 31 European countries through the EU ETS system and in 3 provinces of Canada.
- Feed-in Tariff (FIT): It is a monetary reward for wind-generated electricity at a guaranteed price per kWh. Many countries have implemented this support scheme since the 80s (see Fig. **27**) resulting to a significant increase of the wind power sector. Nowadays, there are 14 countries that have an established FIT system including major markets such as Canada, China, Denmark (for offshore and small WTs), Germany, the UK, and Italy.
- Renewable Portfolio Standards (RPS): Electricity utilities are obliged to use a portion of their electricity supplies from RES. Hence, wind investments are encouraged indirectly. Countries using this incentive include China, Canada, Italy, the UK, USA, Norway and Sweden.
- Green Certificates (GCs): Power plants receive certificates for RES generated electricity. Electricity and certificates are sold in a determined price according to the demand set by the consumer's obligation to purchase an electricity share from RES. The UK, USA, Canada, Norway, and Sweden have implemented this measure.
- Net Metering: Excess electricity from a producer is fed back to the grid at a given price as recorded by a bi-directional electricity meter. Seven IEA countries have implemented this support scheme at the moment including Denmark, Canada, the UK, USA, the Netherlands (for small wind only), and Portugal for micro-generation.
- Income Tax Credits: Expenses associated with wind projects' installation may be deducted from taxable income streams. It is established in China, USA, Canada, Ireland, Norway, and the Netherlands.
- Investment Funds: These funds are provided to private wind investors for promoting wind energy in order to boost financial growth. The UK, Canada, Ireland, and the Netherlands use this support scheme as a tool for business development.
- Incentives for small Wind Turbines: In order to provide support to small wind installations, many countries such as USA, the UK, Canada, Denmark, Italy, and Portugal, have implemented measures that include reduced connection costs, VAT rebate, micro FIT, and accelerated capital subsidies.

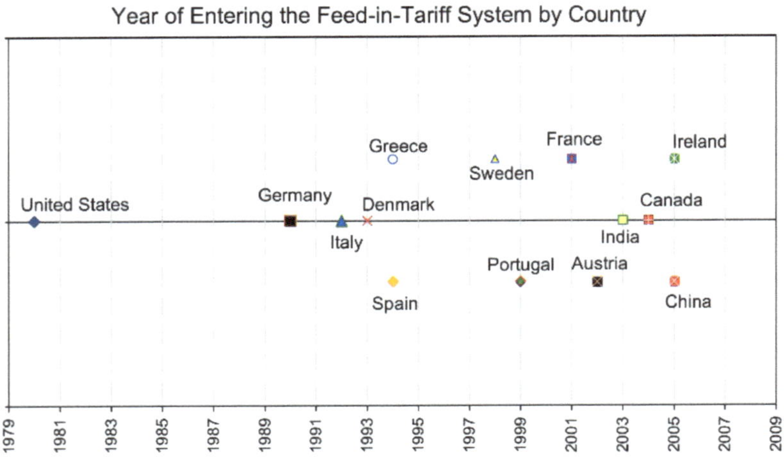

Fig. (27). Countries implementing FIT support scheme [28].

Additionally, there are also some other measures implemented by other countries in order to support wind investments. For example, Capital Subsidies scheme which includes covering a part or the entire wind system cost is established in China and Korea, while in China also a relief from import tax charges is applied to large wind projects. Concerning Greece, in 2014, certain (idle) licences for RES plants are withdrew in order to reduce administrative burden, while RES investments on islands combined with water desalination get priority [34].

Employment Opportunities

In terms of employment, somewhat 280,000 is the number of people employed directly and indirectly in the wind energy field of Europe, respectively. Another 73,000 correspond to the 100 manufacturing plants built in the US, while in China employers reached 470,000 in 2014. These include employment spots in manufacturing companies, in promotion, utilities, engineering and R&D (direct employment) or employment in companies providing services or producing components for wind turbines (indirect relation). Of the leader countries on the basis of people/MW installed, Finland and Denmark employ more than 5 and 7 persons respectively while in terms of absolute numbers Germany employs 140,000 (see Fig. **28**). China although ranks at the third place with around 4 employers per installed MW, it is estimated that in the future this number may reach 15 employers/MW mainly in the manufacturing industry [34, 39].

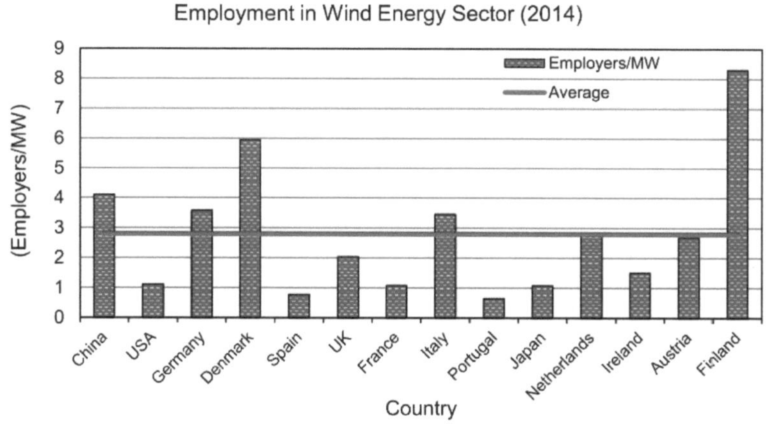

Fig. (28). Wind energy employees per MW installed in major markets [34, 39].

ENVIRONMENTAL ISSUES

Although wind energy is a "green" energy resource, it's exploitation may affect certain environmental and social parameters including visual, noise, land use, electromagnetic interference, ship collisions, embodied energy-LC emissions, and possible impacts on birds, fish, and mammals. Both onshore and offshore wind installations have similar environmental impacts but with significant differences that should be accounted for. Table **2** synopsises the impacts occurred in both cases in order to be able to distinguish the separate effects of wind energy projects to the environment [40]. These concerns about environmental impacts of wind energy development are in many occasions the major hindrance for new wind power projects. However, several studies indicate that the impacts caused by wind-based plants' implementation can be actually perceived as "myths", and do not burden environment in a significant degree [34, 41].

Table 2. Main environmental impacts of wind energy installations [23].

Environmental Parameter	Impact	Construction Phase		Operation Phase		Decommissioning	
		Onshore	Offshore	Onshore	Offshore	Onshore	Offshore
Avian Species	Fatalities, barrier to migration, nest displacement, breeding disturbance due to noise	√	√	√	√		
Wildlife	Disturbance & emigration	√		√		√	

(Table 2) cont.....

Environmental Parameter	Impact	Construction Phase		Operation Phase		Decommissioning	
		Onshore	Offshore	Onshore	Offshore	Onshore	Offshore
Fish	Loss of habitat, disturbance of breeding due to noise-vibrations		√		√		√
Mammals	Reduction of habitat, potential hearing damage	√	√	√	√	√	√
Zoobenthos	Loss of habitat, reduction of abundance		√		√		√
Hydrography - Land	Contamination of pollutant chemicals, land required	√	√			√	√
Humans	Stress symptoms, sleep disturbances due to noise and flickering, aesthetic disruption	√		√		√	
Air Quality	LC emissions	√	√	√	√	√	√

An indicative example of impacts found in both onshore and offshore installations is the one on avian species. These impacts may include risk of collision fatalities, nest displacement and flight routes avoidance, and imposition of a barrier to bird population migration [23]. However, as it is obvious from Fig. (**29**), based on evidence for 10000 bird fatalities only 1 is caused by wind turbines compared to buildings or even communication towers (*i.e.* 5500 and 30 fatalities, respectively).

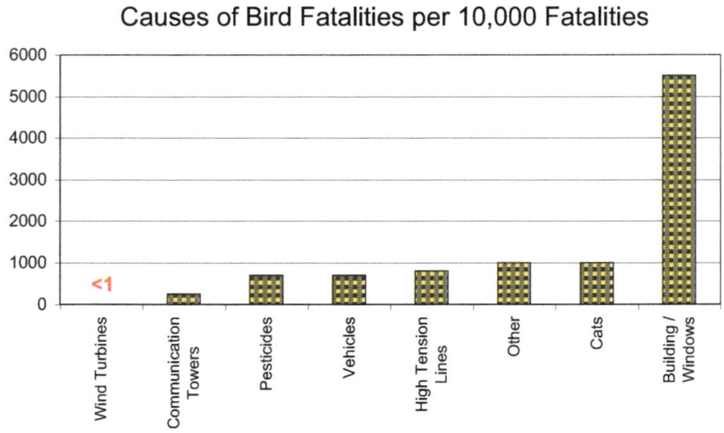

Fig. (29). Bird fatalities caused by different factors [28].

Concerning the average energy intensity in a lifecycle assessment across electricity generation technologies, wind power lies between 0.033kWh$_{pr}$/kWh$_e$ to 0.053kWh$_{pr}$/kWh$_e$ for offshore and onshore, respectively (onshore is higher due to lower capacity factors). These values are by far lower than a conventional power plant where a coal-based power plant ranges between 0.3kWh$_{pr}$/kWh$_e$ to 0.37kWh$_{pr}$/kWh$_e$ (see Fig. **30**) [42].

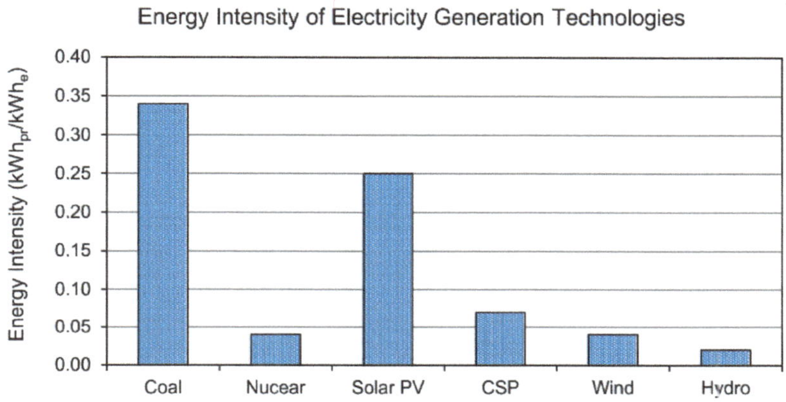

Fig. (30). LCA energy intensity of different electric power production technologies [42].

On top of that, a clear advantage is recorded for wind installations when comparison is attempted with conventional power plants. For example, the LC GHG emissions and the global warming potential of wind energy are much lower than other technologies including other RES with a maximum of 81kg/MWh produced (see Fig. **31**) [43].

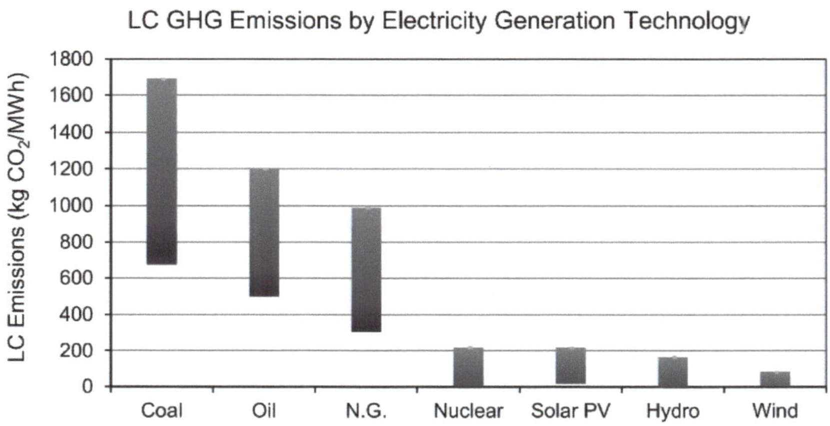

Fig. (31). LC GHG emissions from electricity generation [43].

For these reasons, in general, environmental performance of wind energy is perceived by the majority of people and transformed in to a wide-spread social support that further boosts wind energy developments. According to a study published by European Commission [44], high acceptability seems to correspond to large scale wind power adoption (above 70% in favour) and only solar energy appears to be more socially accepted (80% acceptance) while nuclear meets the greater opposition with more than 35% at the EU level [28]. However, it should be mentioned that although opposed attitude towards wind power installations are only 3%, this share is higher than other RES excluding biomass energy (see Fig. **32**) [44].

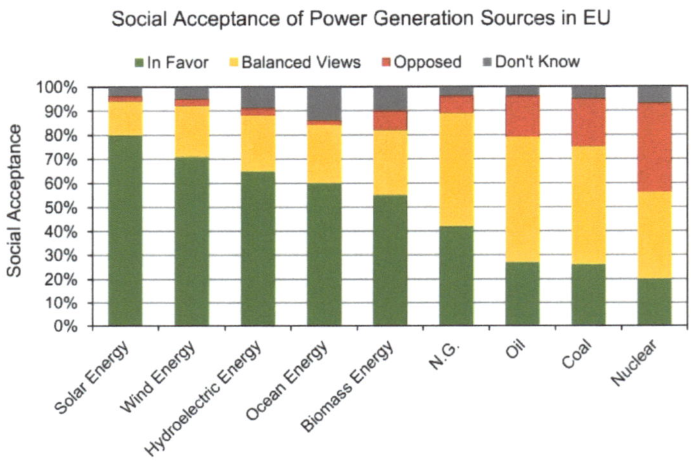

Fig. (32). Social acceptance of various energy technologies [44].

PROSPECTS OF WIND ENERGY

Wind Energy Targets

Most major markets of wind energy have adopted policy frameworks that will enable the further development of wind power investments. EU has released scenarios concerning the targets for 2030 of onshore and offshore installations. According to EWEA, the high (wind energy penetration) scenario includes 392GW of installed capacity: 294GW onshore and 98GW offshore, equivalent to 16.7GW annual installations, *i.e.* 10.9GW onshore (65%) and 5.8GW offshore (3 5%) (see Fig. **33**), while annual investments will reach €39.4billion: €26.5 billion onshore and €12.9 billion offshore. In this context, wind energy in 2030 will produce almost 1000TWh of electricity: 630TWh onshore and 370TWh offshore, equivalent to the consumption of 206 million average EU households [27].

Targets set by many European countries aim in increasing the capacity of wind power in order to reach the goals set by the EU concerning RES contribution to the electricity generation fuel mix in 2020. Specifically, Denmark aims wind energy to reach 50% of its electricity generation by 2020 (at this time contribution equals to 35%), Finland's targets include the installation of 900MW until 2020, and France aims its cumulative installations in 2020 to reach 19GW and 6GW for onshore and offshore respectively. The UK aims to install by 2030 above 25GW, while Spain's targets for 2020 include cumulative wind capacity of 35.7GW. On the other hand, Germany's targets for 2020 include the installation of 12.5GW onshore and 6.5GW offshore wind projects while Poland, the second largest market of Europe in 2015 aims to install 1GW of offshore wind capacity by 2020 [30]. However, estimations indicate that Poland capacity in 2020 may exceed 11GW in total, while another 10GW onshore and 6GW offshore are forecasted by 2030 [45, 46].

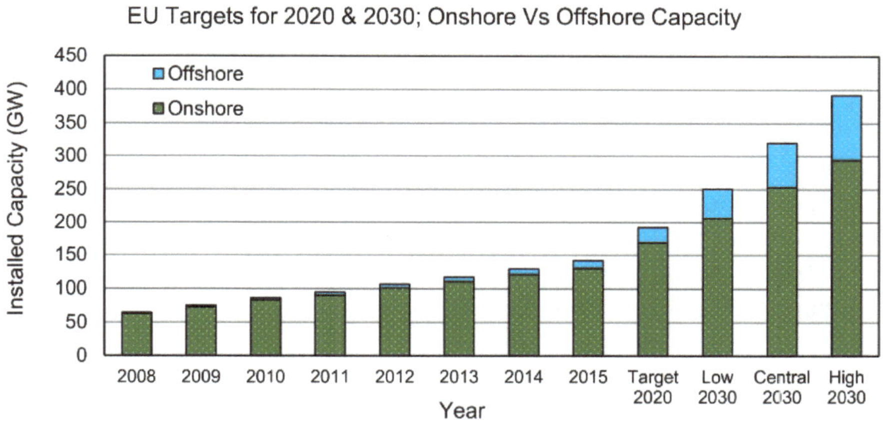

Fig. (33). Wind capacity targets set by EU for 2020 and 2030 [27].

Regarding other major markets in 2015, the US targets for wind energy include 10% by 2020, 20% by 2030 & 35% by 2050 of total electricity needs to be covered by wind energy installations. In this way, more than 380,000 jobs would be supported by 2030. Canada aims to reach 5GW by 2025. In Latin America Brazil aims to reach 15.6GW by 2021, and Mexico wind energy targets for 2030 account of 10GW [12, 30].

In China, domestic production of wind turbines and components captures the entire supply chain. Supply is starting to not only satisfy domestic demand, but also meet international needs, especially for components. At the same time the 2020 target of China is set to a cumulative capacity of 250GW by adding at least 20GW annually. In India, existence of domestic industry and 65-70GW of assessed wind potential along with 10% of RES capacity and 4-5% of RES energy shares by 2012, are the main drivers of wind energy. 4GW will be possibly installed during 2016-2017 due to end of Generation based incentives. Current manufacturing capacity is around 9.5GW, while 2022 and 2030 targets reach 60GW and 125GW, respectively [12, 30].

At the end of 2008, Australia expanded the country's RES target to 20% by 2020. Currently, under construction are 14 projects that are expected to be commissioned until 2017, totalling 1.8GW. Concerning Africa, S. Africa also turns to wind, since 20% of the country's energy demand by 2025 could be provided by wind energy. The capacity target reaches 30GW. Other African countries with wind energy targets include Egypt, Morocco, and Algeria which aim a cumulative capacity of 7.2GW (for 2020), 2GW (for 2020) and 5GW (for 2030) respectively. Middle East countries such as Jordan, Syria, and Kuwait have set significant targets compared to previous years. Jordan government aims to total installations of 1.2GW by 2020, while Syria had set a target to reach 1GW of wind power capacity by 2020. Kuwait on the other hand, aims to fully operational installations by 2030 of 3.1GW [12, 30].

R&D Targets

Since the late 90s, research on wind energy development has been promoted through several funding packages such as the EU's FP5 to FP7 concerning small and large wind power installations. During the last years, research has been turned mostly to offshore installations in Europe where almost 76% of the FP7 resources for wind energy aimed to improve reliability and technology for offshore wind turbines [23].

Nowadays, many governments support research for wind technology and deployment including offshore development and mitigation of possible environmental impacts. Fig. (**34**) indicates the national R&D budgets for wind energy development. According to the figure, in most countries R&D budgets for wind energy in 2014 present a reduction compared to 2013. This is highly obvious in the case of USA, Sweden, and EU in general, where grants have been decreased significantly. In EU, 2014 budget reached 25M€, more than 50% decrease in relation to 2013, while in the USA although the budget had a reduction of around 13% in 2014, total reported national grants are the second highest in the list. On the other hand, Germany was the only European reported country that increased support to wind energy in 2014 compared to 2013 with a budget of 38.5M€ (36.7M€ in 2013). The country which invested significantly in the research of wind technologies with more than 50M€ in 2014, two times the amount that EU disposed, and 100% increase from 2013 budget, was Japan [34].

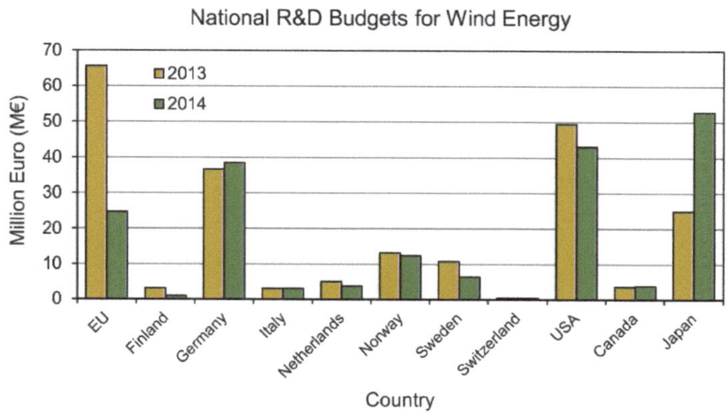

Fig. (34). 2013-2014 R&D national budgets for wind energy by reported countries [34].

The aforementioned research activities are applied to many sectors of wind energy from technical and siting method improvements to financial and environmental issues.

Concerning offshore wind installations, testing and development of turbines up to 10MW is already on the going, and prototypes are now in operation and ready for commercialisation. Additionally, fixed and floating types of foundations are in development in order to implement projects at deeper waters and longer distances from shore, while designs for machines up to 20MW comprise a main part of the research area [34].

In the case of small wind turbines (*i.e.* below 50kW), testing and improvement of components in order to reduce O&M costs and enhance deployment in urban areas are currently under investigation. Technology improvements include also investigation of gearboxes, control systems, two-bladed rotors, upwind and downwind designs, and grid integration through hybrid systems that include wind energy, PVs, hydropower, diesel and energy storage systems [34].

New forecasting techniques and model development for wind resource assessment are also investigated aiming to improve the siting process and therefore increase energy yield of future projects, resulting to financial benefits for future investors. Finally, developing new impact assessment methods, and monitoring procedures such as wildlife impacts and sound propagation will contribute to minimize negative social attitudes and promote positively approval processes [34].

CONCLUSIONS

This chapter investigates the current situation in the wind energy market facts, technology issues, economics, environmental parameters, and future prospects. The undertaken review reveals the dynamics of wind power, the development occurred in major markets, and it's potential for the next 15 years. The role of EU concerning the wind energy development has been analysed, while the growth of wind energy investments in the USA has been reflected to the annual and cumulative 2015 installations. Additionally, the tremendous increase of wind industry during the last six years in China has established the country as the strongest market worldwide. Besides that, development of wind energy at countries where this sector was absent for years, indicates that wind energy comprise one of the main power generating technologies that will substitute in the future a large percentage of conventional carbon containing electricity production plants.

CONSENT FOR PUBLICATION

Not applicable.

CONFLICT OF INTERESTS

The authors declare no conflict of interests.

ACKNOWLEDGEMENT

Declared none.

REFERENCES

[1] P.D. Fleming, and S.D. Probert, "The evolution of wind-turbines: An historical review", *Appl. Energy,* vol. 18, no. 3, pp. 163-177, 1984.

[2] M.J. Pasqualetti, R. Righter, and P. Gipe, Wind Energy, History of A2 - Cleveland, Cutler J.*Encyclopedia of Energy*. Elsevier: New York, 2004, pp. 419-433.

[3] P. Musgrove, *Wind power.* Cambridge University Press: Cambridge, UK; New York, 2010.

[4] T. Ackermann, and L. Söder, "An overview of wind energy-status 2002", *Renew. Sustain. Energy Rev.,* vol. 6, no. 1-2, pp. 67-127, 2002.

[5] N.I. Meyer, "Danish wind power development", *Energy Sustain. Dev.,* vol. 2, no. 1, pp. 18-25, 1995.

[6] J.K. Kaldellis, Wind Energy - Introduction*Comprehensive Renewable Energy - Vol.2 Wind Energy* Elsevier, 2012, pp. 1-10.

[7] G. de Carmoy, "The USA faces the energy challenge", *Energy Policy,* vol. 6, no. 1, pp. 36-52, 1978.

[8] R.L. Thomas, and W.H. Robbins, "Large wind-turbine projects in the United States wind energy program", *ResearchGate,* vol. 5, no. 3, pp. 323-335, 1980.

[9] R.W. Righter, "World Renewable Energy Congress Renewable Energy, Energy Efficiency and the Environment - Pioneering in wind energy: The California experience", *Renew. Energy,* vol. 9, no. 1, pp. 781-784, 1996.

[10] M.D. Dodge, *The illustrated history of wind power development,* 2006. Accessed: 11-Jul-2016 [Onlin e]. Available: http://telosnet.com/wind/early.html

[11] "Wind in Power 2011 - 2015", *WindEurope,* 2016. Accessed: 11-Jul-2016 [Online]. Available: https://windeurope.org/about-wind/statistics/european/

[12] *Global Wind Reports 2006 to 2015,* 2016. Accessed: 11-Jul-2016 [Online]. Available: http://www.gwec.net/publications/global-wind-report-2/

[13] *International Energy Statistics - RES Electricity Production,* 2016. Accessed: 11-Jul-2016 [Online]. Available: http://www.eia.gov/

[14] *30 Years of Policies for Wind Energy: Lessons from Denmark,* 2012. Accessed: 12-Jul-2016 [Online]. Available: https://www.irena.org/documentdownloads/publications/gwec_denmark.pdf

[15] *International Energy Statistics - Electricity Capacity,* 2016. Accessed: 13-Jul-2016 [Online]. Available: http://www.eia.gov/cfapps/ipdbproject/iedindex3.cfm?tid=2&pid=2&aid=7&cid=regions&syid=2000&eyid=2012&unit=MK

[16] Accessed: 13-Jul-2016 [Online]. Available: http://ec.europa.eu/eurostat/web/energy/data/database

[17] J.K. Kaldellis, and D. Zafirakis, "Present situation and future prospects of electricity generation in Aegean Archipelago islands", *Energy Policy,* vol. 35, no. 9, pp. 4623-4639, 2007.

[18] *DAS Monthly Reports - December 2015,* 2016. Accessed: 13-Jul-2016 [Online]. Available: http://www.lagie.gr/en/market/market-analysis/das-monthly-reports/

[19] *HEDNO - Non-Interconnected grid reports -December 2015,* 2016. Accessed: 13-Jul-2016 [Online]. Available: http://www.deddie.gr/en/miniaia-deltia-ape-kai-thermikis-paragwgis-sta-mi-diasunded-emena-nisia/2015

[20] *2015_-HWEA_Statistics_Greece.pdf,* 2016. Accessed: 13-Jul-2016 [Online]. Available: http://eletaen.gr/wp-content/uploads/2016/01/2015_-HWEA_Statistics_Greece.pdf

[21] J.K. Kaldellis, "Critical evaluation of financial supporting schemes for wind-based projects: Case study Greece", *Energy Policy,* vol. 39, no. 5, pp. 2490-2500, 2011.

[22] J.K. Kaldellis, and M. Kapsali, "Shifting towards offshore wind energy—Recent activity and future development", *Energy Policy,* vol. 53, pp. 136-148, 2013.

[23] J.K. Kaldellis, D. Apostolou, M. Kapsali, and E. Kondili, "Environmental and social footprint of offshore wind energy. Comparison with onshore counterpart", *Renew. Energy,* vol. 92, pp. 543-556, 2016.

[24] "The European offshore wind industry- key trends and statistics 2011 to 2016", *WindEurope,* 2016. Accessed: 11-Jul-2016 [Online]. Available: https://windeurope.org/about-wind/statistics/ offshore/

[25] *Wind in our Sails - The coming of Europe's offshore wind energy industry,* 2011.

[26] *Offshore - Wind farms - Database,* 2016. Accessed: 15-Jul-2016 [Online]. Available: http://www.thewindpower.net/store_continent_en.php?id_zone=1006

[27] *EWEA-Wind-energy-scenarios-2030,* 2015. Accessed: 11-Jul-2016 [Online]. Available: http://www.ewea.org/fileadmin/files/library/publications/reports/EWEA-Wind-energy-scenarios-2030.pdf

[28] J.K. Kaldellis, and D. Zafirakis, "The wind energy (r)evolution: A short review of a long history", *Renew. Energy,* vol. 36, no. 7, pp. 1887-1901, 2011.

[29] J. Serrano-González, and R. Lacal-Arántegui, "Technological evolution of onshore wind turbines-a market-based analysis: Technological evolution of onshore wind turbines", In: *Wind Energy*, 2016.

[30] *Renewables 2016 - Global Status Report,* 2016.

[31] H. Smart-Blade Gmb, *Performance Improvement for Wind Turbines*. Accessed: 11-Jul-2016 [Online]. Available: http://smart-blade.com/products-services/vortex-generators.html

[32] *International Energy Statistics - Wind Capacity,* 2016. Accessed: 11-Jul-2016 [Online]. Available: http://www.eia.gov/

[33] J.K. Kaldellis, "Integrated electrification solution for autonomous electrical networks on the basis of RES and energy storage configurations", *Energy Convers. Manage.*, vol. 49, no. 12, pp. 3708-3720, 2008.

[34] *2014 Wind Annual Report,* 2015.

[35] *Renewable Energy Cost Analysis: Wind Power,* 2012.

[36] *Offshore Wind Cost Reduction. Pathways Study,* 2012.

[37] *Annual Energy Outlook 2016,* 2016. Accessed: 11-Jul-2016 [Online]. Available: https://www.eia.gov/forecasts/aeo/electricity_generation.cfm

[38] *Projected costs of generating electricity.* International Energy Agency, Nuclear Energy Agency, Organisation for Economic Co-operation and Development: Paris, 2010.

[39] *A World-Leader in Wind Energy,* 2015. Accessed: 11-Jul-2016 [Online]. Available: http://denmark.dk/en/green-living/wind-energy/

[40] E. Kondili, and J.K. Kaldellis, Environmental-Social Benefits/Impacts of Wind Power*Comprehensive Renewable Energy - Vol.2 Wind Energy,* A. Sayigh, Ed., Elsevier, 2012, pp. 503-539.

[41] J.K. Kaldellis, K.A. Kavadias, and A.G. Paliatsos, "Environmental impacts of wind energy applications: myth or reality?", *Fresenius Environ. Bull.*, vol. 12, pp. 326-337, 2003.

[42] N. Aden, A. Marty, and M. Muller, *Comparative Life-cycle Assessment of Renewable Electricity Generation Technologies,* 2010.

[43] *Life Cycle Assessment Harmonization Results and Findings,* 2014.

[44] *Special Eurobarometer - Energy Technologies: Knowledge, Perception, Measures,* 2007.

[45] TPA Horwath, *Wind Energy in Poland,* 2013.

[46] *Wind Power Poland 2016,* 2016. Accessed: 21-Jul-2016 [Online]. Available: http://www.ewea.org/pwea2016/

CHAPTER 6

Hydrogen Production and Storage

Tryfon Roumpedakis, Petros Vlavakis, Konstantinos Braimakis, Dimitrios Grimekis and **Sotirios Karellas***

Laboratory of Steam Boilers and Thermal Plants, National Technical University of Athens, 9 Heroon Polytechniou St., Athens, 15780, Greece

Abstract: Given the environmental concerns over the use of fossil fuels, the search for non-fossil energy sources to cover the increasing energy demands is gathering a lot of interest. Hydrogen fuel, being CO_2 neutral and non-pollutant, is one of these sources. The main scope of this chapter is to provide a comprehensive review of hydrogen production and storage technologies. Firstly, various hydrogen production methods are discussed, including the utilization of various primary sources. Apart from the conventional natural gas and coal based hydrogen production, more recent technologies, including solar, photobiologic electrolysis and supercritical water gasification are presented. In the second part of the chapter, hydrogen gaseous and liquid storage methods are discussed. Regarding gaseous hydrogen storage, several types of commercially available steel and composite tanks are presented. Furthermore, the option of storage *via* glass microspheres is briefly presented. For liquid hydrogen storage, two of the most common options are analyzed: cryogenic liquid hydrogen and $NaBH_4$ solutions. In the last section of the chapter, the most recent advances in hydrogen storage in solid materials and its storage *via* conversion to other fuels are discussed with an overview of the challenges of these options.

Keywords: Electrolysis, Gas Storage, Hydrogen, POx, Pyrolysis, Steam Methane Reforming, Supercritical Water Gasification.

HYDROGEN PRODUCTION

Hydrogen production has been extensively studied in the last years, due to the numerous advantages associated with its usage. Most importantly, it is a "clean", non-pollutant fuel with zero greenhouse gas emissions.

The main physical properties of hydrogen are summarized in Table **1**.

* **Corresponding Author Sotirios Karellas:** Laboratory of Steam Boilers and Thermal Plants, National Technical University of Athens, 9, Heroon politechniou St., Athens, 15780, Greece; Tel: 0030 2107722810; Fax: 00302107723663; Email: sotokar@mail.ntua.gr

Emmanuel D. Rogdakis & Irene P. Koronaki (Eds.)

Table 1. Summary of physical properties of hydrogen.

Chemical symbol	H_2
Molecular weight	2.02
Specific heat capacity-liquid	9.668 kJ/kgK
Boiling point temperature	-252.8 °C at 1.013 bar
Latent heat of evaporation	446.0 kJ/kg
Gas density	0.08988 kg/m^3
Triple point temperature	-259.2 °C
Triple point pressure	0.07205 bar
Critical temperature	-239.96 °C
Critical pressure	13.15 bar
Critical density	30.12 kg/m^3
Thermal conductivity	0.1869 W/mK at 300 °C
Flammability limits (volume %)	Lower: 4.1 Upper: 74.0
Stoichiometric air/fuel ratio, weight basis	34.3
High heating value	141890 kJ/kg
Low heating value	119972 kJ/kg
Ignition temperature	530 °C

It is estimated that each year, 10-11 million tons of hydrogen are produced in the U.S.A., enough to power 20-30 million vehicles or 5-8 million households. Worldwide hydrogen production amounts to 50 million tons, equivalent to 170 million tons of oil, exhibiting an annual increase of 10% [1]. Hydrogen cannot be found in practically usable quantities in nature, however it can be produced from fossil fuels and biomass or by various other technologies. As of today, the vast majority of hydrogen production takes place by thermochemical conversion of fossil fuels by steam reforming or partial oxidation of methane or the gasification of coal (~95%), while the remaining 5% originates from biomass gasification or water electrolysis [2].

Production from Natural Gas

Hydrogen can be produced from natural gas mainly by three different methods:

- Steam methane reforming (SMR)
- Partial oxidation (POx)
- Autothermal reforming (ATR)

Steam Methane Reforming

Steam methane reforming is the most common method for producing hydrogen, representing around 40% of the worldwide production capacity [1]. This technology is highly developed and is encountered in a wide range of applications, ranging from scale decentralized units (<1 t/h) to larger ammonia production plants (>100 t/h).

The operating principle of this method is the catalytic reaction of methane with steam to produce CO and H_2:

$$CH_4 + H_2O \rightarrow CO + 3H_2 \quad \textbf{(1)}$$

The above reaction is endothermal and requires temperatures between 650-800 °C in order to produce reasonable H_2 yields ranging from 60 to 70% [3, 4]. The catalysts consist of Ni, Pt, Rh or Pd, with Ni catalysts being the most common due to their lower cost. A major issue regarding the use of Ni catalysts, however, is their deactivation that occurs due to char formation at high temperatures, making the use of Pt and Pd more appropriate in such cases owing to their chemical stability. Lastly, when mass transfer phenomena are limited, Rh catalysts are used, since Rh has better catalytic properties than Ni as well as less sensitivity to deactivation due to char formation [4].

The produced CO from the reforming reaction is converted into CO_2 through the water gas shift (WGS) catalytic reaction, at temperatures from 300 to 450 °C:

$$CO + H_2O \rightarrow CO_2 + H_2 \quad \textbf{(2)}$$

Water gas shift reaction catalysts are usually ferrous, copper, molybdenum alloys and Fe-Pd alloys.

The main components of an SMR unit include the natural gas desulfurization unit, the catalytic reforming reactor, the WGS reactor and the hydrogen separation unit, as depicted in Fig. (**1**). Because of the high sensitivity of the reforming and WGS catalysts to sulfur poisoning, intensive desulfurization of the feed gas stream is required. Sulfuric components are firstly converted to H_2S and subsequently react with ZnO in temperatures from 340-390 °C, according to the reaction:

$$H_2S + ZnO \rightarrow ZnS + H_2O \quad \textbf{(3)}$$

After the desulfurization unit, the natural gas is fed into the catalytic reforming reactor, wherein it reacts with steam to produce syngas. If the natural gas contains heavy hydrocarbons, an additional unit is included, the pre-reformer, in which they are converted into CH_4, CO_x and H_2 in an adiabatic reactor at temperatures ranging from 300 to 525 °C.

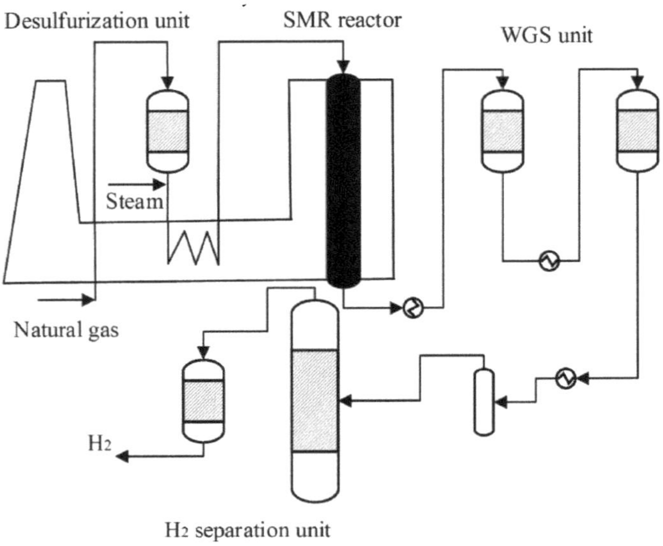

Fig. (1). Overview of the main components of an SMR unit.

The pretreated natural gas is mixed with high pressure stream (at 26 bar) and the produced mixture is preheated to 500 °C before entering the reforming reactor. Inside the reactor, the mixture flows in externally heated tubes filled with Ni catalysts and is converted into CO and H_2. Although reforming is promoted in low pressures, the operating pressure is usually higher than 20 bar, allowing robust reactor design and reduced material costs. The SMR method leads to hydrogen production yields from 60 to 80%, with promising prospects of increasing these values in the future [5].

Partial Oxidation

Partial oxidation of hydrocarbons, including natural gas, is one major route towards the commercial production of hydrogen. The general principle of this process consists in the mixing of the original hydrocarbon fuel with oxygen in such proportions to create CO and H_2.

Depending, on the type of the fuel and the reactor, there are several variations of the process. The process can be carried out in the presence or the absence of a catalyst. The non-catalytic process requires high temperatures (1100-1500 °C) and can take place with any type of original feedstock including coal [6]. The catalytic process, on the other hand, is suited for lighter hydrocarbons and is operating at lower temperatures (600-900 °C).

The catalytic partial oxidation process for methane is a moderately exothermic reaction:

$$CH_4 + 1/2O_2 \rightarrow CO + 2H_2 \quad \text{(4)}$$

In order for the aforementioned reaction to take place, oxygen -or air- and methane pass through a suitable catalyst. If the required oxygen is carried with air, nitrogen has to be removed from the product stream, a process that usually occurs at a separate stage downstream the oxidation reactor [7]. In the catalytic POx process, the air and the fuel are mixed before entering the catalytic bed. Inside the catalytic bed, the hydrocarbons of the fuel are oxidized by a number of heterogeneous reactions, including partial and complete combustion, steam reforming and water gas shift reaction, which are summarized in Table **2**.

Table 2. Summary of the side reactions in a POx reactor [8].

$CH_4 + 2O_2 \rightarrow CO_2 + 2H_2O$	Combustion
$CH_4 + O_2 \rightarrow CO_2 + 2H_2$	
$CO + H_2O \leftrightarrow CO_2 + H_2$	Water Gas Shift
$CH_4 + H_2O \leftrightarrow CO + 3H_2$	Steam Reforming
$CH_4 + CO_2 \leftrightarrow 2CO + 2H_2$	CO_2 Reforming
$CO + H_2 \leftrightarrow C + H_2O$	
$CH_4 \leftrightarrow C + 2H_2$	Methane decomposition
$2CO \leftrightarrow CO_2 + C$	Boudouard
$CO + 1/2\ O_2 \rightarrow CO_2$	
$H_2 + 1/2\ O_2 \rightarrow H_2O$	

Partial oxidation of methane yields higher production rates of hydrogen with temperature, but reaches a peak at approximately 730 °C [9]. Typical POx unit efficiencies lie in the range between 70 and 80%. In general, POx units are smaller than equivalent SMR units, although being inferior regarding their efficiency, due to their higher operating temperatures and non-recoverable waste heat streams. However, contrary to SMR, POx units can be used to utilize a wide range of fuels apart from natural gas. Furthermore, due to lower flame temperatures, they have insignificant NO_x and SO_x emissions.

To minimize energy losses due to the compression of hot gases, the catalytic POx reactor should be operated in the range of 5 to 40 bar [10]. The oxygen to carbon ratio in a typical POx reactor is 1.2 and the gas hourly space velocity ranges from several thousands to several hundred thousand scc/g_{cat}.h [11]. Most commonly used catalysts include transition metals, such as Ni and Co, and noble metal-based catalysts -*e.g.* Pt, Ru, Pd and Rh-. Metallic nickel enhances the production of synthesis gas, while nickel species with an oxidative number higher than 2 tend to catalyze the total combustion of methane. Slagtern *et al.* [12] reported that the activity of Fe- and Co- based catalysts was inferior compared to the Ni-based catalysts. Vernon *et al.* [13] tested several catalysts, including Al_2O_3, 1%Rh/ Al_2O_3 and 1%Ir/ Al_2O_3, and obtained high yields of synthesis gas. Jalowiecki *et al.* [14] reported conversion rates of 75% and hydrogen selectivity of 52% at 600°C with a $CeNi_xO_y$ catalyst. Noble metal catalysts have better stability and activity compared to non-precious metal catalysts, however, due to their high cost and low accessibility are not preferred in industrial applications [15].

POx of methane is a relatively inexpensive alternative to syngas generation and has no need of external heat sources and additional heat carriers such as steam [16]. Furthermore, the POx process is a mildly exothermic reaction in contrary to the endothermic SMR and makes it ideal for integration in high temperature fuel cells. The POx of methane can be carried out under the condition of very high gas hourly space velocity, which makes the process more efficient in terms of production rate and required investment [17, 18]. The absence of water in the feed enables the POx to be used into on board transportation reforming systems [11]. In general, the catalytic POx of hydrocarbons is a complicated process due to the high catalyst surface temperatures that result in hot spots and lack of thermal equilibrium between the solid and the gaseous phases [19]. In addition, even at stoichiometric or sub-stoichiometric conditions, carbon depositing over the catalyst bed is unavoidable, reducing the activity of the catalyst. Catalysts with enhanced thermal conductivity are thought to reduce the hot spots. Experimental results showed that the relative range of carbon deposition, based on the catalysts used, follow the subsequent order Pt=Ir<Ru<Rh<Pd<Ni [20].

Apart from methane, there are alternative fuels that are feasible for POx process, including ethanol. POx of ethanol is a highly exothermal reaction; thus no external heat input is required for the process:

$$C_2H_5OH + 1/2O_2 \rightarrow 3H_2 + 2CO \quad \textbf{(5)}$$

One advantage of the POx of ethanol over other hydrogen production processes is the fact that this reaction is faster than the steam reforming of ethanol, resulting in smaller reactor sizes, while the presence of oxygen is reducing the formation of residues in the catalyst surface prolonging the activity of the catalyst [21]. The main drawback of ethanol POx, as in all partial oxidation processes, is that the corresponding total oxidation is much more favoured thermodynamically, resulting in a low hydrogen selectivity:

$$C_2H_5OH + 3O_2 \rightarrow 3H_2O + 2CO_2 \quad \textbf{(6)}$$

Pirez *et al.* [22] investigated a cerium nickel based oxyhydride catalyst for different O_2/C_2H_5OH ratios and a temperature range from 200 to 480°C and reported that the conversion increases with temperature for a constant O_2/C_2H_5OH ratio, while the increase in the O_2/C_2H_5OH ratio resulted in even more rapid increase in the conversion of ethanol to hydrogen. On the other hand, the ability to break CC bonds, their low selectivity along with their relatively low cost makes Ni-based catalysts promising candidates for enhancing high hydrogen production rates by POx of ethanol [23]. Salge *et al.* [24] reported high hydrogen selectivities-exceeding 80%- using Rh-ceria catalyst at short contact times.

Regarding the non-catalytic POx, the main advantage of this process is the fact that it can utilize a wide range of fuels. The non-catalytic POx for the production of hydrogen by heavy residues is an economically feasible process, that has already been commercially applied [6]. The general exothermic reaction taking place during POx of heavy residues is the following:

$$C_mH_nS_p + m/2O_2 \rightarrow mCO + (n/2 - p)H_2 + pH_2S \quad \textbf{(7)}$$

A number of reactions are taking place simultaneously in the reactor, including incomplete combustion of carbon, coke gasification, water gas shift reaction and methane reforming. Typical flame temperatures of 1300-1500 °C are necessary to ensure complete conversion and to reduce soot formation. Fig. (**2**) shows a basic process flow diagram of hydrogen production by non-catalytic POx, which can roughly be divided into three parts: the pre-treatment equipment, the reactors and the gas cleaning units.

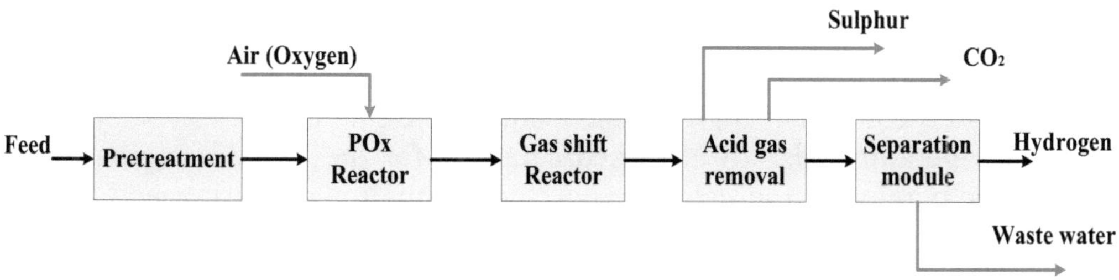

Fig. (2). Flow diagram of hydrogen production *via* POx of sulphurous heavy residue.

Autothermal Reforming

Autothermal reforming is a combination of steam methane reforming and partial oxidation, since the required thermal duty for the formation of syngas is provided by the exothermic partial oxidation of the hydrocarbon feedstock in an adiabatic reactor. The autothermal reformer consists of three zones: the combustion, the thermal and the catalytic zone, as shown in Fig. (**4**). The feed enters first the combustion zone, where is mixed with steam and oxygen of air and is partially oxidized based on the reaction given in Eq. **4**.

In the thermal zone, further conversion occurs by homogeneous gas phase reactions. In this zone, reactions can take place between N_2 and hydrocarbon radicals towards the formation of HCN and NH_3. Proper selection of the ratio O_2/CH_4 provides, *via* the partial oxidation of the fuel, the required heat for the endothermic SMR reaction that takes place in the catalytic zone [19]. Air addition, should be limited, though, to prevent coke deposition on the catalysts' surface. The temperature profile inside the reactor shows a sharp rise in the thermal zone, while in the catalytic zone temperature decreases steadily due to the endothermic reactions [25]. Due to the fact that SMR is highly endothermic, the temperature drop inside the catalytic bed can be up to 300 °C. In general, the composition of steam, methane and oxygen should be adjusted in such a way to ensure that the net enthalpy of the reaction is close to thermal neutrality, ensuring high conversion rates of hydrogen.

One advantage of ATR is that the pressure can be increased compared to SMR, resulting in lower energy requirements for syngas compression and hence to easier heat integration [6]. Furthermore, the autothermal processes can convert gasoline and higher hydrocarbons to hydrogen using suitable catalysts [7]. The ATR process can be stopped and started rapidly, due to its fast reaction rates at start-up, while producing a larger amount of hydrogen compared to POx alone. Additionally, the presence of water in the feed inhibits the flammability and explosion risks of the oxygen-methane mixture, decreases the combustion selectivity and decreases the deposition of residues on the catalyst surface. On the other hand, the presence of steam lowers the reaction temperature and thus the reaction rate [26]. The ATR catalyst is robust with a high thermal stability, however, it is prone to sintering, sulphur poisoning and fouling. The selection of the ATR process catalyst is highly dependent on the fuel characteristics and the operating temperature. Use of Ni based catalysts for the ATR of methane can result in deactivation of the catalyst, hot spot formation, sintering and carbon deposition on the catalysts. To suppress these side effects, the use of catalytic micro-channel reactors is proposed [27]. Akri *et al.* [28] reported that coking phenomenon in Ni based catalysts is limited in case of highly dispersed metal species at the surface of the catalyst. For this reason, Ni based catalysts are combined with promoters to minimize the coke formation. Among these promoters Ce-Zr-O was found to have the higher resistance to coke formation, thanks to its thermal stability, oxygen storage capacity and catalytic activity [29]. Alternatively, the use of rare metals, including La, Ce and Pr, exhibited better performance and reduced the contamination of the catalyst surface, by having higher resistance to coke formation and sintering [30]. For methanol, Cu-based catalysts are used, while for higher hydrocarbons noble metal catalysts are reported -*e.g.* highest selectivity in reforming of isooctane was obtained on Pt catalysis- [31]. Choi *et al.* [32] investigated the use of noble metal catalysts for the ATR of dimethyl ether and found that Ru catalysts performed better under severe reforming conditions. Baruah *et al.* [33] investigated the performance of several catalysts in ATR of ethanol and reported the resulting hydrogen selectivity for a number of them, as shown in Fig. (**3**).

(a)

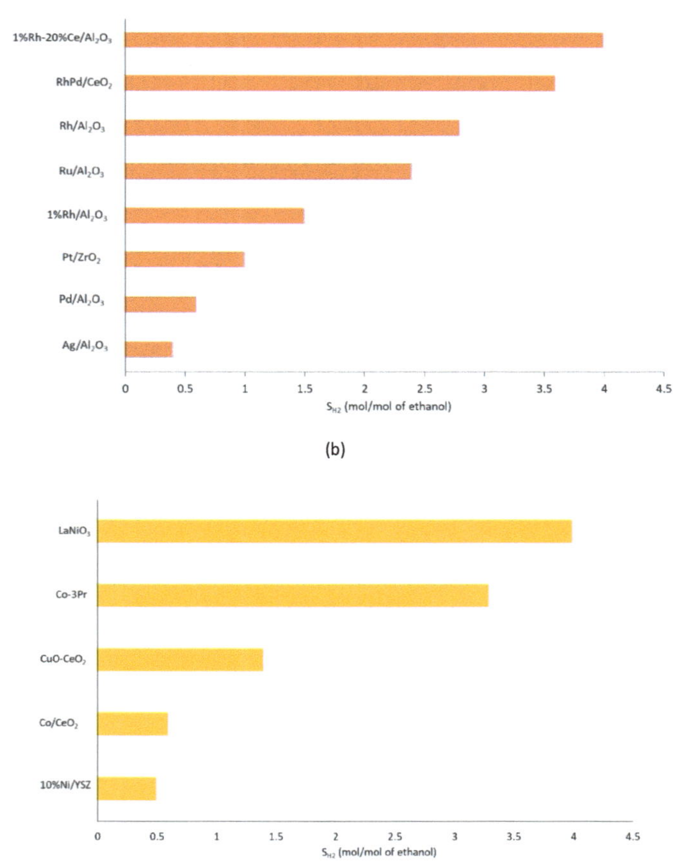

Fig. (3). Hydrogen selectivity for several noble (a) and non-noble (b) metal catalysts [33].

Comparing ATR with catalytic POx, the temperatures in the catalyst bed in ATR are relatively lower and the syngas yield higher. However, premixing and preheating is more straight forward in POx and reaction rates are higher [34].

ATR process is expected to gain interest in the gas-liquid industry thanks to its favourable gas composition for the Fischer-Tropsch synthesis, lower capital cost and potential for economies of scale [35].

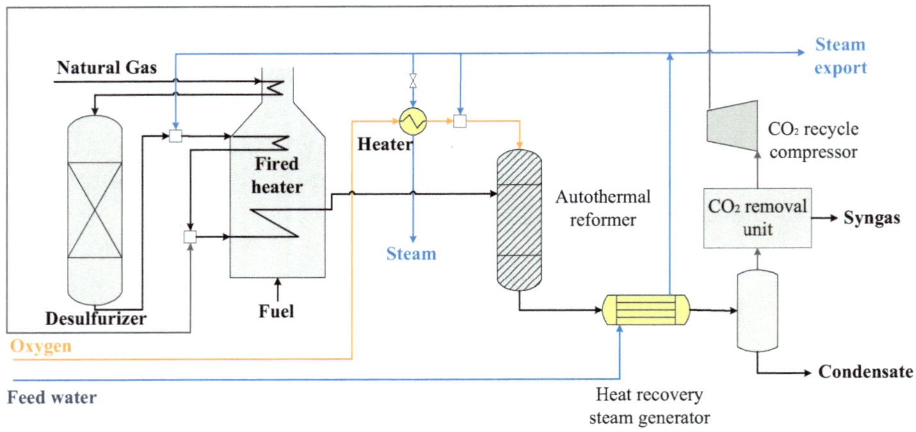

Fig. (4). Simplified scheme of an autothermal reforming plant [36].

Production from Coal

Hydrogen can be produced from coal gasification and separated from the syngas *via* a separation module. The reaction mechanisms have many similarities with the POx of heavy oils, which was described previously [37]. High temperature entrained flow processes are favoured to maximize the carbon conversion to gas. In general, coal is converted to syngas following the reactions [38]:

$$C(s) + (1/a)O_2 \rightarrow \frac{2a-2}{a} CO(g) + \frac{2-a}{a} CO_2(g) \quad \textbf{(8)}$$

$$C(s) + H_2O(g) \rightarrow CO(g) + H_2(g) \quad \textbf{(9)}$$

$$C(s) + CO_2(g) \rightarrow 2CO(g) \quad \textbf{(10)}$$

$$C(s) + 2H_2(g) \rightarrow CH_4(g) \quad \textbf{(11)}$$

In terms of CO_2 capture, there are three main processes: pre-combustion, post-combustion and oxyfuel technologies. In comparison to post-combustion, pre-combustion CO_2 capture removes CO_2 from the stream after gasification. Pre-comb- ustion CO_2 capture enhances the hydrogen yield in syngas. Syngas is then lead to a water gas shift reactor to convert CO to CO_2 which can be then separated from the hydrogen in an amine scrubber column [39, 40]. According to Jordal *et al.* [41], CO_2 capture with solvents cannot achieve sufficiently high hydrogen purity. On the other hand, pressure swing adsorption (PSA) is a very well-known method to purify, downstream to the gasifier, hydrogen at relatively low investment costs. Hydrogen can be separated from the rest gas components in a separation train consisting of four or more beds, resulting in a hydrogen stream with purity higher than 99.9% [42].

Gasification of coal is the oldest known method for production of hydrogen and offers the chance to utilize low grade coal types. Hydrogen production from coal is commercially mature, but is more complex than steam methane reforming [43]. The capital investment costs are higher than steam methane reforming leading to a more expensive product [44]. In principle, all types of coal can be gasified to produce hydrogen, although based on the type of the feedstock certain limitations rise. One of the oldest commercially established processes is the Koppers-Totzek process, operating at atmospheric pressure and purifying the hydrogen by PSA [45]. The resulting hydrogen stream is at a pressure of approximately 28 bar and purity higher than 97.5% [46].

An overview of a coal gasification plant for the production of hydrogen is presented in Fig. (**5**).

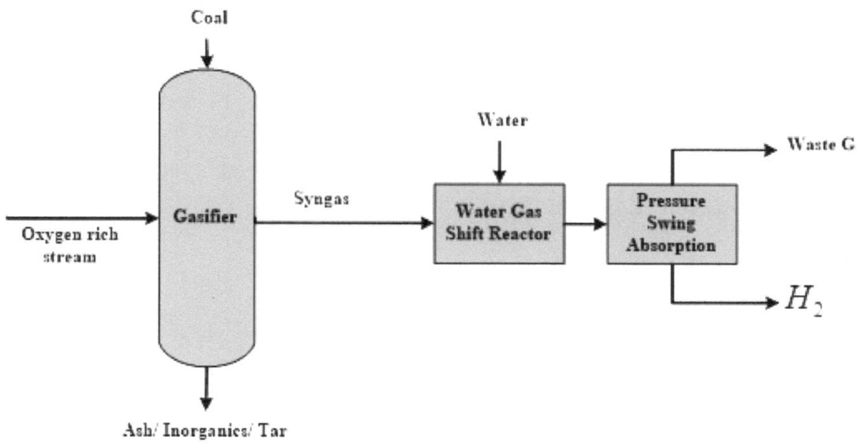

Fig. (5). Simplified flowsheet for the production of hydrogen from coal gasification.

Supercritical Water Gasification

Supercritical water gasification (SWG) is a newly introduced technology, in which the gasification agent is water in supercritical conditions of approximately

220 bar pressure and temperature over 374°C. *Via* SWG, coal can be gasified to a great extent, and the produced gas consists mainly of H_2 and CO_2. The chemical conversion can be presented by the following reaction:

$$CH_xO_y + (2-y)H_2O \rightarrow CO_2 + (2-y+\frac{x}{2})H_2 \quad \textbf{(12)}$$

This reaction can be divided into to two steps: the steam reforming and the water-gas shift reactions, both presented previously. SWG gasification is very efficient on low temperatures -below 700°C-, while it has no moisture content requirement [47]. In addition, supercritical water has the ability to dissolve most organic substances, has excellent mass transfer performance and provides a homogeneous and rapid reacting medium for clean coal conversion [48 - 50]. Another advantage of SCW gasification of coal is its high pressure, which reduces significantly the energy costs for its compression during storage [49]. Furthermore, the solubility of CO_2 in supercritical water has significant difference with the solubility of other gas, allowing CO_2 to be easily separated from the rest of the stream in a high pressure separation system [51]. Jin *et al.* [52] investigated the production of hydrogen from SCW of coal and reported the effects of temperature and pressure on the hydrogen production. It was found that higher temperatures favored the conversion of coal and thus the production of hydrogen, while the pressure was not affecting significantly the gasification characteristics. Ge *et al.* [53] investigated the effect of various alkali catalysts and reported that hydrogen yield was increasing with the use of K_2CO_3, KOH or NaOH compared to Na_2CO_3 and $Ca(OH)_2$. Moreover, it was reported that hydrogen yield was increasing with increasing catalyst loading and decreasing concentration. Li *et al.* [54] reported that supercritical water partial oxidative gasification of coal could improve the hydrogen yield and decrease the production of char and tar.

Integrated Novel Gasification

Recently, a new method has been proposed by Lin *et al.* for the production of hydrogen *via* coal gasification [55 - 57]. This method, called Hydrogen Production by Reaction Integrated Novel Gasification (HyPr-RING) is a combination of water-hydrocarbon reaction, water gas shift reaction and absorption of CO_2 in a single reactor under both sub- and supercritical water conditions that enhances the concentration of hydrogen in the produced syngas. The main reaction is the following:

$$C + 2H_2O + CaO \rightarrow CaCO_3 + 2H_2 \quad \textbf{(13)}$$

This reaction is exothermic and can produce high yield hydrogen at relatively low temperatures, approximately 650-700°C. HyPr-RING advantages over the conventional gasification in the fact that hydrogen separation and gas separation occur in one reactor at a lower temperature [58]. A block diagram of the integrated gasification scheme is depicted in Fig. (**6**).

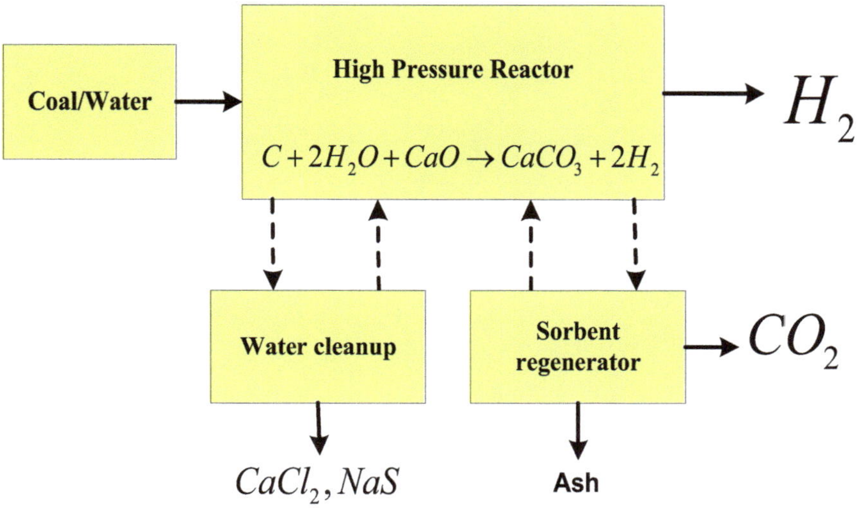

Fig. (6). Simplified flowsheet of the hydrogen production by reaction integrated novel gasification [55].

Production from Biomass with Gasification and Pyrolysis

Since biomass has similar properties to fossil fuels, it can be used as a source for the production of hydrogen as well. One major benefit of utilizing biomass as a source of hydrogen production is the zero CO_2 footprint potential associated with its use [59, 60]. The most common and cost competitive technology for the conversion of biomass to hydrogen is gasification, which is based on the same principles applied for the gasification of coal [61 - 63]. In the first part of the present Section, the most important syngas separation technologies applied for the extraction of hydrogen from the produced syngas are presented. In the second part, a literature review of the integration of pyrolysis in the gasification conversion pathway is included.

Syngas Separation Technologies

Membrane Separators

The most usual hydrogen separation method from syngas is based on the use of selective membranes, which consist of polymers, metals or other ceramic materials. The mechanism behind each membrane type and material is different. A typical schematic of the allocation of a gasification and a membrane separation process is presented in Fig. (**7**).

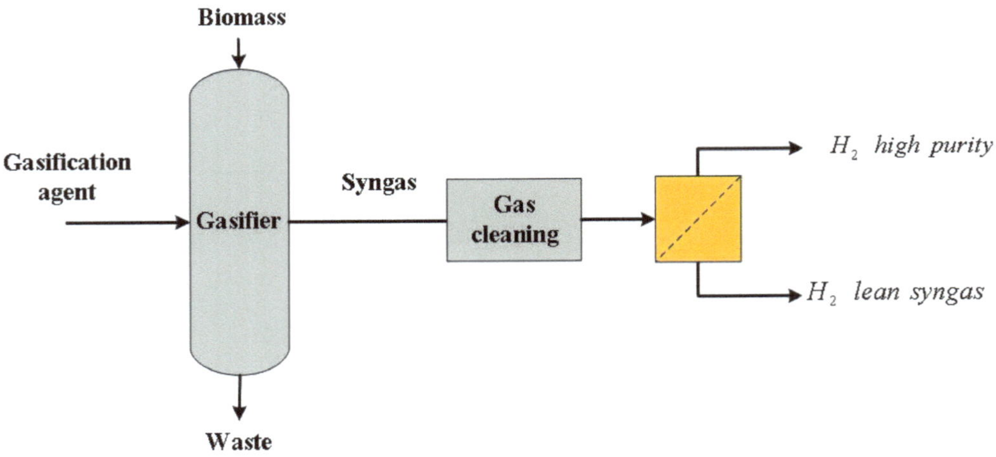

Fig. (7). Gasification and membrane separation process.

In polymer membranes with small pores, the separation is achieved *via* a particle diffusion mechanism which is determined by pore diameter and particle size. In metallic membranes, hydrogen molecules are split in atoms which pass through the membrane and then rejoin to form molecules at its exit. Lastly, in ceramic membranes, the separation is based on transferring hydrogen ions and free electrons through the membrane. Membranes of this type require high operating temperatures in order to achieve sufficient flow rates and attain comparable efficiency to that of the other membranes A typical membrane separator for syngas is depicted in Fig. (**8**).

In order to separate hydrogen from biomass gasification products in pressures above 20 bar and temperatures exceeding 700 °C, Pd-Cu membranes are the most appropriate. Their operating principle lies in the mass transfer due to pressure difference throughout the membrane [64]. One of the main advantages of this technology is the potential for high purity hydrogen production, suitable for use in fuel cells. An important challenge rests in increasing the endurance of these membranes in against pollutants, mainly CO and sulfur, such as their capability of operation in lower temperatures from 250 to 500 °C [65]. If the above improvements are attained, they are expected to reduce the cost of membranes, increase their efficiency and significantly simplify hydrogen separation systems. An overview of the most important membrane separation applications follows.

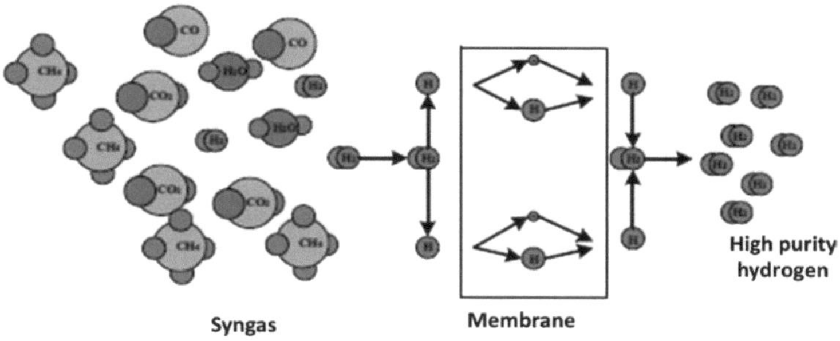

Fig. (8). Hydrogen separation from syngas in a typical membrane.

Membrane Reactor

In the membrane reactor the membrane is located inside the gasifier, allowing the direct removal of hydrogen from the syngas. In its simplified form, the reactor consists of two concentric cylinders, with the membrane being the inner cylinder, as can be seen in Fig. (**9**).

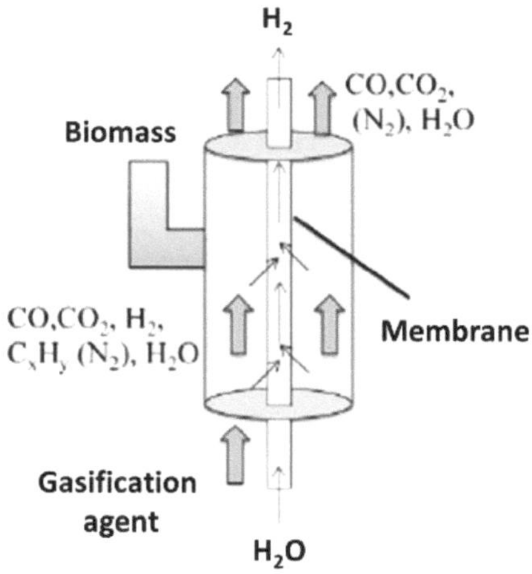

Fig. (9). Membrane gasification reactor.

Multiple smaller internal cylinders can be added in order to increase the membrane surface per volume unit of the reactor and thus increase the hydrogen maximum capacity. Metallic and ceramic membranes are most commonly used in these reactors. A major drawback of this technology concerns operation temperature limitations in order to prevent damage to the membrane due to its sensitivity to mechanical stresses [65].

Reformer-membrane Configuration

Another configuration is the reformer-membrane (Fig. **10**), in which the membrane is located outside the gasifier in order to avoid complications with the reforming catalysts. This configuration thus allows more effective dedicated design of the processes of reforming and separation as well as more simplified mechanical design of the membranes, compared to membrane gasification reactors. On the other hand, the configuration leads to higher membrane surface requirements for the same hydrogen capacity, resulting in increased costs [65].

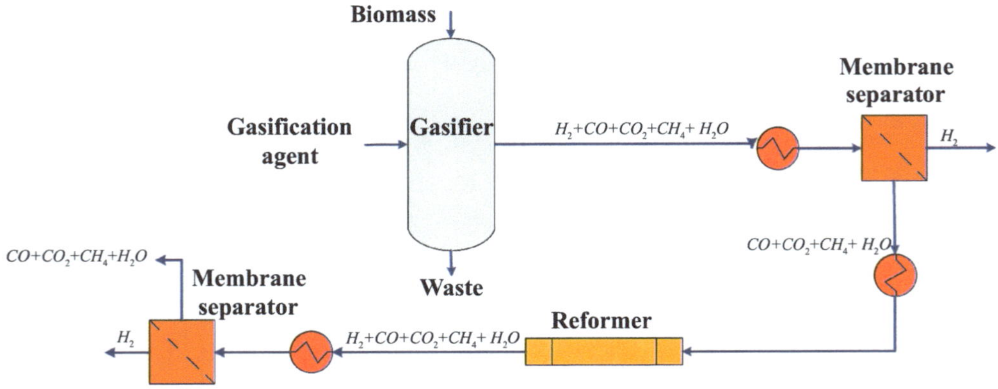

Fig. (10). Reformer-membrane configuration.

Water Gas Shift Reactor and Pressure Swing Absorption (PSA)

The PSA method is used jointly with water gas shift reactors which produce mixtures of CO_2 and H_2. The WGS reaction is carried out in two stages, a high temperature stage from 350 to 500 °C, often with the use of Fe-Cr catalysts and a second low temperature stage at around 200 °C, in which copper catalysts are employed.

The PSA method is used for the separation of gases from pressurized stream with the use of molecular filters with specially designed pores to promote the separation of molecules with different sizes. This method allows significantly high purity levels between 99.0 and 99.9%. The operating pressure varies from 18 to 36 bar. A disadvantage of this method, though, is the requirement of low operating temperatures along with the relatively low hydrogen recovery rate from the original feed stream.

Pyrolysis

Owing to its high thermal efficiency, the gasification of biomass has been widely practiced for the production of syngas and subsequently hydrogen. Nevertheless, this process currently faces obstacles towards large-scale development due to the overall low hydrogen yields and high tar content of the syngas [66]. These obstacles can be partially overcome by the use of catalysts which offer the capability to accelerate biomass conversion rates by lowering the activation energy and thus enhance the catalytic reforming process to obtain higher syngas yields.

Meanwhile, in addition to the direct gasification the pyrolysis of lignocellulosic biomass and the subsequent gasification/reforming of the pyrolysis products (mainly the pyrolysis oils and gases) is one of the considered strategies for the production of hydrogen.

During the pyrolysis process, the biomass feed is heated in the absence of an oxidizing agent and its thermochemical degradation takes place. The produced products are of solid, liquid and gaseous phase, with ratios depending on the reaction conditions (temperature, heat transfer rate, particle size, residence time) and the fuel feed composition. New technologies for the rapid pyrolysis of biomass (known as "fast pyrolysis") can be used for the development of compact and efficient systems to convert biomass into vapors which are largely condensed, with yields as high as 75% by weight of the moisture-free biomass. The resulting liquid bio-oil, also known as "biocrude", is a mixture of aldehydes, alcohols, acids, oligomers from the constitutive carbohydrates, lignin, and water produced during the dehydration reactions. Meanwhile, the non-condensable pyrolysis gases consist of several species, mainly CO, CO_2, H_2 and CH_4. One of the main advantages of the pyrolysis process is that it can be implemented to greatly increase the energy density of the fuel, from 1.5-14.8 GJ/m^3 for solid biomass to 20.4-26 GJ/m^3 (for liquid bio-oil). Thus it can contribute to decreased transportation and storage costs as well as improved handling [67].

Hydrogen can be then produced by reforming the biocrude or the pyrolysis gases with steam. This process has the potential to be cost competitive compared with conventional means of hydrogen production. The use of a renewable resource feedstock is an alternative to conventional means of hydrogen production from petroleum and natural gas. As a result, the environmental benefit of zero net carbon dioxide emissions enables the implementation of CO_2-neutral energy systems [68]. Different pyrolysis-based, biomass-to-hydrogen conversion pathways have been investigated by various authors. Although the majority of these studies are oriented towards steam reforming of bio-oil to produce H_2-rich syngas, some focus on the reforming of the pyrolysis non-condensable gases, while others include the gasification of bio-char.

A number of studies have focused on the experimental investigation of different catalysts for the steam reforming of pyrolysis products in order to produce gas of high H_2 content. Pyrolysis can be carried out in various reactor types, such as entrained flow, fluidized (bubbling or circulating) or microwave reactors. Therefore, most of the existing literature studies include experimental tests aiming to explore the impact of certain reaction conditions and catalysts on the hydrogen yield, considering various reactor types and biomass feedstocks. A literature review presenting some indicative research studies on currently investigated methods follows.

Chen *et al.* [69] experimentally investigated an integrated process for hydrogen production based on the simultaneous (instead of serial) gasification and catalytic conversion (utilizing the commercial catalyst NiO/MgO) of pyrolysis gases and char derived from cotton stalks in an entrained flow bed reactor. The authors observed that the parallel conversion process (gas and solid) resulted in an increase of the carbon conversion from 78.84% to 92.06% compared to the two stage process involving first pyrolysis and subsequent catalytic steam reforming in distinct steps, accompanied by a change of the distribution of the tar components. Finally, the authors estimated that the potential H_2 yield increased from 47.71 to 78.19 g per kg of cotton stalk.

In another study, Chen *et al.* [66] explored the behavior of Ni/CaAlOx catalyst for the production of H_2 rich syngas from the two stage pyrolysis and steam reforming of biomass sawdust. The authors justified the use of Ca to promote the thermochemical conversion of biomass due to its low cost and high availability. The impact of different molar ratios ranging from 1:3 to 3:1 of Ca:Al on the syngas production as well as the coke formation on catalysts' surface was examined. The authors concluded that while it is in fact possible to manipulate the CO/H_2 molar ratio by varying the Ca content in the catalyst within a range from 1.01 to 1.63, the H_2 content remained constant at around 45% vol. They also pointed towards a dependence of the total gas yield on the surface area of the catalyst, while they observed that the $Ni/CaAlO_x$ ratio of 1:3 exhibited the lowest gas production, mainly due to the extreme catalyst deactivation occurring due to coke depositions. On the other hand, a total concentration of 90% vol. syngas (H_2 and CO_2) was obtained with a ratio of 3:1. Consequently, the authors suggested that the investigated catalyst can be used for controlling the composition of the syngas.

Czernik and French [70] investigated the distributed production of hydrogen by auto-thermal reforming of fast pyrolysis-derived bio-oil, using a 0.5% Pt/Al_2O_3 catalyst. The biomass species considered for the production of bio-oil were oak, poplar, pine and lignin-free oak. For a reforming temperature was 800 °C to 850 °C, while the steam-to-carbon ratio was 2.8-4, with an oxygen-to-carbon ratio of 0.9-1.1. The yield of the produced hydrogen was equal to 9-11 g per 100 g of fast pyrolysis oil, corresponding to 70%-83% of the stoichiometric potential. The authors noted that the elemental composition of the bio-oil as well as the conversion ratio of bio-oil carbon to gas, ranging from 70% to 89% had the most important effect on the hydrogen yield. The authors also concluded that the implementation of this process in a large scale plant (1500 kg/day), would result in a production cost of 4.26 $/kg (USD 2013), with the most significant cost component being the bio-oil cost, amounting to 56.3% of this cost.

Xu *et al.* [71] studied the dry and steam reforming of biomass pyrolysis gas for the production of H_2 enriched syngas by using Ni/Fe/Ce/Al_2O_3 catalysts. The effect of different operating conditions, such as the calcination temperature of the catalysts, the reaction temperature, the gas hourly space velocity and the H_2O/C ratio was investigated, regarding the CH_4 and CO_2 conversion as well as the selectivity of hydrogen. The authors identified a reaction temperature of 600 °C, a velocity of 900 h^{-1} and a H_2O/C ratio of 0.92 as the optimal conditions, corresponding to concentrations of 36.80%, 10.48%, 9.61%, 42.62% and 0.49% for the gas species H_2, CO, CH_4, CO_2, and C_2H_n, respectively. The authors noted that although a multitude of reactions occur during the reforming of the pyrolysis gas, the dominant reactions are strongly dependent on the operating conditions.

Li *et al.* [72] carried out an experimental study of microwave pyrolysis for hydrogen production from rice straw biomass. The single and combined effect of the pyrolysis temperature (varied from 400 °C to 700 °C), the microwave power (varied from 1.5 kW to 2.5 kW), the feed moisture content (varied from 7.5% to 61.84%) and the the use of a catalysts (Na_2CO_3, $ZnCl_2$, CaO) on the hydrogen yield was examined. Based on the experimental results, it was observed that the H_2 yield at first increased and subsequently decreased as the pyrolysis temperature and microwave power increased, but monotonously decreased with increasing moisture content. The H_2 yield was maximized at 600 °C (36.50%) when no catalyst was used and when the microwave power was equal to 2.2 kW, with a moisture content of 7.5%. The non-catalytic reaction was found to produce the highest hydrogen yield, equal to 36.14%, while the lowest yield was observed when using $ZnCl_2$, equal to 27.31%.

Zin *et al.* [73] investigated the CO_2 sorption-enhanced hydrogen production potential by steam reforming pyrolysis oils of palm empty fruit bunches (PEFB) and pine wood. In their experiments, a mixture of calcined dolomite and nickel was used as a catalyst., while CaO was used as the sorbent material. The reforming temperature was equal to 600°C at atmospheric pressure. The authors firstly concluded that the sorption enhancement can theoretically lead to a substantial increase in the hydrogen yield and purity of the syngas based on equilibrium calculations considering the bio-oil as acetic acid. From the experimental results, the authors noted that near full sorption enhancements effects were observed at 600 °C with PEFB, while a significant degree of enhancement was attained for pine oil, in the presence of steam-activated calcined natural dolomite and a Ni-based catalyst. For PEFB, the sorption enhancement leaded to a very high H_2 purity of 96% vol and a yield of 10.5% wt of the fuel with the dolomite sorbent, compared to a purity of 67% vol and a yield of 9.51% wt without sorbent. For pine oil, the H_2 purity and yield demonstrated an increase of 60% and 41% respectively compared to the non-sorbent process.

Alvarez *et al.* [74] studied the effect of the addition of different plastic materials inside the reactor during the steam pyrolysis/gasification of wood sawdust with the combined effect of using a Ni/Al_2O_3 on the hydrogen content of the gaseous product stream. The plastics examined were polypropylene, high density polyethylene, polystyrene as well as real world plastics. A sensitivity analysis on the quantity of the added plastic material was carried out for polypropylene, varying its mass ratio in the initial feed stock from 5 to 20% wt. The optimal value was found to be 20%, leading to an overall gas yield of 56.9% wt along with a higher H_2 concentration and production, equal to 36.1% vol and 10.98 mmol g^{-1}, respectively. The authors attributed the improvement in the synergetic effects between intermediate pyrolysis species and the plastics. When the Ni/Al_2O_3 catalyst was used, the hydrogen yield was greatly increased, due to the enhancement of the water gas shift and methane reforming reactions. This result was interpreted by considering the exceptionally high activity for the C-C rupture and H_2 formation. Among the different plastics examined, although high density polyethylene exhibited the highest gas yield, polypropylene leaded to higher overall hydrogen production. The highest bio-oil and char yields were obtained with polystyrene, due to the higher reaction energy associated with cracking polystyrene in the pyrolysis stage.

In another work, Jiang *et al.* [75] studied the atmospheric pyrolysis of six types of biomass in the molten alkali compound NaOH, at temperatures ranging from 350 to 550 °C. The biomass species included fir, birch and redwood sawdust, in addition to rice and cole stalk and rice husk. In the case of NaOH, the produced syngas consisted only of two compounds, H_2 and CH_4. In general, the biomass type had relatively little influence on the H_2 content of the pyrolysis gas. However, it had a strong influence on the overall yield of the gaseous products. Redwood sawdust had the highest H_2 yield of 65.39 g/kg, while stalk and rice husk had lower yields of 40 g/kg. Furthermore, although the H_2 content of the pyrolysis gas was not particularly influenced by the reaction temperature, a temperature shift from 350 to 550 °C resulted in a H_2 yield variation from 30.7% to 66.5%. An additional parameter studied was the N_2 flowrate, which had little influence on both the hydrogen content in the pyrolysis gas as well as the overall hydrogen yield. By comparing the effect of NaOH in conjunction with different catalysts, it was shown that the addition of the molten salt increased the H_2 yield. Among the different catalysts (Ni, Al_2O_3, NaCl and $PdCl_2$), $NiCl_2$, showed the highest increase of the H_2 yield in the case of rice stalk, from 46.33 g/kg (no catalyst) to 67.34 g/kg.

Liu *et al.* [76] considered using a two-stage fixed reactor for the catalytic pyrolysis of biomass in order to produce hydrogen. He studied the effect of the reaction conditions as well as the nickel content in the catalyst. A positive correlation between the hydrogen yield and the catalytic bed temperature, the residence time and the amount of catalyst was observed. The optimal reaction conditions included a reaction temperature for the first-stage pyrolysis around 650-700 °C and a second-stage catalytic bed temperature of 800 °C. The optimal reaction time for the pyrolysis process was 17 minutes, while the optimal nickel content of the catalyst was equal to 9% wt.

Production from Water with Electrolysis

Hydrogen can be produced from splitting water molecules *via* electrolysis. The electrolysis process can in some cases be powered by electricity, while additional cases include photolysis and photobiological electrolysis.

Electrolysis

The electrolysis of water to produce hydrogen and oxygen is a process in which two electrodes in a basic electrolyte are connected to a direct current (DC) supply. If an sufficiently high cell voltage is applied to the cell, the redox reaction takes place producing hydrogen at the cathode and oxygen at the anode. The cathode is the negative electrode and the anode is the positive one (Fig. **11**).

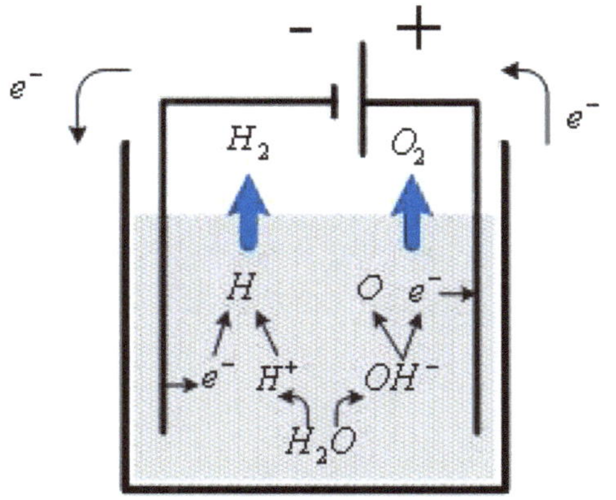

Fig. (11). Principle of water electrolysis.

One of the main advantages of electrolytic hydrogen production is that it yields the highest-purity hydrogen (up to 99.999%). Furthermore, it requires highly available raw materials (electricity and water), has simple system architecture and can be easily scaled economically to serve small and larger applications. The main disadvantage is the price of electricity, which can be seen as an expensive "fuel".

Electrolysis requires energy to take place, it is therefore an endothermic reaction. The overall reaction of water electrolysis can be expressed as:

$$H_2O(l/g) \rightarrow \frac{1}{2}O_2 + H_2 \quad \textbf{(14)}$$

A charge transfer is usually achieved by implementing a basic or acid electrolyte. This is necessary to enable the electron transfer from the DC source to the liquid solvent which is separate to produce hydrogen. Depending on the technology, the charge carriers can be hydroxyl ions (OH^-), protons (H^+) or oxygen anions (O^{-2}).

These carriers correspond to different types of alkaline electrolyzers, PEM electrolyzers and high-temperature steam electrolyzers respectively. Hydrogen and oxygen can either be co-produced or separately generated. In the latter case, separate cells are used for carrying out the electrolysis reactions at the anode and cathode of the electrolyzer.

The half-cell reactions at the cathode and anode for Alkaline electrolyzers can be expressed as:

Anode $$2H_2O + 2e^- \rightarrow H_2 + 2OH^- \quad \textbf{(15)}$$

Cathode $$2OH^- \rightarrow \frac{1}{2}O_2 + 2H^+ + 2e^- \quad \textbf{(16)}$$

The half-cell reactions at the cathode and anode for PEM electrolyzers can be expressed as:

Anode $$2H^+ + 2e^- \rightarrow H_2 \quad \textbf{(17)}$$

Cathode $$2OH^- \rightarrow \frac{1}{2}O_2 + 2H^+ + 2e^- \tag{18}$$

Finally, the half-cell reactions for High-temperature Steam electrolyzers can be expressed as:

Anode $$H_2O + 2e^- \rightarrow H_2 + O^{2-} \tag{19}$$

Cathode $$O^{2-} \rightarrow \frac{1}{2}O_2 + 2e^- \tag{20}$$

Thermodynamics

The amount of energy required for electrolysis is defined by the enthalpy of reaction ΔH_R which is equal to the enthalpy of water formation. The enthalpy of formation is defined as the difference between the enthalpy of formation of products and the enthalpy of formation of reactants as:

$$\Delta H_R = H_{f,H_2} + \frac{1}{2}H_{f,O_2} - H_{f,H_2O(l)} \tag{21}$$

Its value depends on the thermodynamics of the electrochemical reaction. At standard conditions of temperature and pressure (P=1.013 bar, T= 298 K), ΔH_R has a value of 285.8 kJ/mol.

When electrolysis is performed along a reversible path, the Gibbs free energy change for the thermodynamic system can be expressed as:

$$\Delta G_R(T) = \Delta H_R(T) - T\Delta S_R \tag{22}$$

where T (K) is the temperature of the electrochemical reaction and is assumed steady and ΔS_R is the entropy change in J/mol/K. Up to a temperature of 2250 °C, is a positive quantity, since the enthalpy term is predominant over the entropy difference term. At standard conditions, the reversible Gibbs free energy difference for liquid water and water vapor electrolysis are equal to 237.23 kJ/mol and 228.61 kJ/mol, respectively. Moreover, the enthalpy difference between the enthalpies of formation of water and water vapor (difference between HHV and LHV) is equal to the latent heat. The term $T\Delta S_R$ denotes the limit of energy that can be supplied as heat or thermal energy to the process. When water vapor is fed into the electrolyzer, the process energy demand is increased at higher temperatures. In other words, the share of the $T\Delta S_R$ product becomes significant in the equation.

The minimum cell voltage which has to be applied for the electrolysis of water can be derived from the equation

$$\Delta G_R^0 = zFE^0 \tag{23}$$

where E^0 is the standard, reversible, thermodynamic voltage (V), z is the number of electrons transferred per molecule of hydrogen produced and F is the Farraday constant (96485 C/mol). For liquid water at 25 °C, the reversible voltage has a value of 1.229 V. This calculation is based on the fact that two electrons are transferred (z=2) and that ΔG_R^0=237.2 kJ/mol. The reversible voltage depends on both temperature and the thermodynamic activity of each component in the system, as shown by the Nerst equation. The partial pressures can be used for gases and the activity for water is assumed to be close to 1. The Nerst equation can be written in the following form

$$E = E^0 + \frac{RT}{zF} ln \frac{P_{O2}^{\frac{1}{2}} P_{H_2}}{a_{H_2O}} \tag{24}$$

As shown in the Nerst equation, an increase in system pressure yields an increase in the reversible voltage. This may seem as a detrimental effect of applying pressure in the electrolyzer, however, the increase in pressure ultimately compensates for the mechanical compression of hydrogen after the electrolyzer. Moreover, the increase in E_{rev} is minor and therefore, pressurized electrolysis is widely accepted as one of the state-of-the-art options for hydrogen production.

The combined effect of pressure and temperature is shown in Fig. (**12**). Increasing the system pressure from 10 bar to 20 bar has only a minor effect on the total increase of voltage required for water electrolysis to take place.

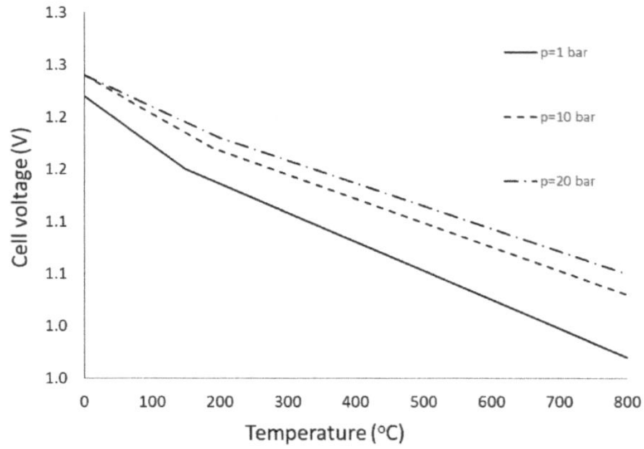

Fig. (12). Effect of temperature (oC) and pressure (bar) on equilibrium cell voltage Erev (V) for water electrolysis [77].

Since the definition of reversible cell voltage only applies for cases in which the energy is supplied as both heat and electricity, another term has been used to refer to the minimum cell voltage when energy is supplied only in the form of electricity. This term is called thermoneutral cell voltage E_{th} and it is calculated by the equation

$$E_{th} = \frac{\Delta H_R}{zF} \tag{25}$$

For water electrolysis at standard conditions (=285.8 kJ/mol), E_{th}=1.481 V.

Losses and Conversion Efficiency

Irreversibilities during hydrogen production by electrolysis cause a drop in voltage supplied in the system, since the reversible voltage which is defined by the aforementioned equations only applies for equilibrium and fully reversible operations. Therefore, additional work in the system in the form of electricity is required in order to carry out the endothermic reaction. Therefore, the non-ideal cell voltage requirement is given by the equation:

$$E_{real,cell} = E_{rev} + |n_{an}| + |n_{cat}| + ir_a \tag{26}$$

In this equation, nan and ncat express the losses at the electrodes of the electrolyzer, while the term ira represents the effect of ohmic losses in the system.

More specifically, the losses can belong to the following categories:

- Losses due to the evolution of hydrogen and oxygen, which disturb the effective area at the electrode area and cause increased current density.
- Ohmic losses due to the ionic resistance of the electrolyte. Moreover, electrical resistance of electrodes, bipolar plates and collectors cause ohmic losses in the electrolyzer.
- Losses due to mass transfer limitations, basically due to charge carrier depletion at the electrodes.

Concerning the efficiency of the electrochemical conversion in the electrolyzer, different definitions can be found in literature. Assuming that hydrogen is the useful output from the electrolyzer and that electrical work is applied for the electrochemical reactions at low temperature electrolyzers, the efficiency of energy conversion can be written as:

$$\eta = \frac{\dot{n}_{H_2} \Delta H_R{}^0}{P_{el}} \tag{27}$$

And in HHV (for liquid water) in the form

$$\eta = \frac{\dot{n}_{H_2} HHV_{H_2}}{P_{el}} \tag{28}$$

In this equation, the term P_{el} denotes the total power supplied to the stack (P_{stack}) and to the auxiliary equipment and devices of the system as well (P_{aux}). Therefore

$$P_{el} = P_{stack} + P_{aux} = V_{stack} I_{stack} + P_{aux} \tag{29}$$

This expression of efficiency assumes that only power is supplied to the electrolyzer to drive the reactions. This means that the equation is fully applicable to low temperature electrolyzer technologies, such as the state-of-the-art alkaline and PEM electrolyzers. When heat is also supplied to the system, the efficiency in terms of energy flow is represented as

$$\eta = \frac{\dot{n}_{H_2} HHV_{H_2}}{P_{el} + \dot{Q}} \tag{30}$$

Another definition of electrolyzer efficiency is the voltage efficiency (η_v). This definition accounts for efficiency as the ratio of minimum theoretical cell voltage at the operating temperature $E_{th}(T)$ to the real cell voltage $E_{cell}(T,i)$. The former can have the value of either reversible cell voltage or the thermoneutral voltage, depending on the operation temperature and ultimately on the electrolyzer type. The latter one is a function of both temperature and current density and the equation has the following form:

$$\eta_V = \frac{E_{th}(T)}{E_{cell}(T,i)} \tag{31}$$

Another common definition of the electrolyzer efficiency is the Faradaic efficiency denoted as η_I. This definition is adopted to describe the losses of energy due to electrolysis of impurities present in the electrolyzer, as well as the effect of stray currents in the stack and diffusion issues in the reactor. The Faradaic efficiency is the ratio of the volume of hydrogen gas produced at real conditions to the theoretical hydrogen volume that can be produced according to the Faraday law at a specified time interval. This definition can be represented as

$$\eta_I = \frac{\dot{V}_{H_2,real}}{\dot{V}_{H_2,ideal}} = \frac{\dot{n}_{H_2,real}}{I(2F)^{-1}} \tag{32}$$

Finally, the overall cell efficiency can be calculated as the product of voltage efficiency and current efficiency:

$$\eta_{cell} = \eta_V \eta_I \tag{33}$$

Types of Electrolyzers

As previously described, the main technologies for hydrogen production through electrolysis include alkaline electrolysis, PEM electrolysis and high-temperature steam electrolysis, employing the respective electrolyzer types. The optimal electrolyzer selection and design point from an economic perspective depends on the current density, the cost of the production materials of the electrolyzer as well as the demand for hydrogen production.

Alkaline electrolyzers operate *via* transport of hydroxide ions (OH^-) through the electrolyte from the cathode to the anode with hydrogen being generated on the cathode side. They operate at temperatures in the range from 40 to 90 °C [78].

A liquid electrolyte, such as a 25% potassium hydroxide (KOH) solution is used. This is because of KOH's high conductivity, and because the oxygen evolution reaction has the least energy loss in this solution. These electrolyzers do not require precious metals and typically use nickel electrodes. The electrolyzer units can be either unipolar or bipolar. A unipolar electrolyzer resembles a tank and has electrodes connected in parallel. This electrolyzer design is a high-current, low-voltage system with a single bus bar connecting all the anodes and another connecting all the cathodes. A membrane is placed between each cathode and anode; this separates the hydrogen and oxygen as the gasses are produced but allows the transfer of ions. The bipolar alkaline electrolyzer resembles a filter press. Electrolysis cells are connected in series; hydrogen is produced on one side of each cell and oxygen on the other side. The bipolar electrolyzer is a high-voltage, lower current device, and a membrane separates the electrodes. Most commercial alkaline systems use the bipolar design [79].

Alkaline electrolyzers are most effective when operating at lower capacities with lower current densities and are best suited for larger systems connected to a power grid. Although they have been commercially available for many years, newer approaches using solid alkaline exchanger membranes as the electrolyte have been recently shown to be promising on the lab scale [80].

Another of electrolysis unit is a solid polymer electrolyte (SPE) electrolyzer, which are also referred to as proton exchange membrane or polymer electrolyte membrane (PEM) electrolyzers. PEM electrolyzers operate at similar temperatures than alkaline, ranging from 20 to 100 °C [78]. In these units the electrolyte is a solid ion conducting membrane and the electrolyzer is fed with pure water. The traditional membrane is Nafion® and consists of a Teflon®-like polymer with attached sulfonic-acid groups. The membrane allows the H^+ ion to transfer from the anode side of the membrane-where oxygen is produced-to the cathode side where it forms hydrogen. The SPE membrane then also serves to separate the hydrogen and oxygen gases. This is effectively an acid environment, therefore significant precious metal (Pt, Ir, Ru) loadings are used. PEM electrolyzers typically are configured in the bipolar mode.

One of the main advantages of PEM electrolyzers is their capability to operate at high current densities, in contrast to alkaline ones [81], thus having lower operational costs, especially when coupled with variable electricity inputs that originates for example from renewable energy (such as solar and wind).

High temperature steam electrolysis (HTSE) is in principle similar to room temperature water electrolysis. The main difference is that part of the required reaction energy is provided in the form of heat, which is less expensive than electricity. Moreover, the electrolysis reaction is more efficient at higher temperatures, thus producing higher yields of hydrogen. Steam electrolyzers use ceramic ion-conducting electrolytes.

Owing to material durability limitations, there is an optimal temperature range that is suitable for this type of electrolysis, which is around 700 to 900 °C [82]. HTSE units consist of a stack of several single repeat units (SRUs), as seen in Fig. (**13**). SRUs are in turn comprised of the electrochemical cell, (also known as solid oxide electrolysis cell or SOEC), which is the main component of the electrolyzer in which the electrochemical reaction takes place. The SOEC is composed of three ceramic layers: a dense electrolyte and two porous electrodes (cathode and anode) on its sides. In addition to this, the stack includes connections to carry the electric current that ensure sufficient fluid distribution for every cell and separate the anode and cathode of adjacent cells. Lastly, seals are used for tight-proofing the stack from the external environment.

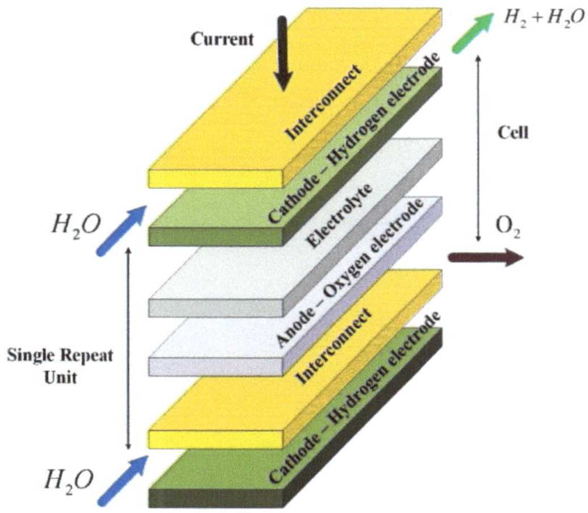

Fig. (13). Schematic representation of an SRU.
Reference for caption: The AGRAL Project, Hydrogen and Fuel Cell application, link: https://www.agral-project.com/the-agral-project/hydrogen-and-fuel-cell-application/ (accessed 06/09/2018).

As previously mentioned, energy needs for HTSE are lower compared to low temperature water electrolytes, enabling the operation under more favourable *i-V* regimes and leading to improved efficiency. HTSE is also efficient with respect to hydrogen production rates, due to increased current density. For example, while for alkaline and PEM water electrolysis the efficiencies are 62% and 68% respectively, they can reach values of 96% in the case of HTSE. Of course, HTSE is not without disadvantages. These include the heat losses that can occur at system level, as well as considerable degradation rates when long duration operation is considered (above 25000 h).

Due to their higher cost compared to alkaline and PEM electrolysis, no commercial HTSE systems are still available. However, cost of HTSE hydrogen is expected to drop as the technology becomes more commercialised and units are mass produced.

Solar Electrolysis and Photolysis

Solar Electrolysis

Because of their limited nature and non-homogenous distribution, fossil fuels are not expected to keep up with the increasing global energy demand, while at the same time their reserves are getting less and less easily accessible. Based on the above considerations, the synthesis of hydrogen is becoming significantly attractive as a means to produce a CO_2 neutral fuel with certain advantages such as high energy conversion efficiency, production from water without any emissions, abundance, flexibility of storage, effective long distance transportation, high heating value and ease of conversion to other forms of energy [83]. Electrolysis systems for hydrogen production can be coupled with photovoltaic panels in order to utilize solar energy for powering the process. These systems are capable of more flexible operation, since they can alternate between electricity generation or hydrogen production, depending on the availability of the solar radiation and the electricity demand. A literature review of such applications follows.

Koumi Ngoh *et al.* [84] carried out a techno-economic analysis of a hybrid hydrogen production system utilizing solar photovoltaic and thermal energy. The system combined parabolic trough collectors (PTC) used to generate hot pressurized steam at 320 °C, which was subsequently electrolyzed to H_2 and O_2 by electricity produced by photovoltaic panels. The production cost of the system was estimated at 5235.45 $/t of hydrogen.

Ahmadi *et al.* [85] proposed a novel proton exchange membrane (PEM) system for hydrogen production powered by an ORC utilizing solar and ocean thermal energy. The net power output of the system was 102 kW with a hydrogen production rate of 1.2 kg/h. The exergetic efficiency of the system was estimated at around 22%. The components causing the highest exergy destruction were the condenser and the turbine of the ORC. Moreover, he showed that the increase of the solar radiation intensity leads to the increase of the overall exergy efficiency. Meanwhile, the increase of the ambient temperature increases the efficiency when it is lower than 292 K and it decreases it when it has higher values.

Ratlamwala and Dincer [86] assessed two solar driven systems for the production of hydrogen by carrying out energy and exergy analyses. The first system included a solar heliostat field with a Cu-Cl and a Kalina cycle. The second system was similar to the first, with the exception of the inclusion of a photocatalytic reactor for producing hydrogen. The Cu-Cl cycle was considered by the authors due to its capability of low temperature operation (around 530 °C) and the reduced material and maintenance costs. The total heliostat field had an area of 2000 m^2 and an efficiency of 75% and the low and high operating pressures of the Kalina cycle were equal to 2.5 and 4 bar respectively. The authors concluded that when the solar radiation intensity increases, the hydrogen production increases from 126.9 L/s to 289.4 L/s and 154.1 L/s to 343.9 L/s. The exergetic efficiency of the systems increased from 45.6% to 47.79% for the first one and from 54.94% to 56.41% for the second. Moreover, although the ambient temperature has no impact on the energy efficiencies of the systems, its increase leads to an increase of their exergy efficiency from 47.98% to 50.82% and 56.87% to 59.64%.

Ozcan and Dincer [87] extended the scope of the previous study by considering a solar driven Mg-Cl hybrid system for producing electricity and hydrogen. A concentrating solar thermal power cycle with energy storage, a steam power plant with reheating and regeneration and a Mg-Cl hydrogen synthesis cycle constitute the overall system proposed by the authors. Considering solar energy as the heat source of the system, the overall system energy and exergy efficiencies were estimated equal to 18.8% and 19.9%, which are significantly lower compared to a heat input system consisting of molten salts. The highest exergy destruction of the system takes place in the solar field, accounting for 79% of the total system irreversibility.

Hacatoglu *et al.* [88] performed an exergetic assessment of a hybrid solar hydrogen system with activated carbon storage. The system includes hybrid PV/T cells for making use of the solar radiation. The hydrogen is produced by electrolysis and is stored in a cluster of tanks containing carbon. A simple "base case" system was considered and it was subsequently compared to an enhanced system with multiple outputs. The authors recognized that solar photovoltaic-based systems have the lowest exergetic performance, with efficiency values in the range between 14% and 18%. They therefore consider these systems as having the highest potential for further improvement. From the comparison of the base case and the modified systems, the authors concluded that the exergy efficiency values significantly differ, being equal to 4% and 11%, respectively. They also point out the environmental benefits of the hybrid system, which, as they state, may counter balance the economic obstacles for its adoption.

Photolysis

Another discussed method for the production of hydrogen by utilizing solar energy is the solar photolysis of water [89 - 91]. Photolysis constitutes an alternative to electrolysis, in which light is directly used for splitting water to hydrogen and oxygen. Special semiconductors called photoelectrodes are used to absorb solar radiation and simultaneously generate the required voltage for the process. Hence, if the photoelectrode is submerged inside an aquatic electrolyte exposed to solar radiation, electricity is produced to power the reactions of hydrogen and oxygen.

Contrary to the applications involving electrolysis powered by solar-derived electricity from intermediate PV panels, the aim of solar photolysis systems is the direct chemical conversion of water into O_2 and H_2 by employing photoelectrode cells, in which light harvesting and the generation of electric charge carriers and migration towards the external circuit for water cleavage take place [92]. It is estimated that for the production of 1 kg of hydrogen an energy input of 39.4 kWh is required [93].

The reaction is directly affected by the type of the photoelectrode used and the radiation intensity, which produces currents from 10 to 30 mA/cm^2. For these intensity levels the required electrolysis voltages is around 1.35 V [94]. These systems offer the future potential for greatly reducing the cost of electrolytic hydrogen compared to conventional technologies. Although still at a research level, photolysis presents promising efficiencies and competitive costs. Some of the major challenges currently that it is currently facing are expected to be overcome with progress in material science, bringing about the discovery of more efficient materials.

Conibeer and Richards [95] compared a PV/electrolyser with various proposed photoelectrolytic technologies for solar hydrogen production. Although they pointed out the simplicity, reliability and applicability of photoelectrolytic systems, they concluded that for the time being, the PV/electrolysis pathway is more appropriate having a higher efficiency, due to its maturity and due to the major challenges faced by photoelectrolytic systems. Owing to its promising perspectives, photoelectrolysis as a means to produce hydrogen is currently under extensive basic research and critique [89], with many different strategies involving various photoelectrode types and arrays being proposed [92]. Most current investigations are focused on the search of new semiconductor materials that are capable of absorbing visible light radiation that are necessary for the application of the technology, with TiO_2 being the most widely used [91, 96]. Other photoelectrodes reported include SiO_2, Al_2O_3, In_2O_3 and RuO_2 [97]. Varghese and Grimes [92] stress out the importance of standardizing the evaluation procedures for determining the efficiency of different cells. Furthermore, they concluded that realistic solar photoconversion efficiencies can be predicted by using the incident photon to electron conversion efficiency (IPCE) values in the energetic analysis of such systems. For the time being, most studies are proof-of-concept oriented and focus on the experimental testing of different cells and integrated energy systems have not been considered. Reported photoelectrolytic cell efficiency values (solar to H_2 efficiency) range between 1.5, 4.6, 7 and 15% [95, 98 - 102]. Nevertheless, so far there is scarcity of reported energetic and especially exergetic efficiencies of such systems, although they may seem promising for a sustainable "hydrogen economy".

Photobiological Electrolysis

Photobiological hydrogen production is achieved from microorganisms that convert water and organic matter into hydrogen through their biological processes. The microorganisms are usually green microalgae or cyanobacteria and use sunlight to split the water molecules to atoms of oxygen and hydrogen. A major obstacle of the technology at its current phase is the fact that the production of hydrogen is impossible to maintain for longer than a few minutes [103]. Another issue is the production of oxygen which inhibits the reaction of producing hydrogen and can pose a safety risk when mixed with hydrogen at certain concentrations [104].

The technology is currently not competitive compared to others because of its significantly high investment cost and low efficiency. A lot of research is still required before its successful commercial implementation. Nevertheless, it could be ultimately a reliable method for renewable hydrogen production.

Hydrogen Storage

Gas Storage

The most common method of hydrogen storage in gaseous phase is in steel tanks, although lately there is shown an increasing interest towards the composite tanks, which are structured from such materials to withstand the heavy pressure loads. Cooling of hydrogen down to temperatures close to cryogenic is another alternative, enhancing the energy density of gaseous hydrogen. A more innovative method for the storage of hydrogen in gaseous phase is the use of glass microspheres.

Steel and Composite Tanks

Use of steel tanks for hydrogen storage, as already mentioned, is a well proven method. Most suitable materials for hydrogen storage applications are mainly austenitic stainless steel. In steel tanks, hydrogen is usually compressed to pressures of 200 to 350 bar. With the use of composite tanks, the pressure ranges has increased up to 700 bar. The tendency of hydrogen to adsorb and dissociate at the material surfaces can result in embrittlement and diffusion, which limits the range of materials that can be used for these applications.

Tanks of made of steel or as called type I containers offer a satisfying solution in terms of strength and safety, but disadvantage due to their high weight. In order to reduce the weight problem, steel has been replaced by composite containers, while to ensure gas tightness a thin metal layer has been added. Type II tanks include a mesh partially made of carbon fibers to provide stability, while Type III tanks have a mesh fully made out of carbon fibers. Regarding Type I and II tanks, the mass of metals that is required to preserve the pressure limits the hydrogen storage to 1-2% of the vessel's mass, making this types impractical for mobile applications [105]. Finally, Type IV tanks (Fig. **14**) have a synthetic liner. In general, composite tanks achieve lower weights at the expense of increasing cost. In terms of comparing type III and IV tanks, type III tanks lead to lower gas temperature [106, 107].

Fig. (14). Compressed gaseous hydrogen vessel.
Wikipedia, Hydrogen tanks for Toyota Mirai, link: https://en.wikipedia.org/wiki/Hydrogen_tank#/media/File: Hydrogen_tanks_for_Toyota_Mirai.png (accessed 10.10.2018).

Use of composite tanks for hydrogen storage poses several advantages. Having been field of extended research over the past years, they are now in commercial level. There is no requirement for heat transfer on their interior and thus can be used down to cryogenic temperatures as well. Among their main disadvantages are their large volume requirement, which even at such high pressures is approximately two times higher than that for the storage of the liquid hydrogen; the high cost which is up to 150 €/kWh for type IV tanks and the energy costs that are a consequence for the compression to such high pressure levels [108]. Furthermore, there are still several safety issues that have to be resolved, including the rapid loss of hydrogen during a breakdown and the lack of knowledge considering the impact of hydrogen in the properties of the tank materials in cryogenic conditions. This rapid loss of hydrogen increases the risk of explosion, posing a huge drawback for the use of this method for on board storage applications [109].

Glass Microspheres

Glass microspheres are produced from a glass powder, mixed with a blowing agent, which by increasing rapidly temperature decomposes causing expansion from the interior resulting in the formation of the microspheres. Storage of hydrogen in glass microspheres consists of three main steps: charging, filling and discharging. Initially, the empty glass microspheres are filled with hydrogen under high pressure (350-700 bar) and high temperature, of approximately 300 °C, through permeation in a high pressure tank. Afterwards, the glass microspheres are cooled down to ambient temperatures and transferred in a low pressure tank. In the final step, microspheres are reheated in temperatures of 200-300 °C in order to control the hydrogen losses.

Glass microspheres offer a safe mean of hydrogen storage at relatively low pressures for automotive applications. Their hydrogen volumetric capacity is reported to be up to 15 g/L and their gravimetric capacity of 17% [111]. Main disadvantages of this technology are the low density of the microspheres and the high pressure that is required for their filling. Furthermore, microspheres tend to have extended hydrogen leakages at ambient temperatures. However, the main challenge of this technology is the demand for heat at temperatures higher than the ones that a PEM fuel cells can provide, while the stability of the system is also difficult to control at high temperatures [43].

Liquid Storage

Most common method for hydrogen storage in liquid phase is by cooling hydrogen down to cryogenic temperatures (-253 °C). Alternatively, hydrogen can be dissolved and stored into liquid solutions, such as $NaBH_4$, rechargeable organic liquids or in anhydrous NH_3.

Cryogenic Liquid Hydrogen

In principle, the mass concentration in this method can reach 100%, however, the existing applications are limited only up to 20%. The corresponding values in terms of density are 80 kg/m^3 and 30 kg/m^3, respectively. Jorgensen [112] reported the optimum value for the liquid hydrogen storage of automotive applications to be equal to 15%. Approximately 30-40% of the reported losses are occurring during the production of liquid hydrogen [113]. Heat transfer has to be minimized in the tank; thus the highly insulated tanks consist of an inner tank and an outer container with an insulated vacuum layer in the space between. The requirement of extensively insulated cryogenic tanks -an average insulation thickness of 3 cm- is a main drawback of this method, especially in terms of long term thermal fatigue. A low surface to volume ratio for the tank is key to minimize heat transfer and thus cylindrical shaped tanks are preferred.

Boil-off of hydrogen is a main challenge for this technology [7]. The inevitable heat losses of hydrogen result in evaporation phenomena in the tank and increase in the pressure. Therefore, liquid hydrogen storage tanks should be equipped with pressure relief system and safety valves. Losses due to evaporation range between 0.3% and 3% in today's applications [105, 114]. On the other hand, liquid hydrogen provides high storage densities achieved in relatively low pressures, usually only slightly above the atmospheric. Liquid hydrogen has already found application in few commercial vehicles, while in the future is expected to be used as an aircraft fuel. The storage density is generally expected to increase in pressure vessels since liquid hydrogen is slightly compressible [115]. A liquid nitrogen storage tank installed at NASA's facilities is shown in Fig. (**15**).

Fig. (15). Liquid hydrogen tank at NASA's Kennedy Space Center [116].
Liquid Hydrogen tank at Cape Canaveral, link: https://www.flickr.com/photos/warriorwoman531/8155235271/in/photostream/ (accessed 10.10.2018).

An alternative that reduces the effect of boil-off is the cryo-compressed tank, which operates at cryogenic temperatures under high pressures. Aceves *et al.* [117] investigated the performance of a cryogenic capable pressure vessel and reported a hydrogen mass fraction of 7.38%. Cryo-compressed tanks are light and capable of fast refueling. An advantage of this method is that is less energy intensive compared to liquid hydrogen. Their main drawback is their significantly high cost, which for the moment is 3-4 times higher than the compressed hydrogen tanks [118].

NαBH$_4$ Solutions

Storage of hydrogen in $NaBH_4$ solutions is based on the following catalytic hydrolysis reaction:

$$NaBH_4(l) + 2H_2O(l) \rightarrow 4H_2(g) + NaBO_2(s) \quad \textbf{(34)}$$

Theoretical maximum energy storage density is limited to 10.9 wt.% in hydrogen [43]. This method provides a safe and controllable way to produce hydrogen in a vehicle. Recovery of $NaBH_4$ is an issue, since the resulting $NaBO_2$ can be converted back to $NaBH_4$ only under certain conditions, resulting in a high cost for the recovery of $NaBH_4$. As long as no other method is found for the recovery of the solvent, this method is considered highly unlike to be used in large scale.

Storage in Solid Materials

Hydrogen storage in metal hydrides has the potential to become a safe and efficient energy storage system, for both stationary and mobile applications [119]. There are at least three main types of metal hydrides: the ionic hydrides, the covalent hydrides and the metallic hydrides. In ionic hydrides the metal atoms exist as cations and the hydrogen exists as H^- anions. They are highly exothermic and have salt-like properties [42]. Covalent hydrides involve relatively weak, shared-electron bonds. Most of them are quite unstable. Metallic hydrides are formed by transition metals, in which hydrogen acts as a metal and forms metallic bonds with them [120]. They present higher storage density than gas and liquid phase hydrogen storage methods. One challenge for the use of metallic hydrides as a hydrogen storage method is the fact that their surfaces are usually covered with an oxide layer that does not allow hydrogen to access the metal and form bonds. Further challenges of these technologies include the existence of impurities in the hydrogen stream which can lead to poisoning, agglomeration of particles and phase change [108].

Carbon nanotubes, which were introduced in 1991 by Ijima [121], are an alternative for achieving high energy densities. Single-walled carbon nanotubes (SWNTs) include a single graphene layer and have a diameter of few nanometers and length in the scale of microns [122]. Due to van der Waals interactions between single tubular nanostructures, SWNTs typically are used in bundles of tubes [108]. Multi-walled carbon nanotubes (MWNTs) consist of a maximum of 50 tubes. On the other hand, carbon nanofibers consist of graphite platelets stacked together with an interlayer spacing in various orientations [123]. They are formed from the catalytic thermal decomposition of hydrocarbons.

During the previous decade carbon nanotubes and nanofibers attracted a lot of interest due to several reports claiming to achieve very high storage capacities, Fig. (**16**). As it was proven, these results were a product of experimental errors [110]. At extreme conditions of 500, bar Jordá-Beneyto *et al.* [124] reported a maximum storage capacity of 2.7 wt.%.

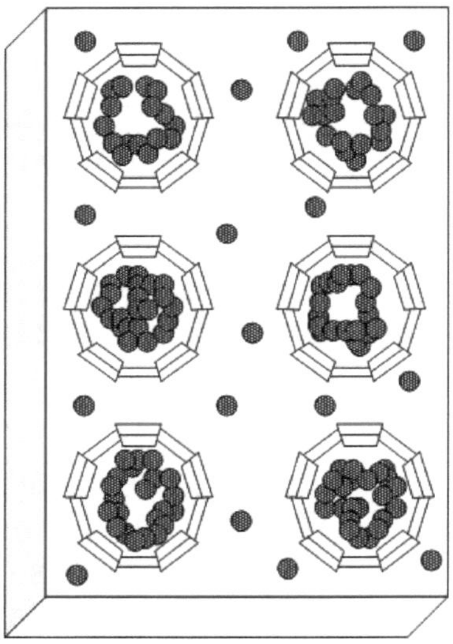

Fig. (16). A configuration of H_2 molecules adsorbed on a triangular array of carbon nanotubes [125].

Methanation Storage

A different way to store hydrogen in addition to conventional technologies is the conversion of hydrogen to other fuels like methane, substitute natural gas (SNG), ethanol, methanol, *etc.* The conversion of hydrogen to methane can be achieved through the methanation process *via* the Sabatier reaction, which can be expressed as:

$$CO_2 + 4H_2 \rightarrow CH_4 + 2H_2O \quad \textbf{(35)}$$

During this process, amounts of hydrogen and carbon dioxide are firstly converted into methane and subsequently to SNG. It should be noted that the process requires large amounts of hydrogen to take place. Unlike to the methanation of synthesis gas, wherein the required hydrogen is available from the gasification of biomass, in this case the required hydrogen is applied *via* the process of Power-to -Gas (PtG) where hydrogen is produced from electrolysis of water using excess electricity from renewable sources. The methanation technologies used for the conversion of H_2 and CO_2 to methane are the thermochemical (catalytic) and biological methanation.

A key advantage of this storage technology is the possibility of channelling the amount of hydrogen that is converted to an existing gas distribution network. In this way, the energy can be stored for a long period of time and transported with low costs, as no capital investment is required for dispensing. Another advantage of this process is the parallel storage of carbon dioxide (CO_2), which is required for the conversion of hydrogen.

Methanation Process

Methanation is an exothermic process. In its general form, a gas with a low methane content converted into a gas of high methane content. For the methanation of synthesis gas from the gasification of solid fuels, such as biomass and coal, the process can be both, thermochemical (catalytic) and biological, with the catalytic methanation to be the process that is mainly used up to now for the methanation of synthesis gas. Biological methanation mainly applied to gases from other processes such as biogas, originating from digestion of organic material, separated carbon dioxide from flue gases, hydrogen generated by electrolysis and others. A brief overview of the catalytic and biological methanation processes follows.

Catalytic Methanation

Thermochemical methanation is defined as the conversion of hydrogen and carbon monoxide or carbon dioxide into methane in the presence of catalyst (usually nickel based). Typical process conditions are temperatures in the range from 300 to 550°C and pressures from 1 to 100 bar.

On a molecular level, the pathway of the reactant molecules is illustrated by Fig. (**17**). When the reactants are in the particle boundary layer of the catalyst (1) they diffuse (2) into the catalyst because the concentration of the reactants inside the catalyst is lower than outside. Reactants are adsorbed on an active site (3) directly with strong bonds, or *via* weaker bonds followed by strong bonds. Binding with strong bonds is called chemisorption and binding with weak bonds is called physisorption. When four hydrogen molecules and one carbon dioxide molecule have been adsorbed (4) they undergo the Sabatier reaction and desorb (5) from the catalyst surface as methane molecules and then diffuse out (6) to the boundary layer (7) and leave the catalyst [126].

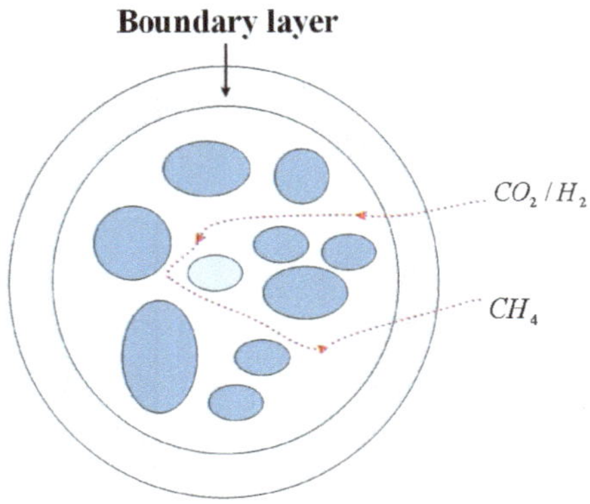

Fig. (17). Catalysis process where reactant gases diffuse into active sites where the Sabatier reaction occurs [126].

The process of catalytic methanation is carried out in reactors which contain appropriate catalysts as mentioned previously. The reactors which are applicable for such processes are fixed-bed reactors, fluidized bed reactors and three phase reactors. The fluidized bed reactor and the fixed bed reactors are known and already applicable technologies, while the other reactor technologies are still in the development stage.

Biological Methanation

In this process, microorganisms serve as biocatalysts, converting the hydrogen and carbon dioxide to methane. Such microorganisms are called methanogens, which produce methane as a metabolic byproduct in anoxic conditions. These biological reactions are known since 1906 [127], however, the technical implementation still faces technical difficulties. Biological methanation (BM) is performed at low temperatures of about 40 °C to 70 °C, which makes the process simple and therefore ideal for small-scale facilities. Moreover, the micro-organisms have a high tolerance to impurities such as hydrogen sulfide (H_2S), organic acids and ammonium (NH_4^+) which are commonly found in gases used for methanation [128]. The sudden change of load is also not an issue for the biological methanation process, something very important for the utilization of electricity from renewable sources.

The biological methanation is a part of the process of biogas production. The biogas production process shown in Fig. (**18**) consists of different reactions numbered in the figure. The degradation of the substrate is performed in several steps and the methanation is to be performed throughout the reactions 4 to 7. The methane formation is performed from methanogenic archaea that convert carbon dioxide and hydrogen or acetic acid (CH_3COOH) to methane. The various steps are carried out by organisms that use the energy released when the longer carbon chains are broken down into smaller ones.

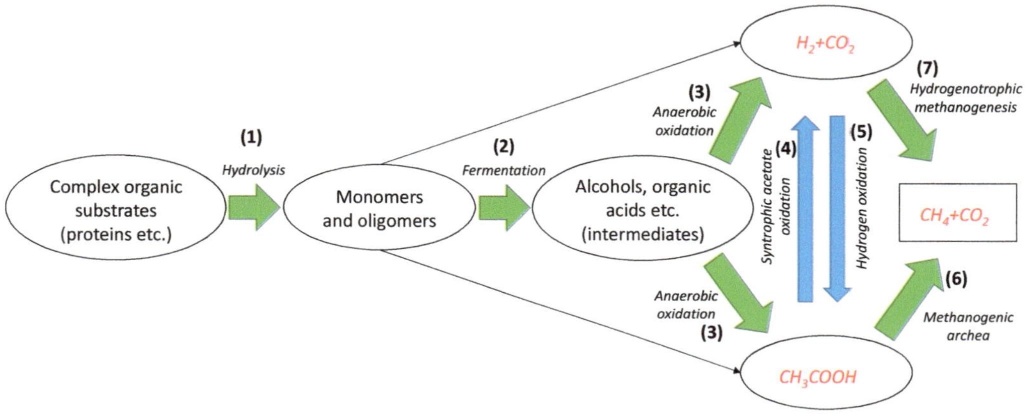

Fig. (18). Decomposition of organic matter into methane.

The addition of hydrogen to an existing reactor affects the biogas process in step 3. Several compounds are decomposed by anaerobic oxidation to carbon dioxide, hydrogen and acetic acid. If the concentration of hydrogen is very high, the reaction becomes thermodynamically unfavorable. Under normal conditions, the concentration of hydrogen is low as it is consumed from archaea to produce methane. In Fig. (**19**), the importance of hydrogen for anaerobic oxidation of propionate is showcased.

Propionic is a molecule that forms acetate salt (ionic form of the acetic acid) by the oxidation process. As seen in Fig. (**19**), only a small portion in the middle of the diagram receives the propionate oxidizing bacteria and methanogenic archaea enough energy to grow. The propionates oxidizing organisms prefer a low hydrogen pressure, and the methanogens high hydrogen pressure. This means that, if hydrogen is added to an existing biogas plant, this addition should be done in such a way that does not occur any significant effect on the anaerobic oxidation.

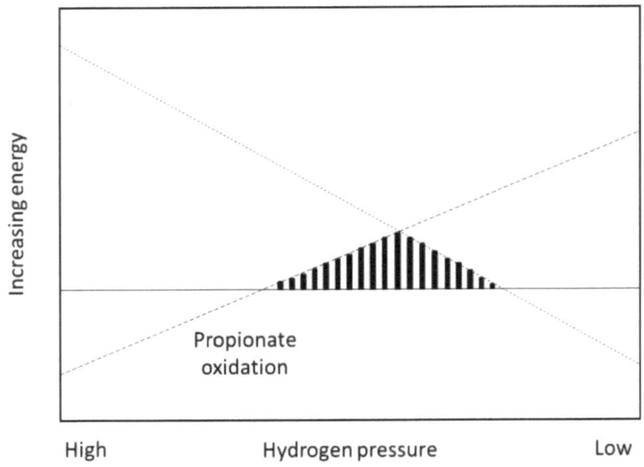

Fig. (19). Hydrogen concentration requirement for anaerobic digestion and formation of CH_4.

The products of the anaerobic oxidation are used by methane-producing organisms. Acetate salt can either be directly decomposed into carbon dioxide and methane (6) or into carbon dioxide and hydrogen (4) and then into methane (7). The path shown with the number 4 and followed by the number 7 can to be achieved through reactions in two stages, according to the following reactions:

$$CH_3COO^- + 4H_2O \rightarrow 2HCO_3^- + 4H_2 + H^+ \quad \textbf{(36)}$$

$$4H_2 + HCO_3^- + H^+ \rightarrow CH_4 + 3H_2O \quad \textbf{(37)}$$

The addition of hydrogen to an existing biogas plant complicates the metabolic pathway for the production of methane by syntrophic acetate oxidation. A high hydrogen partial pressure results in a concentration that is difficult for the microbes that are responsible for carrying out the reaction in Eq. **36** to manage. Thus, the microorganisms that perform the reaction in Eq. **37** must have time to grow in number as more hydrogen is added. Reaction 4 in Fig. (**18**) can also be performed in the opposite direction. When this happens, hydrogen oxidation takes place, in which bacteria produce acetic acid from hydrogen and carbon dioxide, as shown by the reaction:

$$4H_2 + 2CO_2 \rightarrow CH_3COOH + 2H_2O \quad \textbf{(38)}$$

Then the acetic acid can be decomposed into carbon dioxide and methane, according to the reaction:

$$CH_3COOH \rightarrow CH_4 + CO_2 \quad \textbf{(39)}$$

Acetic acid can also be produced directly by fermentation and anaerobic oxidation. The decomposition of acetate by methanogens contributes to about 70% of the total methane produced in a biogas process [129]. Hydrogen and carbon dioxide can also be converted directly into methane through route 7. The formation of methane to reduce the carbon dioxide takes place according to the reaction [130]:

$$4H_2 + HCO_3^- + H^+ \rightarrow CH_4 + 3H_2O \quad \textbf{(40)}$$

CONCLUSIONS

In this chapter, the state-of-the-art and R&D directions towards H_2 production and storage were presented. Firstly, hydrogen can be produced from fossil fuel sources *via* processes such as steam methane reforming (SMR), which is widely considered due its lower costs. Meanwhile, substantial research is being conducted on the H_2 production by the utilization of renewable energy sources, such as the thermochemical conversion of biomass or water electrolysis powered by wind and solar energy. The latter options can also serve as power-to-fuel energy storage schemes and thus provide a means to overcome grid stabilization issues that occur due to the intermittent nature of these energy sources. However, water electrolysis is still facing increased production costs and further research is required to improve its cost competitiveness, otherwise significant incentives are required to promote its development. Concerning hydrogen storage, different options were reviewed such as direct hydrogen storage in gas and liquid tanks and storage in solid sorbents and liquid solutions. Safety issues, storage cost and achievable volumetric and gravimetric densities are of great importance considering also the application for fuel cell operation in the transportation sector. Furthermore, the conversion of hydrogen to fuels can be a pathway for carbon dioxide recycling and utilization at lower fuel storage costs due to the use of existing distribution infrastructure, as in the case of the natural gas grid (for hydrogen conversion to methane).

CONSENT FOR PUBLICATION

Not applicable.

CONFLICT OF INTERESTS

The authors declare no conflict of interests.

ACKNOWLEDGEMENT

Declared none.

REFERENCES

[1] R.B. Gubta, *Hydrogen Fuel: Production, transport and Storage.* CRC Press, 2009.

[2] J.M. Ogden, "PROSPECTS FOR BUILDING A HYDROGEN ENERGY INFRASTRUCTURE", *Annu. Rev. Energy Environ.,* vol. 24, no. 1, pp. 227-279, 1999. [http://dx.doi.org/10.1146/annurev.energy.24.1.227]

[3] X. Zhai, S. Ding, Z. Liu, Y. Jin, and Y. Cheng, "Catalytic performance of Ni catalysts for steam reforming of methane at high space velocity", *Int. J. Hydrogen Energy,* vol. 36, no. 1, pp. 482-489, 2011. [http://dx.doi.org/10.1016/j.ijhydene.2010.10.053]

[4] A. Iulianelli, G. Manzolini, M. De Falco, S. Campanari, T. Longo, and S. Liguori, "H2 production by low pressure methane steam reforming in a Pd-Ag membrane reactor over a Ni-based catalyst: Experimental and modeling", *Int. J. Hydrogen Energy,* vol. 35, no. 20, pp. 11514-11524, 2010. [http://dx.doi.org/10.1016/j.ijhydene.2010.06.049]

[5] P. Dodds, and W. McDowall, *A review of hydrogen production technologies for energy system models.* UCL Energy Institute, University College London, 2012.

[6] Z.M. Nazim, *Production of Hydrogen from Hydrocarbons. Hydrogen Fuel.* CRC Press, 2008, pp. 33-101.

[7] B. Sørensen, *Hydrogen and fuel cells: emerging technologies and applications.* 1st ed.. Elsevier Academic Press: Amsterdam, 2005.

[8] B. Christian Enger, R. Lødeng, and A. Holmen, "A review of catalytic partial oxidation of methane to synthesis gas with emphasis on reaction mechanisms over transition metal catalysts", *Appl. Catal. A Gen.,* vol. 346, no. 1-2, pp. 1-27, 2008. [http://dx.doi.org/10.1016/j.apcata.2008.05.018]

[9] S. Fukada, N. Nakamura, and J. Monden, "Effects of temperature, oxygen-to-methane molar ratio and superficial gas velocity on partial oxidation of methane for hydrogen production", *Int. J. Hydrogen Energy,* vol. 29, no. 6, pp. 619-625, 2004. [http://dx.doi.org/10.1016/j.ijhydene.2003.08.002]

[10] M. Lyubovsky, S. Roychoudhury, and R. LaPierre, "Catalytic partial "oxidation of methane to syngas" at elevated pressures", *Catal. Lett.,* vol. 99, no. 3-4, pp. 113-117, 2005. [http://dx.doi.org/10.1007/s10562-005-2103-y]

[11] M.W. Smith, and D. Shekhawat, *Catalytic Partial Oxidation. Fuel Cells: Technologies for Fuel Processing.* Elsevier: Amsterdam, 2011, pp. 73-128. [http://dx.doi.org/10.1016/B978-0-444-53563-4.10005-7]

[12] Å. Slagtern, H.M. Swaan, U. Olsbye, I.M. Dahl, and C. Mirodatos, "Catalytic partial oxidation of methane over Ni-, Co- and Fe-based catalysts", *Catal. Today,* vol. 46, no. 2-3, pp. 107-115, 1998. [http://dx.doi.org/10.1016/S0920-5861(98)00332-0]

[13] P.D.F. Vernon, M.L.H. Green, A.K. Cheetham, and A.T. Ashcroft, "Partial oxidation of methane to synthesis gas, and carbon dioxide as an oxidising agent for methane conversion", *Catal. Today,* vol. 13, no. 2-3, pp. 417-426, 1992. [http://dx.doi.org/10.1016/0920-5861(92)80167-L]

[14] L. Jalowiecki-Duhamel, H. Zarrou, and A. D'Huysser, "Hydrogen production at low temperature from methane on cerium and nickel based mixed oxides", *Int. J. Hydrogen Energy,* vol. 33, no. 20, pp. 5527-5534, 2008.

[http://dx.doi.org/10.1016/j.ijhydene.2008.07.031]

[15] H.E. Figen, and S.Z. Baykara, "Hydrogen production by partial oxidation of methane over Co based, Ni and Ru monolithic catalysts", *Int. J. Hydrogen Energy,* vol. 40, no. 24, pp. 7439-7451, 2015. [http://dx.doi.org/10.1016/j.ijhydene.2015.02.109]

[16] Z. Al-Hamamre, and M.A. Hararah, "Hydrogen production by thermal partial oxidation of ethanol: Thermodynamics and kinetics study", *Int. J. Hydrogen Energy,* vol. 35, no. 11, pp. 5367-5377, 2010. [http://dx.doi.org/10.1016/j.ijhydene.2010.03.018]

[17] Q. Zhu, X. Zhao, and Y. Deng, "Advances in the partial oxidation of methane to synthesis gas", *J. Nat. Gas Chem.,* vol. 13, no. 4, pp. 191-203, 2004.

[18] S. Freni, G. Calogero, and S. Cavallaro, "Hydrogen production from methane through catalytic partial oxidation reactions", *J. Power Sources,* vol. 87, no. 1-2, pp. 28-38, 2000. [http://dx.doi.org/10.1016/S0378-7753(99)00357-2]

[19] F. Joensen, and J.R. Rostrup-Nielsen, "Conversion of hydrocarbons and alcohols for fuel cells", *J. Power Sources,* vol. 105, no. 2, pp. 195-201, 2002. [http://dx.doi.org/10.1016/S0378-7753(01)00939-9]

[20] J.A. Velasco, C. Fernandez, L. Lopez, S. Cabrera, M. Boutonnet, and S. Järås, "Catalytic partial oxidation of methane over nickel and ruthenium based catalysts under low O2/CH4 ratios and with addition of steam", *Fuel,* vol. 153, pp. 192-201, 2015. [http://dx.doi.org/10.1016/j.fuel.2015.03.009]

[21] J. Llorca, V.C. Corberán, N.J. Divins, O. Fraile Raquel, and E. Taboada, *Hydrogen from Bioethanol A2 - Diéguez, Luis M. GandíaGurutze ArzamendiPedro M. Renewable Hydrogen Technologies.* Elsevier: Amsterdam, 2013, pp. 135-169.

[22] C Pirez, W Fang, M Capron, S Paul, H Jobic, and F Dumeignil, "Steam reforming, partial oxidation and oxidative steam reforming for hydrogen production from ethanol over cerium nickel based oxyhydride catalyst", *Applied Catalysis A: General..* [http://dx.doi.org/10.1016/j.apcata.2015.10.035]

[23] E. Kraleva, M-M. Pohl, A. Jürgensen, and H. Ehrich, "Hydrogen production by bioethanol partial oxidation over Ni based catalysts", *Appl. Catal. B,* vol. 179, pp. 509-520, 2015. [http://dx.doi.org/10.1016/j.apcatb.2015.06.004]

[24] J.R. Salge, G.A. Deluga, and L.D. Schmidt, "Catalytic partial oxidation of ethanol over noble metal catalysts", *J. Catal.,* vol. 235, no. 1, pp. 69-78, 2005. [http://dx.doi.org/10.1016/j.jcat.2005.07.021]

[25] J.D. Holladay, J. Hu, D.L. King, and Y. Wang, "An overview of hydrogen production technologies", *Catal. Today,* vol. 139, no. 4, pp. 244-260, 2009. [http://dx.doi.org/10.1016/j.cattod.2008.08.039]

[26] W-H. Chen, C-T. Shen, B-J. Lin, and S-C. Liu, "Hydrogen production from methanol partial oxidation over Pt/Al2O3 catalyst with low Pt content", *Energy,* vol. 88, pp. 399-407, 2015. [http://dx.doi.org/10.1016/j.energy.2015.05.055]

[27] M. Karakaya, Z. Ilsen Onsan, and A.K. Avci, "Microchannel Autothermal Reforming of Methane to Synthesis Gas", *Top. Catal.,* vol. 56, no. 18-20, pp. 1716-1723, 2013. [http://dx.doi.org/10.1007/s11244-013-0107-1]

[28] M. Akri, T. Chafik, P. Granger, P. Ayrault, and C. Batiot-Dupeyrat, "Novel nickel promoted illite clay based catalyst for autothermal dry reforming of methane", *Fuel,* vol. 178, pp. 139-147, 2016. [http://dx.doi.org/10.1016/j.fuel.2016.03.018]

[29] M.V. Phanikrishna Sharma, J.F. Akyurtlu, and A. Akyurtlu, "Autothermal reforming of isobutanol over promoted nickel xerogel catalysts for hydrogen production", *Int. J. Hydrogen Energy,* vol. 40, no. 39, pp. 13368-13378, 2015. [http://dx.doi.org/10.1016/j.ijhydene.2015.07.113]

[30] S.H. Park, W.L. Yoon, and S.H. Kim, "Autothermal reforming of methane to syngas using co-precipitated Ni-(La2O3)x-(ZrO2)1-x catalyst", *Res. Chem. Intermed.,* vol. 34, no. 8-9, pp. 781-786, 2008.
[http://dx.doi.org/10.1007/BF03036938]

[31] A. Godula-Jopek, and W. Jehle, *Wellnitz Jr. Hydrogen storage technologies new materials, transport, and infrastructure.* Wiley-VCH: Weinheim, 2012.
[http://dx.doi.org/10.1002/9783527649921]

[32] S. Choi, and J. Bae, "Autothermal reforming of dimethyl ether with CGO-based precious metal catalysts", *J. Power Sources,* vol. 307, pp. 351-357, 2016.
[http://dx.doi.org/10.1016/j.jpowsour.2015.12.068]

[33] R. Baruah, M. Dixit, P. Basarkar, D. Parikh, and A. Bhargav, "Advances in ethanol autothermal reforming", *Renew. Sustain. Energy Rev.,* vol. 51, pp. 1345-1353, 2015.
[http://dx.doi.org/10.1016/j.rser.2015.07.060]

[34] P.M. Biesheuvel, and G.J. Kramer, "Two-section reactor model for autothermal reforming of methane to synthesis gas", *AIChE J.,* vol. 49, no. 7, pp. 1827-1837, 2003.
[http://dx.doi.org/10.1002/aic.690490719]

[35] D.J. Wilhelm, D.R. Simbeck, A.D. Karp, and R.L. Dickenson, "Syngas production for gas-to-liquids applications: technologies, issues and outlook", *Fuel Process. Technol.,* vol. 71, no. 1-3, pp. 139-148, 2001.
[http://dx.doi.org/10.1016/S0378-3820(01)00140-0]

[36] T.S. Christensen, and I.I. Primdahl, "Improve syngas production using autothermal reforming", *Journal Name: Hydrocarbon Processing; (United States),* vol. 73, no. 3, pp. 39-46, 1994. Medium: X

[37] R. Kothari, D. Buddhi, and R.L. Sawhney, "Comparison of environmental and economic aspects of various hydrogen production methods", *Renew. Sustain. Energy Rev.,* vol. 12, no. 2, pp. 553-563, 2008.
[http://dx.doi.org/10.1016/j.rser.2006.07.012]

[38] Wd. Jong, *Ommen JRv. Biomass as a sustainable energy source for the future: fundamentals of conversion processes.* Wiley: Hoboken, New Jersey, 2014.

[39] E. Mostafavi, N. Mahinpey, M. Rahman, M.H. Sedghkerdar, and R. Gupta, "High-purity hydrogen production from ash-free coal by catalytic steam gasification integrated with dry-sorption CO2 capture", *Fuel..*
[http://dx.doi.org/10.1016/j.fuel.2016.03.026]

[40] G.J. Stiegel, and M. Ramezan, "Hydrogen from coal gasification: An economical pathway to a sustainable energy future", *Int. J. Coal Geol.,* vol. 65, no. 3-4, pp. 173-190, 2006.
[http://dx.doi.org/10.1016/j.coal.2005.05.002]

[41] K. Jordal, R. Anantharaman, T.A. Peters, D. Berstad, J. Morud, and P. Nekså, "High-purity H2 production with CO2 capture based on coal gasification", *Energy,* vol. 88, pp. 9-17, 2015.
[http://dx.doi.org/10.1016/j.energy.2015.03.129]

[42] Y. Yürüm, Hydrogen Production Methods.*Hydrogen Energy System: Production and Utilization of Hydrogen and Future Aspects.,* Y. Yürüm, Ed., Springer Netherlands: Dordrecht, 1995, pp. 15-30.
[http://dx.doi.org/10.1007/978-94-011-0111-0_2]

[43] I.E. Agency, *Hydrogen Production and Storage: R&D Priorities and Gaps.*Paris, 2006.

[44] P.E. Dodds, and W. McDowall, *A review of hydrogen production technologies for energy system models.* UCL Energy Institute, University College London, 2012.

[45] S.T. Mirabal, *An Economic Analysis of Hydrogen Production Technologies Using Renewable Energy Resources.* University of Florida, 2003.

[46] M. Steinberg, and H.C. Cheng, "Modern and prospective technologies for hydrogen production from fossil fuels", *Int. J. Hydrogen Energy,* vol. 14, no. 11, pp. 797-820, 1989.

[http://dx.doi.org/10.1016/0360-3199(89)90018-9]

[47] M. Steinberg, and H.C. Cheng, "Modern and prospective technologies for hydrogen production from fossil fuels", *Int. J. Hydrogen Energy,* vol. 14, no. 11, pp. 797-820, 1989. [http://dx.doi.org/10.1016/0360-3199(89)90018-9]

[48] S.N. Reddy, S. Nanda, A.K. Dalai, and J.A. Kozinski, "Supercritical water gasification of biomass for hydrogen production", *Int. J. Hydrogen Energy,* vol. 39, no. 13, pp. 6912-6926, 2014. [http://dx.doi.org/10.1016/j.ijhydene.2014.02.125]

[49] S. Li, Y. Lu, L. Guo, and X. Zhang, "Hydrogen production by biomass gasification in supercritical water with bimetallic Ni-M/γAl2O3 catalysts (M = Cu, Co and Sn)", *Int. J. Hydrogen Energy,* vol. 36, no. 22, pp. 14391-14400, 2011. [http://dx.doi.org/10.1016/j.ijhydene.2011.07.144]

[50] H. Jin, Y. Chen, Z. Ge, S. Liu, C. Ren, and L. Guo, "Hydrogen production by Zhundong coal gasification in supercritical water", *Int. J. Hydrogen Energy,* vol. 40, no. 46, pp. 16096-16103, 2015. [http://dx.doi.org/10.1016/j.ijhydene.2015.09.003]

[51] L. Guo, and H. Jin, "Boiling coal in water: Hydrogen production and power generation system with zero net CO2 emission based on coal and supercritical water gasification", *Int. J. Hydrogen Energy,* vol. 38, no. 29, pp. 12953-12967, 2013. [http://dx.doi.org/10.1016/j.ijhydene.2013.04.089]

[52] H. Jin, Y. Lu, B. Liao, L. Guo, and X. Zhang, "Hydrogen production by coal gasification in supercritical water with a fluidized bed reactor", *Int. J. Hydrogen Energy,* vol. 35, no. 13, pp. 7151-7160, 2010. [http://dx.doi.org/10.1016/j.ijhydene.2010.01.099]

[53] Z. Ge, H. Jin, and L. Guo, "Hydrogen production by catalytic gasification of coal in supercritical water with alkaline catalysts: Explore the way to complete gasification of coal", *Int. J. Hydrogen Energy,* vol. 39, no. 34, pp. 19583-19592, 2014. [http://dx.doi.org/10.1016/j.ijhydene.2014.09.119]

[54] Y. Li, L. Guo, X. Zhang, H. Jin, and Y. Lu, "Hydrogen production from coal gasification in supercritical water with a continuous flowing system", *Int. J. Hydrogen Energy,* vol. 35, no. 7, pp. 3036-3045, 2010. [http://dx.doi.org/10.1016/j.ijhydene.2009.07.023]

[55] M. Ni, D.Y. Leung, M.K. Leung, and K. Sumathy, "An overview of hydrogen production from biomass", *Fuel Process. Technol.,* vol. 87, no. 5, pp. 461-472, 2006. [http://dx.doi.org/10.1016/j.fuproc.2005.11.003]

[56] S. Lin, M. Harada, Y. Suzuki, and H. Hatano, "Process analysis for hydrogen production by reaction integrated novel gasification (HyPr-RING)", *Energy Convers. Manage.,* vol. 46, no. 6, pp. 869-880, 2005. [http://dx.doi.org/10.1016/j.enconman.2004.06.008]

[57] S-Y. Lin, Y. Suzuki, H. Hatano, and M. Harada, "Hydrogen Production from Hydrocarbon by Integration of Water−Carbon Reaction and Carbon Dioxide Removal (HyPr−RING Method)", *Energy Fuels,* vol. 15, no. 2, pp. 339-343, 2001. [http://dx.doi.org/10.1021/ef000089u]

[58] S-Y. Lin, Y. Suzuki, H. Hatano, and M. Harada, "Developing an innovative method, HyPr-RING, to produce hydrogen from hydrocarbons", *Energy Convers. Manage.,* vol. 43, no. 9-12, pp. 1283-1290, 2002. [http://dx.doi.org/10.1016/S0196-8904(02)00014-6]

[59] L.J. Guo, Y.J. Lu, X.M. Zhang, C.M. Ji, Y. Guan, and A.X. Pei, "Hydrogen production by biomass gasification in supercritical water: A systematic experimental and analytical study", *Catal. Today,* vol. 129, no. 3-4, pp. 275-286, 2007. [http://dx.doi.org/10.1016/j.cattod.2007.05.027]

[60] "Direct Hydrogen production from Biomass Gasifier using Hydrogen-Selective Membrane", *Direct Hydrogen production from Biomass Gasifier using Hydrogen-Selective Membrane,* 2007.

[61] J. Rezaiyan, and N.P. Cheremisinoff, *Gasification Technologies: A primer for engineers and scientists.* CRC Press, 2005.

[62] "Review of Technologies for Gasification of Biomass and Wastes", *E4Tech,* 2009.

[63] H. Knoef, *Handbook of biomass gasification: BTG*, 2005.

[64] S. Karellas, E. Kakaras, T. Papadopoulos, C. Schäfer, and J. Karl, "Hydrogen production from allothermal biomass gasification by means of palladium membranes", *Fuel Process. Technol.,* vol. 89, no. 6, pp. 582-588, 2008.
[http://dx.doi.org/10.1016/j.fuproc.2007.11.033]

[65] M. Falco, G. Iaquaniello, G. Cucchiella, and L. Marelli, *Membrane reactors for hydrogen production processes.* Springer, 2011.
[http://dx.doi.org/10.1007/978-0-85729-151-6]

[66] F. Chen, C. Wu, L. Dong, A. Vassallo, P.T. Williams, and J. Huang, "Characteristics and catalytic properties of Ni/CaAlOx catalyst for hydrogen-enriched syngas production from pyrolysis-steam reforming of biomass sawdust", *Appl. Catal. B,* vol. 183, pp. 168-175, 2016.
[http://dx.doi.org/10.1016/j.apcatb.2015.10.028]

[67] K. Braimakis, K. Atsonios, K.D. Panopoulos, S. Karellas, and E. Kakaras, "Economic evaluation of decentralized pyrolysis for the production of bio-oil as an energy carrier for improved logistics towards a large centralized gasification plant", *Renew. Sustain. Energy Rev.,* vol. 35, pp. 57-72, 2014.
[http://dx.doi.org/10.1016/j.rser.2014.03.052]

[68] K. Braimakis, K. Atsonios, K.D. Panopoulos, S. Karellas, and E. Kakaras, "Economic evaluation of decentralized pyrolysis for the production of bio-oil as an energy carrier for improved logistics towards a large centralized gasification plant", *Renew. Sustain. Energy Rev.,* vol. 35, pp. 57-72, 2014.
[http://dx.doi.org/10.1016/j.rser.2014.03.052]

[69] Z. Chen, S. Zhang, Z. Chen, and D. Ding, "An integrated process for hydrogen-rich gas production from cotton stalks: The simultaneous gasification of pyrolysis gases and char in an entrained flow bed reactor", *Bioresour. Technol.,* vol. 198, pp. 586-592, 2015.
[http://dx.doi.org/10.1016/j.biortech.2015.09.015] [PMID: 26433156]

[70] S. Czernik, and R. French, "Distributed production of hydrogen by auto-thermal reforming of fast pyrolysis bio-oil", *Int. J. Hydrogen Energy,* vol. 39, no. 2, pp. 744-750, 2014.
[http://dx.doi.org/10.1016/j.ijhydene.2013.10.134]

[71] X. Xu, E. Jiang, M. Wang, and Y. Xu, "Dry and steam reforming of biomass pyrolysis gas for rich hydrogen gas", *Biomass Bioenergy,* vol. 78, pp. 6-16, 2015.
[http://dx.doi.org/10.1016/j.biombioe.2015.03.015]

[72] H. Li, X. Li, L. Liu, K. Li, X. Wang, and H. Li, "Experimental study of microwave-assisted pyrolysis of rice straw for hydrogen production", *Int. J. Hydrogen Energy,* vol. 41, no. 4, pp. 2263-2267, 2016.
[http://dx.doi.org/10.1016/j.ijhydene.2015.11.137]

[73] R.M. Zin, A. Lea-Langton, V. Dupont, and M.V. Twigg, "High hydrogen yield and purity from palm empty fruit bunch and pine pyrolysis oils", *Int. J. Hydrogen Energy,* vol. 37, no. 14, pp. 10627-10638, 2012.
[http://dx.doi.org/10.1016/j.ijhydene.2012.04.064]

[74] J. Alvarez, S. Kumagai, C. Wu, T. Yoshioka, J. Bilbao, and M. Olazar, "Hydrogen production from biomass and plastic mixtures by pyrolysis-gasification", *Int. J. Hydrogen Energy,* vol. 39, no. 21, pp. 10883-10891, 2014.
[http://dx.doi.org/10.1016/j.ijhydene.2014.04.189]

[75] H. Jiang, Y. Wu, H. Fan, and J. Ji, "Hydrogen Production from Biomass Pyrolysis in Molten Alkali",

AASRI Procedia., vol. 3, pp. 217-223, 2012.
[http://dx.doi.org/10.1016/j.aasri.2012.11.036]

[76] S. Liu, J. Zhu, M. Chen, W. Xin, Z. Yang, and L. Kong, "Hydrogen production via catalytic pyrolysis of biomass in a two-stage fixed bed reactor system", *Int. J. Hydrogen Energy*, vol. 39, no. 25, pp. 13128-13135, 2014.
[http://dx.doi.org/10.1016/j.ijhydene.2014.06.158]

[77] M. Götz, J. Lefebvre, F. Mörs, A. McDaniel Koch, F. Graf, and S. Bajohr, "Renewable Power-to-Gas: A technological and economic review", *Renew. Energy*, vol. 85, pp. 1371-1390, 2016.
[http://dx.doi.org/10.1016/j.renene.2015.07.066]

[78] T. Smolinka, E. Ojong, and J. Garche, "Hydrogen production from renewable energies - electrolyzer technologies", *Freiburg, Germany: Fraunhofer Institute for Solar Energy Systems ISE.*.

[79] "Current State-of-the-Art Hydrogen Production Cost Estimate Using Water Electrolysis", *National Renewable Energy Laboratory*, 2009.

[80] "HYDROGEN PRODUCTION: ELECTROLYSIS", *energygov.*.

[81] M. Carmo, D.L. Fritz, J. Mergel, and D. Stolten, "A comprehensive review on PEM water electrolysis", *Int. J. Hydrogen Energy*, vol. 38, no. 12, pp. 4901-4934, 2013.
[http://dx.doi.org/10.1016/j.ijhydene.2013.01.151]

[82] J. Mougin, *Hydrogen production by high-temperature steam electrolysis*. CEA-Liten: Grenoble, France.
[http://dx.doi.org/10.1016/B978-1-78242-361-4.00008-X]

[83] I Dincer, and C Acar, "Review and evaluation of hydrogen production methods for better sustainability", *International Journal of Hydrogen Energy.*.

[84] S. Koumi Ngoh, L.M. Ayina Ohandja, A. Kemajou, and L. Monkam, "Design and simulation of hybrid solar high-temperature hydrogen production system using both solar photovoltaic and thermal energy", *Sustainable Energy Technologies and Assessments.*, vol. 7, no. 0, pp. 279-293, 2014.
[http://dx.doi.org/10.1016/j.seta.2014.05.002]

[85] P. Ahmadi, I. Dincer, and M.A. Rosen, "Energy and exergy analyses of hydrogen production via solar-boosted ocean thermal energy conversion and PEM electrolysis", *Int. J. Hydrogen Energy*, vol. 38, no. 4, pp. 1795-1805, 2013.
[http://dx.doi.org/10.1016/j.ijhydene.2012.11.025]

[86] T.A.H. Ratlamwala, and I. Dincer, "Comparative energy and exergy analyses of two solar-based integrated hydrogen production systems", *Int. J. Hydrogen Energy*, vol. 40, no. 24, pp. 7568-7578, 2015.
[http://dx.doi.org/10.1016/j.ijhydene.2014.10.123]

[87] H. Ozcan, and I. Dincer, "Energy and exergy analyses of a solar driven Mg-Cl hybrid thermochemical cycle for co-production of power and hydrogen", *Int. J. Hydrogen Energy*, vol. 39, no. 28, pp. 15330-15341, 2014.
[http://dx.doi.org/10.1016/j.ijhydene.2014.07.089]

[88] K. Hacatoglu, I. Dincer, and M.A. Rosen, "Exergy analysis of a hybrid solar hydrogen system with activated carbon storage", *Int. J. Hydrogen Energy*, vol. 36, no. 5, pp. 3273-3282, 2011.
[http://dx.doi.org/10.1016/j.ijhydene.2010.12.034]

[89] A. Melvin, "The impracticability of large-scale generation of hydrogen from water photolysis by utilization of solar radiation", *Int. J. Hydrogen Energy*, vol. 4, no. 3, pp. 223-224, 1979.
[http://dx.doi.org/10.1016/0360-3199(79)90027-2]

[90] A.J. Bard, and M.A. Fox, "Artificial Photosynthesis: Solar Splitting of Water to Hydrogen and Oxygen", *Acc. Chem. Res.*, vol. 28, no. 3, pp. 141-145, 1995.
[http://dx.doi.org/10.1021/ar00051a007]

[91] J.L. Ropero-Vega, J.A. Pedraza-Avella, and M.E. Niño-Gómez, "Hydrogen production by photoelectrolysis of aqueous solutions of phenol using mixed oxide semiconductor films of Bi-Nb-M-O (M = Al, Fe, Ga, In) as photoanodes", *Catal. Today,* vol. 252, pp. 150-156, 2015. [http://dx.doi.org/10.1016/j.cattod.2014.11.007]

[92] O.K. Varghese, and C.A. Grimes, "Appropriate strategies for determining the photoconversion efficiency of water photoelectrolysis cells: A review with examples using titania nanotube array photoanodes", *Sol. Energy Mater. Sol. Cells,* vol. 92, no. 4, pp. 374-384, 2008. [http://dx.doi.org/10.1016/j.solmat.2007.11.006]

[93] M. Grätzel, and A.J. McEvoy, "Hydrogen Production by Solar Photolysis of Water", *American Physical Society Symposium Basic Research for the Hydrogen Economy.,* 2004 Montreal, Canada

[94] C.M. Kalamaras, and A.M. Efstathiou, *Hydrogen Production Technologies: Current State and Future Developments.* Power Options for Eastern Medditerranean Region: Limassol, Cyprus, 2013.

[95] G.J. Conibeer, and B.S. Richards, "A comparison of PV/electrolyser and photoelectrolytic technologies for use in solar to hydrogen energy storage systems", *Int. J. Hydrogen Energy,* vol. 32, no. 14, pp. 2703-2711, 2007. [http://dx.doi.org/10.1016/j.ijhydene.2006.09.012]

[96] T. Uchida, H. Sugimura, A. Sekiguchi, N. Kitamura, N. Shimo, and H. Masuhara, "An International Journal Devoted to all Aspects of Electrode Kinetics, Interfacial Structure, Properties of Electrolytes, Colloid and Biological ElectrochemistryPhotoelectrolysis of water on a titanium dioxide/platinum microelectrode array", *J. Electroanal. Chem.,* vol. 351, no. 1, pp. 343-348, 1993. [http://dx.doi.org/10.1016/0022-0728(93)80245-D]

[97] M.P. Nair, K.V.C. Rao, and C.G.R. Nair, "Investigation of the mixed oxide materials—TiO2□SiO2, TiO2□SiO2□Al2O3, TiO2□SiO2□In2O3 and TiO2□SiO2□RuO2—In regard to the photoelectrolysis of water", *Int. J. Hydrogen Energy,* vol. 16, no. 7, pp. 449-459, 1991. [http://dx.doi.org/10.1016/0360-3199(91)90097-3]

[98] M. Grätzel, "Photoelectrochemical cells", *Nature,* vol. 414, no. 6861, pp. 338-344, 2001. [http://dx.doi.org/10.1038/35104607] [PMID: 11713540]

[99] T. Bak, J. Nowotny, M. Rekas, and C.C. Sorrell, "Photo-electrochemical hydrogen generation from water using solar energy. Materials-related aspects", *Int. J. Hydrogen Energy,* vol. 27, no. 10, pp. 991-1022, 2002. [http://dx.doi.org/10.1016/S0360-3199(02)00022-8]

[100] O. Khaselev, A. Bansal, and J.A. Turner, "High-efficiency integrated multijunction photovoltaic/electrolysis systems for hydrogen production", *Int. J. Hydrogen Energy,* vol. 26, no. 2, pp. 127-132, 2001. [http://dx.doi.org/10.1016/S0360-3199(00)00039-2]

[101] M.K. Nowotny, T. Bak, J. Nowotny, and C.C. Sorrell, "Titanium vacancies in nonstoichiometric TiO2 single crystal", *Physica Status Solidi (b).,* vol. 242, no. 11, pp. R88-R90, 2005.

[102] C. Santato, M. Ulmann, and J. Augustynski, "Photoelectrochemical Properties of Nanostructured Tungsten Trioxide Films", *J. Phys. Chem. B,* vol. 105, no. 5, pp. 936-940, 2001. [http://dx.doi.org/10.1021/jp002232q]

[103] M.L. Girardi, "Biological Systems for Hydrogen Photoproduction", *DOE Hydrogen and Fuel cells Program,* 2000.

[104] "HYDROGEN PRODUCTION: PHOTOBIOLOGICAL", *energygov.*

[105] N.T. Stetson, S. McWhorter, and C.C. Ahn, *1 - Introduction to hydrogen storage A2 - Veziroğlu, Ram B. GuptaAngelo BasileT. Nejat. Compendium of Hydrogen Energy.* Woodhead Publishing, 2015, pp. 3-25.

[106] N. de Miguel, R. Ortiz Cebolla, B. Acosta, P. Moretto, F. Harskamp, and C. Bonato, "Compressed

hydrogen tanks for on-board application: Thermal behaviour during cycling", *Int. J. Hydrogen Energy,* vol. 40, no. 19, pp. 6449-6458, 2015. [http://dx.doi.org/10.1016/j.ijhydene.2015.03.035]

[107] T.Q. Hua, R.K. Ahluwalia, J.K. Peng, M. Kromer, S. Lasher, and K. McKenney, "Technical assessment of compressed hydrogen storage tank systems for automotive applications", *Int. J. Hydrogen Energy,* vol. 36, no. 4, pp. 3037-3049, 2011. [http://dx.doi.org/10.1016/j.ijhydene.2010.11.090]

[108] M. Hirscher, *Handbook of hydrogen storage: new materials for future energy storage.* Wiley-VCH: Weinheim, 2010. [http://dx.doi.org/10.1002/9783527629800]

[109] C. Chilev, and F.D. Lamari, "Hydrogen storage at low temperature and high pressure for application in automobile manufacturing", *Int. J. Hydrogen Energy,* vol. 41, no. 3, pp. 1744-1758, 2016. [http://dx.doi.org/10.1016/j.ijhydene.2015.11.099]

[110] M. Felderhoff, C. Weidenthaler, R. von Helmolt, and U. Eberle, "Hydrogen storage: the remaining scientific and technological challenges", *Phys. Chem. Chem. Phys.,* vol. 9, no. 21, pp. 2643-2653, 2007. [http://dx.doi.org/10.1039/b701563c] [PMID: 17627309]

[111] N.K. Zhevago, *8 - Other methods for the physical storage of hydrogen A2 - Veziroğlu, Ram B. GuptaAngelo BasileT. Nejat. Compendium of Hydrogen Energy.* Woodhead Publishing, 2016, pp. 189-218.

[112] S.W. Jorgensen, "Hydrogen storage tanks for vehicles: Recent progress and current status", *Curr. Opin. Solid State Mater. Sci.,* vol. 15, no. 2, pp. 39-43, 2011. [http://dx.doi.org/10.1016/j.cossms.2010.09.004]

[113] R.A. Huggins, *Hydrogen Storage. Energy Storage.* Springer US: Boston, MA, 2010, pp. 95-117.

[114] A. Léon, Hydrogen Storage.*Hydrogen Technology: Mobile and Portable Applications.,* A. Léon, Ed., Springer Berlin Heidelberg: Berlin, Heidelberg, 2008, pp. 81-128. [http://dx.doi.org/10.1007/978-3-540-69925-5_3]

[115] R.K. Ahluwalia, T.Q. Hua, J.K. Peng, S. Lasher, K. McKenney, and J. Sinha, "Technical assessment of cryo-compressed hydrogen storage tank systems for automotive applications", *Int. J. Hydrogen Energy,* vol. 35, no. 9, pp. 4171-4184, 2010. [http://dx.doi.org/10.1016/j.ijhydene.2010.02.074]

[116] B. Granath, Liquid Hydrogen--the Fuel of Choice for Space Exploration. 2015.

[117] S.M. Aceves, F. Espinosa-Loza, E. Ledesma-Orozco, T.O. Ross, A.H. Weisberg, and T.C. Brunner, "High-density automotive hydrogen storage with cryogenic capable pressure vessels", *Int. J. Hydrogen Energy,* vol. 35, no. 3, pp. 1219-1226, 2010. [http://dx.doi.org/10.1016/j.ijhydene.2009.11.069]

[118] S. Lasher, K. McKenney, J. Sinha, and J. Rosenfeld, *Analyses of hydrogen storage materials and on-board systems.* DOE Merit Review, 2007.

[119] S. Mellouli, H. Dhaou, F. Askri, A. Jemni, and S. Ben Nasrallah, "Hydrogen storage in metal hydride tanks equipped with metal foam heat exchanger", *Int. J. Hydrogen Energy,* vol. 34, no. 23, pp. 9393-9401, 2009. [http://dx.doi.org/10.1016/j.ijhydene.2009.09.043]

[120] S. Satyapal, J. Petrovic, C. Read, G. Thomas, and G. Ordaz, "The U.S. Department of Energy's National Hydrogen Storage Project: Progress towards meeting hydrogen-powered vehicle requirements", *Catal. Today,* vol. 120, no. 3-4, pp. 246-256, 2007. [http://dx.doi.org/10.1016/j.cattod.2006.09.022]

[121] S. Iijima, "Helical microtubules of graphitic carbon", *nature,* vol. 354, no. 6348, pp. 56-8, 1991.

[122] F.L. Darkrim, P. Malbrunot, and G.P. Tartaglia, "Review of hydrogen storage by adsorption in carbon nanotubes", *Int. J. Hydrogen Energy,* vol. 27, no. 2, pp. 193-202, 2002. [http://dx.doi.org/10.1016/S0360-3199(01)00103-3]

[123] A. Züttel, P. Sudan, P. Mauron, T. Kiyobayashi, C. Emmenegger, and L. Schlapbach, "Hydrogen storage in carbon nanostructures", *Int. J. Hydrogen Energy,* vol. 27, no. 2, pp. 203-212, 2002. [http://dx.doi.org/10.1016/S0360-3199(01)00108-2]

[124] M. Jordá-Beneyto, F. Suárez-García, D. Lozano-Castelló, D. Cazorla-Amorós, and A. Linares-Solano, "Hydrogen storage on chemically activated carbons and carbon nanomaterials at high pressures", *Carbon,* vol. 45, no. 2, pp. 293-303, 2007. [http://dx.doi.org/10.1016/j.carbon.2006.09.022]

[125] H.M. Cheng, Q.H. Yang, and C. Liu, "Hydrogen storage in carbon nanotubes", *Carbon,* vol. 39, no. 10, pp. 1447-1454, 2001. [http://dx.doi.org/10.1016/S0008-6223(00)00306-7]

[126] G. Benjaminsson, J. Benjaminsson, and B.R. Rudberg, *Power-to-Gas - A technical review.* Svenskt Gastekniskt Center AB, 2013.

[127] N.L. Söhngen, Het ontstaan en verdwijnen av waterstof en methan, 1906.

[128] M. Burkhardt, T. Koschack, and G. Busch, "Biocatalytic methanation of hydrogen and carbon dioxide in an anaerobic three-phase system", *Bioresour. Technol.,* vol. 178, pp. 330-333, 2015. [http://dx.doi.org/10.1016/j.biortech.2014.08.023] [PMID: 25193088]

[129] A. Schnürer, and Å. Jarvis, Mikrobiologisk handbok för biogasanläggningar, 2009.

[130] J.G. Ferry, "The chemical biology of methanogenesis", *Planet. Space Sci.,* vol. 58, no. 14-15, pp. 1775-1783, 2010. [http://dx.doi.org/10.1016/j.pss.2010.08.014]

CHAPTER 7

Exploitation of Normal Shallow Geothermal Energy for Heating and Cooling Applications

Michalis Gr. Vrachopoulos*, Maria K. Koukou and **Nikolaos P. Tsolakoglou**

Energy and Environmental Research Laboratory, Mechanical Engineering Department, Technological Education Institute of Sterea Ellada, Psachna, 34400, Euboea, Greece

Abstract: The main objective of this chapter is to provide a fully and comprehensive introduction to geothermal systems that exploit normal (or swallow) geothermal energy including construction and designing guidelines. The necessity to decrease power consumption and consequently cost led to the integration of these systems to cover heating, cooling and domestic hot water loads in domestic or even industrial facilities. Basic operational and construction guidelines concerning all possible geothermal plants (closed or open loop) are provided along with representing drawings. The chapter is divided into various sections focusing on certain issues. It begins with basic geological information regarding the ability of the subsoil to absorb or supply heat *via* the integrated heat exchanger. Subsequently, a summary on the operation of heat pumps in general and the advantages of geothermal pumps is provided. Then, the two main categories of geothermal loops are analyzed: closed loop horizontal or vertical geothermal exchangers and open loop system installations. Finally, the proposed methodology to follow when such a system should be designed is discussed, containing cost elements issues of such systems.

Keywords: Closed Loop, Design, Geothermal Heat Pump, Horizontal System, Methodology, Normal Geothermy, Open Loop, Swallow Geothermy, Vertical System.

INTRODUCTION

Normal or Swallow Geothermal energy is called the heat of the geological formations and the water, surface water and groundwater, which are not classified as geothermal potential [1 - 6]. It refers normally [7 - 12] to a depth of 400 m, while in most real cases, depths do not exceed 100 m. It may also be defined as the heat that exists below each subsurface, heat of lakes, rivers and the sea, with temperature less than the annual average temperature of the air in the area

* **Corresponding Author Michalis Gr. Vrachopoulos:** Energy and Environmental Research Laboratory, Technological Education Institute of Sterea Ellada, Psachna, 34400, Euboea, Greece; Tel: +302228099661; Fax: +302228099660; Email: mgrvrachop@gmail.com, mvrachop@teiste.gr

Emmanuel D. Rogdakis & Irene P. Koronaki (Eds.)

normally equivalent below 25°C.

The temperature in the underground remains nearly constant throughout the year, without sharp fluctuations. The heat capacity of the subsurface materials is extremely high, and if it is compared with the heat capacity of the air, which is the main heat source for heat pumps used in air conditioning devices, it is considered to tend to infinity [1]. Actually, this means that it provides the possibility of pumping or absorbing large quantities of heat in the subsoil, without significantly altering the temperature conditions. In Europe, there are different conditions in soil temperatures, ranging from 2-3°C near the polar circle, to 20°C in southern Europe, having an impact on the choice and design of normal geothermy installations [2 - 4, 9, 10]. Applications of normal geothermy for heating and cooling buildings consist of three main parts [1, 9, 10]: heat pump, connection with the earth through geothermal heat devices and the heating, cooling and hot water distribution system inside the building.

The exploitation of normal geothermal energy takes place through open and closed loop Geothermal Heat Pumps (Ground Source Heat Pumps, GSHP) or/and underground thermal energy storage (Underground Thermal Energy Storage, UTES) [8, 12, 13]. Normal geothermal energy is a renewable energy source, which has great potential for saving energy and reducing emissions of greenhouse gases. Applications mainly include heating and air conditioning of buildings as in connection with conventional heating and cooling systems (which consume oil and natural gas) they can achieve energy savings up to 70%. It is considered as one of the promising technologies that can contribute to energy savings in heating and air conditioning, helping to achieve the 20-20-20 targets of the 2020 EU energy policy [14 - 16]. GSHP systems currently hold the largest share in the installed capacity and the energy use of all geothermal energy uses. The number of installed 12 kW power units (standard power for homes in the US and Western Europe) is about 4.16 million. The size of individual units, however, ranges from 5,5 kW, for domestic use, up to more than 150 kW for commercial facilities, schools, hospitals, *etc*.

GEOLOGY INFORMATION

General

For the installation of heating energy systems or summer air conditioning -and cooling- by exploiting the heat and thermal properties of geological formations and water, surface water and groundwater, not qualifying as geothermal potential, it is necessary to determine their thermal characteristics [2, 5, 12, 13, 17, 18]. The soil surface temperature and the temperature near to the surface may be determined by a balance between sunlight, geothermal heat flux and the thermal inertia of the material making up the subsoil and are influenced by these two factors. In GSHP, it is desired soil has a high heat transfer capacity. Geological parameters and thermal properties that are taken into account in the calculation of the thermal characteristics of soils and rocks include [2, 5, 12, 13, 17 - 24] the physical characteristics of soils and of the rocks such as:

- the actual density or specific gravity for each of the materials that consist the formation and the apparent density or the weight per unit volume
- the soil porosity, moisture and the volumetric water content
- the average specific heat and thermal conductivity as a resultant of various materials
- the average special thermal diffusivity as a resultant of different materials

For the calculation of the specific heat capacity both the solid rock and the fluid must be taken into account (Table **1**) [12].

Table 1. Indicative values for density, thermal conductivity and specific heat capacity per volume of different soils at about 20 °C.

Rock (type)	Density ρ	Thermal Conductivity λ		Special heat capacity
	$*10^3$ kg/m^3	Typical range W/(mK)	Typical value W/(mK)	MJ/(m^{3}K)
Magmatic rocks				
Basalt (volcanic)	2,6 - 3,2	1,3 - 2,3	1,7	2,3 - 2,6
Diorite (volcanic)	2,9 - 3,0	2,0 - 2,9	2,6	2,9
Granite (plutonic)	2,4 - 3,0	2,1 - 4,1	3,4	2,1 - 3,0
Peridotite (plutonic)	3,0	3,8 - 5,3	4,0	2,7
Metamorphous rocks				
Marble	2,5 - 2,8	1,3 - 3,1	2,1	2,0
Gneiss	2,4-2,7	1,9-4,0	2,9	1,8 - 2,4
Argillaceous schists	2,7	1,5 - 2,6	2,1	2,2 - 2,5
Sedimentary rocks				
Limestone	2,6 - 2,7	2,5 - 4,0	2,8	2,1 - 2,4
Marl	2,5 - 2,6	1,5 - 3,5	2,1	2,2 - 2,3
Quartzite	about 2,7	3,6 - 6,6	6,0	2,1 - 2,2
Sandstone	2,2 - 2,7	1,3 - 5,1	2,3	1,5 - 2,8
Claystone	2,5 - 2,6	1,1 - 3,5	2,2	2,1 - 2,4
Unconsolidated rocks				

(Table 1) cont.....

Rock (type)	Density ρ	Thermal Conductivity λ		Special heat capacity
Gravel, dry	2,7 - 2,8	0,4 - 0,5	0,4	1,4 - 1,6
Gravel, saturated with water	about 2,7	about 1,8	1,8	about 2,4
Sand, dry	2,6 - 2,7	0,3 - 0,8	0,4	1,3 - 1,6
Sand saturated with water	2,6 - 2,7	1,7 - 5,0	2,4	2,2 - 2,9
Clay / mud dry	-	0,4 - 1,0	0,5	1,5 - 1,6
Clay / mud saturated with water	-	0,9 - 2,3	1,7	1,6 - 3,4
Peat	-	0,2 - 07	0,4	0,5 - 3,8
Other substances				
Bentonite	-	0,5 - 0,8	0,6	about 3,9
Concrete	about 2,0	0,9 - 2,0	1,6	about 1,8

Evaluation of Thermal Potential

The use of geological maps is indicated for the calculation of the thermal potential in each region with the aim to investigate the possibility of installing open or closed cooling and heating system, taking advantage of the existing heat at surface water or groundwater and geological formations, which are not classified as geothermal potential. Actually, of all the minerals that make up the petrology, those that have the greatest spread and form the basis of the project area are selected. It is considered that each rock or rock unit is isotropic in the thermal conductivity and thermal capacity throughout the volume of the heat exchanger influence. Each petrological section is considered to have a single value of thermal conductivity and thermal capacity with isotropic character. The temperature distribution in the subsoil plays an important role in the geothermal field.

In a wide range of facilities or in large power plants it is useful to utilize the thermal response method results (Thermal Response Test, TRT), mainly because in the analysis of subsurface materials thermal properties, accurate assessment of their size is not possible. In that method, the network is connected with conditioned water production apparatus and is introduced into the heat exchanger by continuously measuring the outlet temperature and the heat flow from 24 to 48 hours. The measurements are performed with a mobile unit that includes a heat pump for the production of hot or cold water, measuring instruments (two heat flow meters and two temperature probes) and data logger. The device (Fig. **1**), is directly connected to the geothermal heat exchanger and therefore it must be located very close to drilling so that the system is not influenced by other parameters.

Fig. (1). Thermal response test device.

TRT enables to identify and predict the thermal behavior of the subsoil around the geothermal exchanger over time and the average specific heat, the mean specific thermal conductivity and the average thermal diffusion of materials are calculated. The thermal response tests are performed in the area of the subsoil considered, so that the geothermal system has better performance. The method measures: the average temperature across the heat exchanger prior to system operation, the average thermal conductivity of the subsoil formation, which includes the conductivity of the subsoil water, the thermal resistance of the geothermal exchanger.

Types of Normal Geothermal Energy Exploitation Systems

There are two basic technologies for the exploitation of normal geothermal potential:

- **Open loop geothermal systems or open systems:** These are installations that operate using pumping of underground fluids (especially water). They utilize the capacity of the constant temperature that have subsurface fluids and they use surface water or groundwater (by the lake, well, river, drilling) for thermal use. At the end, they reinject water to the underground or they dispose it to surface water available in the area. They draw water from the underground reservoir using drilling and usually through the use of an intermediate water/water exchanger between the GSHP and the open circuit impart or absorb energy in the system before the water returns to the reservoir. This type of system is appropriate in areas with a shallow depth of aquifer.
- **Closed loop geothermal systems or closed systems:** These are installations that use a closed piping circuit buried in the ground and they are divided into two categories:

- Vertical closed loop systems or Borehole Heat Exchangers (BHEs) or Vertical Earth Heat Exchangers (VEHEs). In that case, there is requirement for drilling one or more deep wells usually up to 100m, where usually double U type tubes are placed vertically forming a closed circuit. The initial cost is high, mainly due to drilling, their lifespan is large, and they require a small area for their development.
- Horizontal closed loop systems with different variations in their configuration. The initial cost is relatively small compared to other systems, their installation is easy, while their life is long. However, they require large surface area for their development, which is a crucial point for the final selection.

In most cases, both systems can be used for heating and for cooling, since it is permitted by the distribution of heating and cooling in the building. Detailed references to these systems is given in the following sections. There are also other geometries (*e.g.* systems in conical form or inclined drilling) or other forms that are not tackled here.

Along with the above systems, research is conducted on the ability to store heat or cooling energy in underground thermal energy storage systems (Underground Thermal Energy Storage, UTES). The latter is based on the concentration of heat in the subsurface mass or in a part of the underground with rate greater than the rate of diffusion. The storage capability is defined by the difference between the heat deposition speed in a part of the subsurface mass, generating local temperature increase and the heat diffusion speed, which tends to damp this difference. This phenomenon generates a temporally finite storage which can be exploited since its time of use is less than the time of full dissipation of heat underground. It is completely transitory and it cannot be exploited under all conditions. These systems are divided into [12, 22]: Aquifer Thermal Energy Storage (ATES) and Borehole Thermal Energy Storage (BTES). Thermal energy storage is used in power supply systems to compensate for any difference in the size of the thermal efficiency of an energy source and the thermal efficiency of an energy producer and to bridge the time between supply and demand for thermal energy.

The thermal energy storage system must be capable of absorbing energy rapidly, to store it for a relatively shorter period of time, and then to attribute it quickly. A typical example is the storage of hot water in a district heating or air conditioning system [22]. If the key objective of storage is to bridge the time between supply and demand, the storage system should receive a relatively large amount of heat energy, to keep it in storage for an extended period (*e.g.* few hours, days or a few months) and then to attribute it when needed. The storage cycle extends from the start of the charging process until the end of the discharge process. These systems are suitable for seasonal storage.

In temperate latitudes this type of storage can be used to balance the changing demand in winter and summer, since the stored thermal energy is used at the minimum within such a time, from the end of the charging process, able to not be restored temperature field. If the system temperature is less than the natural subsurface temperature, then we refer to cold storage. The heat is transferred through heat geothermal heat exchangers or *via* groundwater *e.g.* aquifer storage. Ground water can be pumped through wells from aquifers or from other provisions [22].

The systems that prevail are those of the closed loop (Figs. **3 & 4**), about 80% of all systems in the European Union, and mainly vertical geothermal heat exchangers. The horizontal system is a specialized application that in Denmark is still used, in France its use decreases despite the fact that it holds a significant share a few years ago. The open-circuit systems, (Fig. **2**) are implemented in different countries, depending on the existing legal and environmental context of the country and the availability of groundwater, and their permitted uses.

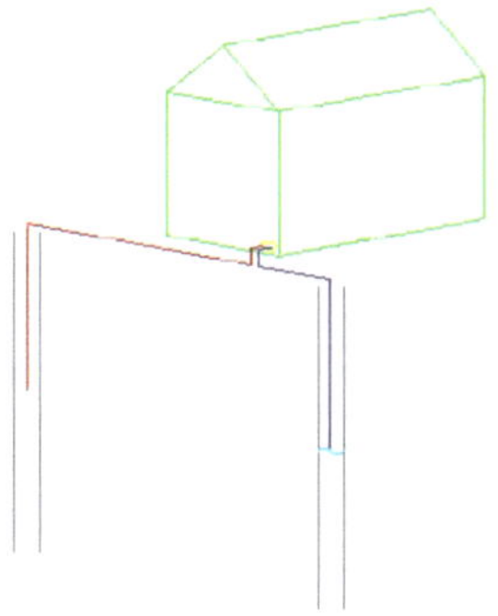

Fig. (2). Typical GSHP open circuit layout.

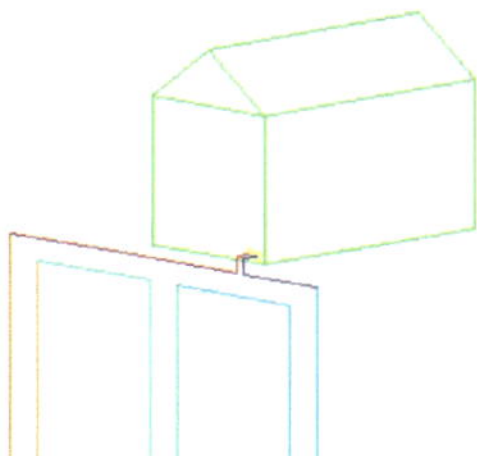

Fig. (3). Typical layout for closed loop GSHP with vertical geothermal system.

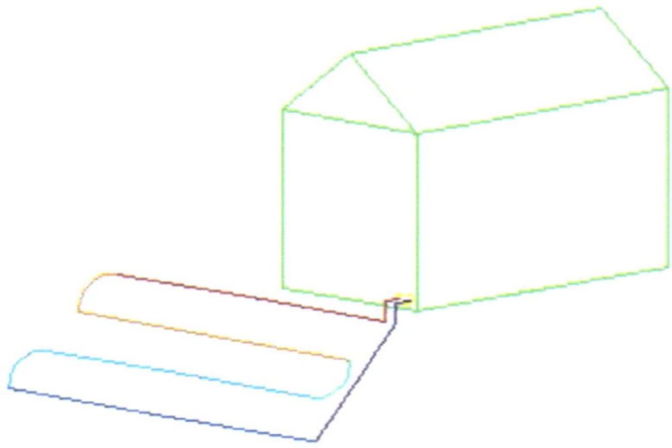

Fig. (4). Typical layout f or closed loop GSHP with horizontal geothermal system.

HEAT PUMPS TECHNOLOGY

Basic Theory

Heat pumps are machines that produce cooling or heating by consuming electrical energy based on a thermodynamic cycle (vapor compression refrigerant cycle) [25]. Heat is pumped from low temperature reservoirs and transferred to the working refrigerant which after vapor compression provides heat to the system. The most common heat reservoirs are air and water (lake, sea, river). Lately, the use of soil (geothermal energy) and sun radiation (solar energy) is considered as promising outcome.

The main feature of heat pumps is the ability to produce heating or cooling depending on the desired conditions. In cooling mode, heat is extracted from a high temperature environment to a lower one (which is always performed in all changes in nature – heat flows from higher temperature means to lower). Heat pump has the ability to pump heat even in heating mode (from lower temperature reservoirs to higher – environment to conditioned space) [25].

The operational performance of a heat pump is defined in both cooling or heating mode as the ratio of the load supplied (in both modes) over the power consumed (electric energy). **Coefficient of Performance (COP)** is the performance factor in heating mode and **Energy Efficiency Ratio (EER)** in cooling mode. Both factors are affected by reservoir temperature. These factors are calculated in preset conditions for all manufacturers (EER is measured in environment temperature of 35°C and COP with 7°C).

In general, heat pump performance is described by the following factors:

- **EER/COP**, describe the overall performance of a heat pump based on a set of environmental and working temperatures (in both modes).

- **Seasonal COP (SCOP)**, describes the yearly performance of the heat pump given by the ratio of heating produced by the heat pump over the energy consumed (electrical) over a year. Seasonal COP is divided into three regional zones for Europe (warmer, average and colder) based on the average environment temperature. Seasonal COP is greater than COP as it includes the environment temperature variation during winter in relation to the preset temperature of 7° C used in the calculation of COP. Cooling mode operation is described by **Seasonal Energy Efficiency Ratio (SEER)** which is the ratio of cooling produced by the heat pump over the energy consumed (electrical) over a year.
- **European seasonal COP and EER (ESCOP & ESEER)** are factors describing the annual performance of heat pumps used by Eurovent Certification Company. The calculation combines partial and peak loads in both operations (cooling and heating) by load ratios and weighting factors. The equation used is: ESEER=0.03x100% load+0.33x75% load+ 0.41x50%+0.23x25%.

Geothermal Heat Pumps

Geothermal heat pumps are based on the exploitation of geothermal energy originated by the subsoil or subsoil water to provide heating, cooling and domestic hot water production *via* vapor compression cycle with ecofriendly refrigerants.

The main components of a geothermal site are the heat pump, the geothermal exchanger (vertical/ horizontal closed loop or water from a drilling) and the internal installation (underfloor system, fan coils, and radiators) to supply heating or cooling to the interior [26 - 28]. The main principle is based on subsoil's constant temperature through a year that reduces consumption (electrical energy) and contributes in the elimination of greenhouse effect.

Geothermal applications maybe integrated in sites for domestic or industrial cooling, heating and domestic hot water preparation. They also maybe integrated in agricultural plants to provide heat in winter exploiting drilling water. The performance of geothermal installations is the best achievable among the known systems nowadays.

The main components of geothermal heat pumps, Fig. (**5**), are the compressor, the expansion valve, and the heat exchangers to perform energy transaction. Two heat exchangers are integrated into the refrigerant cycle. One is extracting energy from the subsoil and the other provides energy (upgraded by the compressor) to the internal installation. The working refrigerant is always chosen to be nontoxic, nonflammable and with low depletion factors (usually used refrigerants are R410a, R407C and R134a).

Fig. (5). View of GSHP system machinery in a house.

CLOSED LOOP GEOTHERMAL SYSTEMS

Vertical Geothermal Systems

In these systems, the geothermal heat exchanger is integrated in vertical positioning within the underground to depths till 100 m acting as the mean to absorb or reject heat to the soil [2, 5, 29]. According to DIN 4640, the depth of the exchanger may rise up to 400 m in depth but in these occasions the pipes have to be reinforced to withstand the pressure developed within their cavity [12, 13, 21, 22]. The basic working criteria when designing the geothermal exchanger are the heat transfer fluid within the pipes not to exceed a temperature variation of ±11K (mean weekly operation) and ±17K in peak loads. These working conditions are difficult to overcome as the heat transfer ratio to the soil is rather slow.

Vertical closed loop systems take advantage of the geothermal heat flux. The geothermal heat flux may be neglected when the exchanger in working in depths from 0 to 30 m as the main source of energy is solar radiation and the interaction of rain water depending on the soil. The main advantage of the vertical closed loop heat exchanger is the ability of the system to be integrated in sites where there is lack of space (retrofit actions in buildings without spare space) (Fig. **6**). It is the optional choice when there is no sufficient supply of heat transfer fluid to integrate an open loop system or not enough surrounding space to install a closed loop horizontal system.

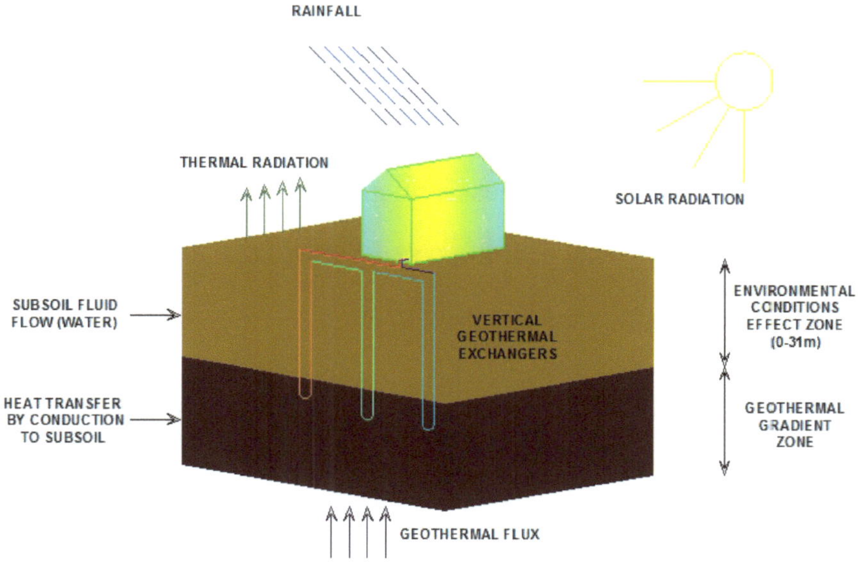

Fig. (6). Heat transfer mechanisms underground.

A geothermal heat pump takes advantage of the heat transfer fluid's ability to absorb or reject heat *via* the vertical closed loop to the soil hence the main design criterion is the temperature variation of the soil in depth. Geothermal heat flux depends on depth and the type of soil which the loop is integrated within. Different soil layers alter the heat transfer phenomena and the problem becomes more complex when water is present (there is a change of heat transfer coefficient in depth and time). When designing a geothermal heat exchanger, the main difficulty is to define the soil temperature variation with depth for different substances. A lot of effort is being put from researchers worldwide to create a mathematical equation to approach this phenomenon. One of these equations, developed by Kasuda and Archenbach, is [29 - 31]:

$$T = T_{mean} - T_{amp} * e^{-Depth*\sqrt{\frac{\pi}{365 \over a}}} * \cos\left\{\frac{2\pi}{365} * \left[t_{now} - t_{shift} - \frac{Depth}{2} * \sqrt{\frac{365 \over \pi}{a}}\right]\right\} \quad (1)$$

where:

T: soil temperature (K)

T_{mean}: mean surface temperature (K)

T_{amp}: yearly peak temperature surface variation (K)

Depth: depth (from surface) (m)

t_{now}: surface temperature (real time measurement) for a certain day

t_{shift}: minimum surface temperature (for a certain day of the year)

α: soil thermal diffusion (m^2/s), originating from:

$$a = \frac{k}{\rho c} \quad \textbf{(2)}$$

where:

k: thermal conductivity (W/mK)

ρ: density (kg/m^3)

c: soil thermal capacity (J/kgK)

When setting the values for thermal diffusion, specific density and soil specific thermal capacity, it is taken into consideration that the soil act as a uniform mean hence the values used are the mean.

Design Approach

When it comes to the designing of such systems for heating, cooling and domestic hot water preparation there are a lot of methods to calculate the geothermal exchanger regarding the load required [32 - 42]. For small systems (load up to 30 kW) the design of the system is based on assessments. Geological maps or technical geological studies are used to define soil characteristics. If the system is used to cover only heating loads the calculations may be based on the values of specific heat pumping rate (W/m) given in Table **2** [12, 13]. If the working hours of the system are extensive besides the specific heat pumping rate values a safety factor has to be considered that defines the impact of the loop to the soil and must not exceed a value of 100 to 150 kWh/m year. This safety factor is for heating operation only and if the system covers cooling loads then this factor may vary. The overall length of the geothermal exchanger (L) originates from the simplified relation [31]:

$$L_{BHE} = \frac{P_{ground}}{N*P_{BHE}} \quad (3)$$

Where N: is the number of vertical geothermal loops, P_{BHE}: is the specific heat pumping rate and P_{ground}: is total amount of energy that maybe extracted from soil, given by:

$$P_{ground} = q_{h,HP,out}\left(1 - \frac{1}{SPF}\right) \quad (4)$$

Where $q_{h,HP,out}$ is heat pump heating output and SPF is heat pump seasonal power factor.

The outcomes of this relation are valid only when:

- The system operates in heating mode and domestic hot production only (cooling is neglected).
- The overall length of the vertical exchanger is no less than 40 m and no more than 100m.
- The pipes are coaxial or doubled U.
- The distance among the vertical loops has to be more than 5m for loops less than 50 m in depth and 6 m for loops more than 50 m in depth.

The values given in the table below may be used to assess the design of a system and need to be verified before the final outcomes. They represent the ability of a vertical closed loop heat exchanger to absorb heat from soil for central Europe regions. Statistical results for loops working in Germany (small systems) showed that the average heat pumping ratio from soil is 50 W/m [34].

When it comes to the design of large systems for working hours more than 2400 hrs, covering of cooling loads and demanding load more than 30 kW the geothermal loops have to be calculated in detail. It is proven that it is better to use the mean monthly desirable heating and cooling loads values when designing the geothermal closed loop exchanger and in parallel the short term maximum loads [12, 13]. The length of the vertical closed loop exchanger is calculated based on technical manuals and national standards [12, 13, 21, 22, 31 - 42].

For larger systems besides the standards, a computational software has to be used to approach the temperature variation in depth (soil characteristics). For example, German Norm VDI 4640 [12, 13, 21, 22] calculates the length of the geothermal exchanger by taking into consideration the values from Table **2**. Representing the specific heat pumping rates from soil and yearly working hours (1800 – 2400 hrs) and ASHRAE suggests the use of Kavanaugh and Rafferty equation [33 - 37, 42] which is analyzed later on this chapter. European countries that have developed sufficient manuals and guidance are those that have a well-established market for geothermal systems such as Germany, Austria, Switzerland [32] and lately Italy [36]. These countries base their standards on ASHRAE's guidance [37]. Other countries [35] provide legal notices concerning the distance of the loops from water drillings, pollutant sources and other public service installations but no technical guidance on the design of such systems.

In general, the design of vertical closed loop systems have to take into consideration a lot of factors such as heating/cooling loads, geothermal heat pump manual, soil characteristics, type of geothermal exchanger (material, filling mean, heat transfer fluid). Most empirical and simplified calculations use the specific heat pumping rate (W/m) which involves soil thermal conductivity, working hours, drilling diameter, filling mean and the impact between the loops. A method for a simplified calculation of the temperature variation along a geothermal vertical closed loop exchanger was originated in the USA in 1950. Based on Kelvin's linear source theory [13] and Ingersoll 's equation [41] which is based on Kelvin's theory, the temperatures based on soil thermal conductivity, heat pumping rate and loop distance along the heat exchanger may be calculated.

Table 2. Indicative specific heat pumping rate values [12, 13].

Underground	**Specific Heat Pumping Rate**	
	For 1800 Hours Operation	***For 2400 Hours Operation***
Typical indicative values:		
Poor underground (Dry sediment) (λ<1,5 W/(m·K))	25 W/m	20 W/m
Normal rocky underground and saturated with water sediment (λ< 1,5 -3,0 W/(m·K))	60 W/m	50 W/m
Consolidated rock (λ>3,0 W/(m·K))	84 W/m	70 W/m
Individual rocks:		
Gravel, sand, dry	< 25 W/m	< 20 W/m
Strong water flow in grave lands and for individual systems	80-100 W/m	80-100 W/m
Clay, loam	35-50 W/m	30-40 W/m
Limestone	55-70 W/m	45-60 W/m
Sandstone	65-80 W/m	55-65 W/m

(Table 2) cont.....

Underground	Specific Heat Pumping Rate	
	For 1800 Hours Operation	*For 2400 Hours Operation*
Siliceous magmatite	65-85 W/m	55-70 W/m
Basic magmatite	40-65 W/m	35-55 W/m
Values can vary significantly due to the tissue of the rock, such schisms, leaf formation, weathering etc.		

Guernsey *et al.* [40] came up with an empirical equation based on linear source theory that may be used for calculations (valid when αt/r2>1):

$$\Delta T = \frac{0{,}1833Q}{\lambda}\left(log_{10}\frac{\alpha t}{r^2} + 0{,}106\frac{r^2}{\alpha t} + 0{,}351\right) \quad \textbf{(5)}$$

Where:

ΔT: temperature difference (K)

Q, heat flux per pipe length (W)

λ, thermal conduction factor (W/mK)

r, distance from pipe center (m)

t, time (s)

α, thermal diffusion

ρ, soil density (kg/m^3)

c_p, specific thermal capacity (kJ/kgK)

Kavanaugh and Rafferty [37, 42], came up with a more accurate method to calculate the overall length of the exchanger by transforming the steady state equation in order to supply the variant transfer ratio of a geothermal loop by using a series of steady heat transfer pulses. The steady thermal resistance of the soil per unit length is calculated related to a time step which provides a steady heat transfer pulse. A wall pipe factor is added to calculate the thermal resistance between the soil and the pipe and the pipe with the heat transfer fluid. The equations for the overall length in cooling (L_c) and in heating mode (L_h) are:

$$L_c = q_a R_{ga} + (q_{lc} - 3{,}14W_c)(R_b + PLF_m R_{gm} + R_{gd}F_{sc})/(t_g - (t_{wi} + t_{wo})/2 - t_p) \quad \textbf{(6)}$$

$$L_h = q_a R_{ga} + (q_{lh} - 3{,}14W_h)(R_b + PLF_m R_{gm} + R_{gd} F_{sc})/(t_g - (t_{wi} + t_{wo})/2 - t_p) \quad \textbf{(7)}$$

Where:

F_{sc}, thermal losses factor (small systems)

PLF_m, partial load factor (per designed month)

q_a, yearly mean heat rejected to the ground (Btu/h)

q_{lh} total yearly thermal load (Btu/h)

q_{lc}, total yearly cooling load (Btu/h)

R_{ga}, soil thermal resistance, yearly pulse (h.ft.°F/Btu)

R_{gd}, soil thermal resistance, daily pulse (h.ft.°F/Btu)

R_{gm}, soil thermal resistance, monthly pulse (h.ft.°F/Btu)

R_b, drilling thermal resistance (h.ft.°F/Btu)

t_g, non disturbed soil temperature (°F)

t_p, close loop impact correction factor

t_{wi}, heat transfer fluid temperature to geothermal heat pump input

t_{wo}, heat transfer fluid temperature to geothermal heat pump output

W_c, input power at the design cooling load

W_h, input power at the design heating load

Heat transfer loads, designing loads and temperature correction factors are positive values for heating and negative for cooling. The aforementioned equations take into consideration three different values of soil thermal resistance pulse a) daily, b) monthly and c) yearly. The installed overall vertical closed loop geothermal exchanger length is based on the biggest value (in heating or cooling) originating from the calculations. In many buildings the cooling load is larger than the one for heating which leads to a longer heat exchanger. This means that during winter excess heat will be available for rejection. In order to keep the cost low a cooling tower may be integrated to reject heat to the surroundings instead to the soil.

Equations **6**, **7** may be further simplified as:

$$L_c = q_a R_{ga} + (C_{fc}\cdot q_{lc})(R_b + PLF_m R_{gm} + R_{gd}F_{sc})/(t_g - (t_{wi} + t_{wo})/2 - t_p) \quad \textbf{(8)}$$

$$L_h = q_a R_{ga} + (C_{fh}\cdot q_{lh})(R_b + PLF_m R_{gm} + R_{gd}F_{sc})/(t_g - (t_{wi} + t_{wo})/2 - t_p) \quad \textbf{(9)}$$

Correction factors,, refer to the total thermal load absorbed or rejected by the geothermal heat pumps and are related to the EER and COP of the units. In order to calculate correction factors values technical information by the heat pump manufacturer have to be supplied (EER and COP). The mean annual heat rejected ratio to the soil q_α may be calculated by using the annual equivalent maximum load working hours in heating or cooling mode:

$$q_a = (C_{fc} \cdot q_{lc} \cdot EFL_{hoursc} + C_{fh} \cdot q_{lh} \cdot EFL_{hoursh})/T_a \quad \textbf{(10)}$$

where EFL_{hoursc}, EFL_{hoursh}, are the equivalent working hours in total cooling or heating mode annually and T_α, is the annual working hours (8760h). Equivalent working hours in total load for heating are the working hours in total heating load per yearly operation and in cooling are the working hours in total cooling load per yearly operation:

$$EFL_{hoursc} = \frac{E_c}{q_{lc}} \kappa\alpha\iota EFL_{hoursh} = \frac{E_h}{q_{lh}} \quad \textbf{(11)}$$

E_cκαι E_h, annual energy consumptions in heating and cooling mode.

C_{fc}, C_{fh} are given in the Table **3** below:

Table 3. Geothermal heat pump correction factors [32, 38].

Cooling EER	**C_{fc}**	**Heating COP**	**C_{fh}**
11 (3,22)	1,31	3,0	0,75
13 (3,81)	1,26	3,5	0,77
15 (4,40)	1,23	4,0	0,80
17 (4,98)	1,20	4,5	0,82

Construction Issues

General Guidelines

Vertical closed loop geothermal exchangers are used when there is lack of surrounding space regardless soil composition. The minimum distance of the loop with any external wall of the building has to be 2 m and the static stability of the construction must not be disturbed in any way. For large systems (>30kW) a testing drill may be performed to verify the geological subsoil composition and then this testing drill may be used to integrate a vertical geothermal exchanger [13, 31].

Geothermal exchangers are installed into the holes from the drilling contractor and needs extra attention to avoid mechanical wear, Fig. (**7**). The installation steps to be fulfilled are [13, 31]:

- The heat exchanger has to be inserted into the hole filled with water so as to add weight and avoid elevation when filling of the drilling will be performed. It is also preferable to add weight at the bottom (to avoid friction wear of the pipes).
- The filling tube has to be installed attached to the geothermal loop at the same time (usually strapped on the heat exchanger).
- Before filling, a pressure test has to be performed (air may be used to ensure to damage has been done to the pipes).
- The hole is filled with conductive medium (grout).
- After filling the pipes with water, a pressure of 6 bar is applied (for at least 1 hour) and the pressure change has to be less than 0,2 bar. If after the pressure test the weather is cold (risk of frost) the loops need to be emptied of water before they are connected.
- Geothermal loop pipes are sealed to avoid dust, rocks or any other impurity that may insert their cavity and block the heat transfer mean flow.

Vertical closed loop geothermal exchangers are indicated to be connected into parallel loops and the following general instructions are suggested [13, 31, 34]:

a. Drilling diameter has to be 6-8" and the depth may vary from 60 to 100 m.
b. The distance between the vertical loops has to be greater than 5-6m to avoid thermal saturation of the subsoil.
c. The preferable construction material of the geothermal loop pipes is crosslinked polyethylene (PeX-A or any other crosslinked pipe) to ensure mechanical properties. When the pipes are installed into the drilling hole a conductive filling medium is applied, namely the grout (cement, bentonite, quartz sand *etc.*).
d. The average heat pumping rate of a vertical closed loop geothermal exchanger

from the subsoil is 35 to 65 W/m (depending on subsoil composition and the working hours in maximum load demand). These values may vary in special occasions.

Fig. (7). Stages for vertical geothermal heat exchanger construction.

Vertical closed loop geothermal exchangers may be classified depending on the heat transfer direction from the subsoil to the heat transfer fluid circulating within the pipes cavity. Fig. (**8**) represents the two basic operating designs:

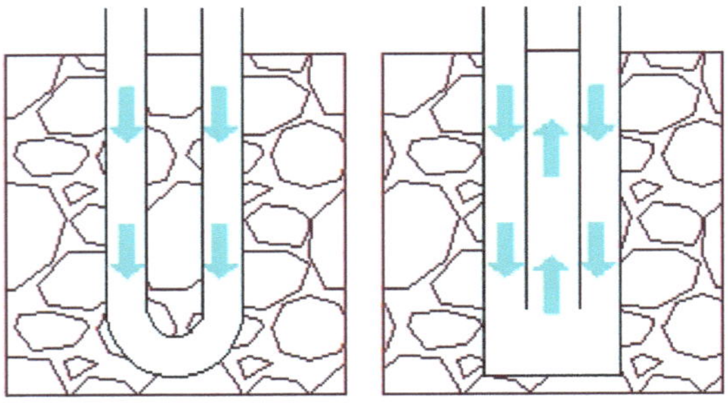

Fig. (8). Vertical geothermal heat exchangers types based on the pipeline layout [31].

- U type geothermal exchanger (left)
- Coaxial geothermal exchanger (right)

Types of Vertical Geothermal Heat Exchangers

Geothermal Heat Exchanger with U-Tube

The most common pipe arrangement within a geothermal well is the integration of a U-shaped, Fig. (**9**). polyethylene exchanger to achieve heat transfer from the subsoil to the circulating media. Single U shaped configuration is widely used in Northern America and Northern Europe when double U shaped configuration is used in Southern Europe applications. A common feature for drilling wells in Northern Europe is the presence of water in swallow depths. In central Europe and the U.S., drilling holes (after the integration of the exchanger) are filled with a conductive medium made of cement, bentonite and quartz sand. New filling media with improved heat transfer coefficient are available today in the market.

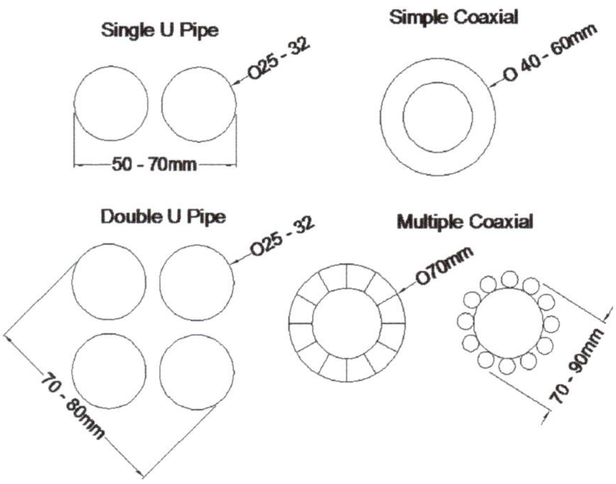

Fig. (9). Types of vertical geothermal exchangers based on the pipeline layout [12, 31].

Single U Shaped Geothermal Exchanger with Spacers

Single U shaped exchangers were widely used in the early years. U shape configuration assures heat transfer in downwards and upwards movement of the working fluid with the pipes. Main advantages are the designing simplicity and the ease of transportation and integration in the geothermal drills. The main disadvantages of this configuration are that if the pipe is damaged (during installation or due to high working conditions) then the geothermal hole is useless and the flow has to be turbulent otherwise the heat transfer phenomena is limited. The performance of single U shape configuration may be upgraded by applying spacers between the pipes and placing them close the drilling walls.

Multiple U Shaped Geothermal Heat Exchanger

Double, triple and in general multiple U shaped pipe configurations are the constructive extensions of single U shaped pipes. The main advantage of such configurations is the increment of heat exchange surface and the decrease in value of thermal resistance of polyethylene pipes. Conduction is also decreased hence the flow does not need to be turbulent unless maximum loads are required. Reduced flow leads to pressure drop (manometric head) within the pipes. Factors affecting the thermal resistance of the drilling are the flow rate of the heat transfer fluid, the construction material of the exchanger and the number of the loops and finally the thermal characteristics of the grout.

Coaxial Geothermal Heat Exchanger

The main characteristic of the coaxial pipe configuration is that heat transfer may be performed by the working fluid either downwards or upwards depending of the flow (inner or outer diameter). In some occasions the inner tube is insulated to avoid heat transfer between the two flows.

Coaxial Geothermal Heat Exchanger without Jacket (Standing Column Well)

The simplest configuration regarding vertical loop geothermal exchanger is the integration of a single pipe in the drilling well and the supply of the working fluid in its cavity. The fluid is then released at the bottom of the well and is supplied to the surface by filling the gap with the drilling walls. The heat transfer is the maximum achievable (working fluid in direct contact with the drilling walls). This type of configuration is known as standing column well. If the subsoil is permeable leaking working fluid increases heat transfer ability.

Coaxial Geothermal Heat Exchanger Tube-in-Tube Type

Most drilling wells have unstable walls so a closed loop is required (geochemical interactions are also a preventing factor). The outer tube may be constructed inserted in the well and then fill the gap with the walls with cement or any other filling mean. The construction material may be plastic or even metallic but this will increase cost. This configuration is easy to be applied to shallow drills or when the subsoil is firm. This method is suitable for clayey subsoil (the gap is filled with clay and no other material is required).

Coaxial Geothermal Heat Exchanger with Soft Jacket

The soft jacket is inserted in the drilling well when the hole is full with liquid to ensure sealing between the walls and the internal cavity. Unfortunately, in most occasions the material was torn and this method is no longer used.

Multi-Channel Geothermal Heat Exchanger

The outer cavity of the external wall may be divided into smaller chambers. Due to increased thermal resistance of the grout between the outer surface of the tube and the drill walls this configuration is the same as multiple U shaped pipes.

Multitube Geothermal Heat Exchanger

The outer flow channel may be consisted of multiple tubes (with smaller diameter that the internal tube). There are no constructed configurations in the market hence the designing and the testing needs further development but some raw experimental data have shown promising outcomes.

Materials and Grout Filling

Due to the presence of concentrated water geothermal exchanger must be resistive to corrosion [12, 30 - 34] and no toxic. All pipes exposed to solar radiation must be fitted with UV resistant layers and furthermore must withstand the designing pressure and temperature conditions. Polymer based pipes (polyethylene, polypropylene and polybutylene) pipes are suitable for such installations. In order to avoid freezing problems, glycol in various ratios is added to the heat transfer fluid to lower working temperature (down to -20 °C) [12, 30, 32 - 34]. When the working conditions are not extreme the use of antifreeze agent is useless as the temperature of the heat transfer fluid does not drop below 0°C (the lack of glycol leads to improved performances in both cooling and heating modes).

Pex-A are cross-linked polyethylene pipes (high pressure peroxide method) which may withstand high burst strength and high pressure loadings and are the most suitable for vertical closed loop geothermal exchangers. PE-100 polyethylene pipes are also used but due to low resistance to single point stress are more appropriate for horizontal closed loop geothermal exchangers (usually protected with sand).

Grout or filling medium is used for two major reasons:

- To ensure thermal transfer from the subsoil to the heat transfer fluid circulating in the geothermal heat exchanger (absorbed or extracted)
- To seal the drilling well (especially onto the surface) from contaminants (fluids, rocks) and to prevent water from aquifer to float (if present).

Grout is inserted to the geothermal well from a pipe attached to the heat exchanger (the pipe and the exchanger are inserted at the same time). When the heat exchanger is set to its final position, grout is pumped *via* the additional pipe to the bottom of the well. The pipe is lifted to the surface in a slow rate to ensure that the end is always submerged in grout and that no air is trapped in between. If the depth of the well is more than 60m the lift of the pipe is difficult and it is better to insert the grout without lifting the pipe (actually the pipe is filled and left in the well). To ensure that grout is equally distributed in the well a pipe may be attached till the depth of the hole and a second one to the middle. Filling is performed to the longest pipe first and when the level of the grout reaches the second pipe then it is inserted though that pipe. An easy way to confirm if the bottom of the filling pipe is emerged into grout is to place a manometer and check when pressure is risen (pumping in the grout).

The most common grout is consisted of bentonite, water and cement. Mixture of bentonite and water must be avoided due to its low thermal conductivity (less than 0,7 W/mK) and low resistance to freezing temperatures. To face these two problems, cement is added to the mixture to allow usage with temperatures down to -15° C, and quartz sand to improve thermal conductivity (>0,8 W/mK). The volumetric composition of the mixture is approximately 25% bentonite, 25% cement, 25% quartz sand and the rest is water. If the ratio of cement is risen, then thermal expansion of the grout is lowered. Other expandable clay rocks may be used instead of bentonite. The main disadvantage of bentonite based grout is its sensitivity to frost. When the dissolved water in the mixture freezes and then liquefies it may not be absorbed by bentonite. The remaining water in the mixture (liquid phase) may lead to rise in pressure within grout's cavity when it is about to freeze once more. The added pressure may lead to fracture of the geothermal heat exchanger.

Horizontal Type Geothermal Systems

Design Approach

The basic considerations that need to be clarified before a horizontal closed loop geothermal heat exchanger is applied are:

- Maximum required cooling and thermal loads through the year.
- Soil composition.
- Available surrounding area to integrate the geothermal heat exchanger.

The basic designing steps are [2, 5, 12, 13, 21, 22, 43 - 47]:

- Selection of the heat exchanger arrangement and orientation.
- Total length of the heat exchanger.
- Flow velocity in the heat exchanger.

A basic designing concept to calculate the total length of the horizontal closed loop geothermal heat exchanger is the use of tables that contain values of available specific heat pumping rate (W/m^2) from the soil depending on its composition. These tables are for heating mode only and the operating hours are 1800 to 2400.

Table **4** shows specific heat pumping rates for horizontal closed loop geothermal heat exchangers in heating mode. The followings requirements must be met to use the Table **4** [13]:

- Heating mode only (including domestic hot water production).
- Running hours for domestic hot water production *via* the heat pump must be added to the overall working hours.

If the working hours are more than those of Table **4** (cooling is included) then the annually specific heat pumping rate has to be calculated by taking into consideration the long-term influence factor for the soil. For heating mode only the values are from 50 to 70 kWh/(m^2. year) [13]. These values alter when cooling mode is performed as well (recharging of the soil). According to DIN 4640 [13], for both cooling and heating mode the temperature of the heat transfer fluid must not exceed a value of ±12K, (mean weekly operation value) in relevance to the soil temperature (non-disturbed soil temperature) and ±18K (peak operation valu e). In general, fine grained sediment soils with high water content such as muddy or clay soils are best suitable for horizontal closed loops geothermal applications and on the other hand sandy or pebble soils must be avoided. These assumptions are valid for geothermal heat pump with working temperature difference of ±5 έως ±7K.

The basic techno-economical criterion to operate a heat pump with a geothermal based heat exchanger is the payback period. Geothermal systems work with the best achievable performance, in both cooling and heating mode than any other available system nowadays. The initial cost though is a limiting factor for the final user to adopt such a system and in order to overcome this barrier the payback period must not exceed 8 operational years.

Table 4. Specific heat pumping rates for horizontal closed loop geothermal heat exchangers in heating mode [13].

Underground	Specific Heat Pumping Rates	
	For 1800h(W/m²)	*For 2400h(W/m²)*
Dry non-firm soil	10	8
Firm soil	20-30	16-24
Water saturated sand, pebble	40	32

Construction Issues

Types of Horizontal Geothermal Systems

Horizontal closed loop geothermal heat exchangers are applied to sites where there is enough surrounding place to integrate the pipes regarding soil composition. Heat transfer fluid (water or mixture of water with glycol) is circulated to the exchanger and rejects or absorbs heat transferred to the heat pump. The surface occupied by the geothermal closed loop horizontal heat exchanger depends on soil composition and the maximum loads that need to be covered while a significant factor is also the depth of installation (Fig. **10**).

Soil temperatures may reach freezing points to depth lower than 1 m, even without the use of geothermal heat pump. In depth lower than 2m, soil temperature may not go beneath 5°C even in polar conditions. The temperature is always risen moving deeper to the subsoil due to geothermal gradient but heat flux is reduced. In order to avoid presence of ice, even in spring periods, the horizontal pipes of the geothermal loop must be placed in distance so as to eliminate the overlapping formation of ice to the soil. The mean distances vary between 0,3 to 0,8m [13]. Besides frost the distance must be in a range to avoid thermal overloading of the soil (the exchanger must be able to cover heating and cooling loads throughout its operation). These limitations are not considered to countries in the Mediterranean basin as mean winter temperatures are high and heating operation hours are low. These conditions allow the integration of the horizontal loop to depth close to 0,8 m.

The minimum distance of the closed loop geothermal heat exchanger with water supplies or sewage pipes is 0,7 m [12, 30, 32 - 34]. The heat exchanger may be integrated to the soil by either removing the upper layer to the installation depth or by digging ditches and install pipes in slinky mode. The upper surface of the pipes needs to be protected by a sandy layer and in order to avoid damage a warning strip must be placed 0,3 m above the pipes. The surface above the horizontal closed loop geothermal heat exchanger must remain non-constructible and breathable.

Fig. (10). Indicative form of horizontal geothermal heat exchanger in the installation phase.

Figs. (**11**) to (**13**) represent selected available modes for the installation of the horizontal closed loop geothermal heat exchanger. Modes may vary from single pipe loops to multiple pipes loops, from horizontal installation to vertical or even from plain pipes to slinky formation. The final choice of the geothermal heat exchanger is based on its advantages and disadvantages depending on the available space, soil composition, yearly raindrop and depth of installation [12, 30, 32 - 34].

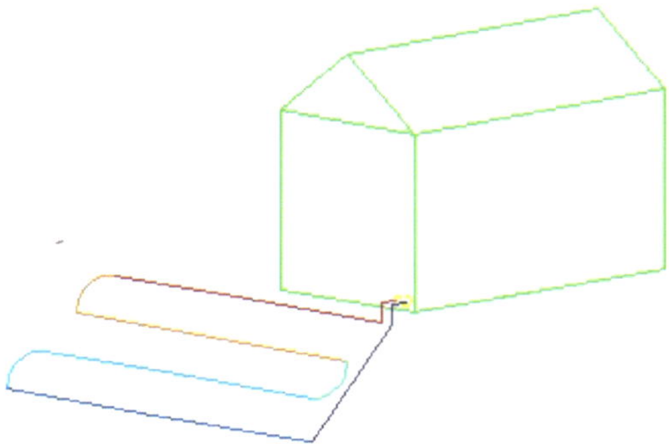

Fig. (11). Horizontal Geothermal System of Simple form in Series.

Systems in Series

The basic advantage of such a formation is the high values of specific heat pumping rate from the soil as the flow is simple and large diameter pipes may be used (compared to other formations). The basic disadvantages are the high installation cost due to the required excavation works, and due to larger pipes in diameter there is an increase in cost and heat transfer fluid volume. Heat transfer is performed in one dimension and the decrease in performance is linear (depending on the length of the pipe).

Parallel Systems

The main advantage of parallel formation is the use of small diameter pipes hence lower cost and less anti freezing agent. The main disadvantage is the difficulty in purging air during operation and the achievement of designed operating conditions (flow adjustment).

Single loop in Ditch System

During this formation, a single loop is integrated to a ditch. The orientation is usually vertical so as to increase available area of installation. The main advantage of this formation is the increased performance compared to series or parallel loops due to its larger area of interaction (heat transfer) with the soil.

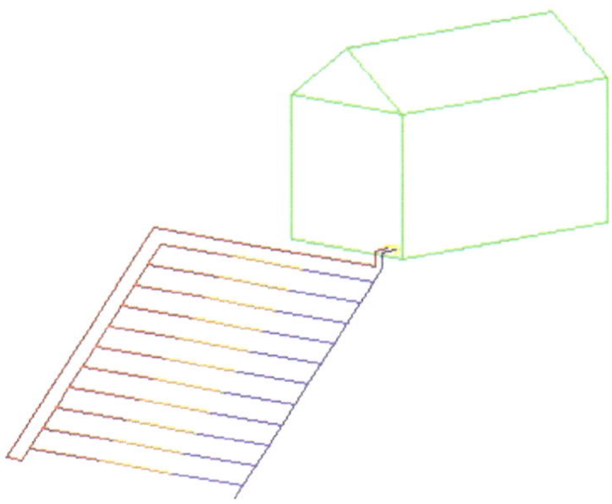

Fig. (12). Horizontal geothermal system of simple form in parallel.

Spiral Layout Systems (Slinky)

Slinky loop formation is used when the available area for integrating the heat exchanger is limited. Compared to single loops in ditches, slinky loops need less installation space and volume and contain larger heat transfer mean volume. The main designing concept is soil composition and operational loads through the year. Installation depth varies from 1.2 to 2.5 m.

Slinky loop systems may be constructed horizontally or vertically. In horizontal orientation, the length of the ditch depends on the spiral diameter of the pipe whence in vertical positioning the ditch is no more than 0.3 m. In both occasions the filling on the gaps after the exchanger is installed is of great importance to achieve the designed conditions. The cost of installation is decreased due to less excavation work but slinky pipes are larger in length (and cost) and so the overall balance is equal.

Fig. (13). Horizontal geothermal system with spiral layout (slinky) – Horizontal or Vertical installation.

Materials

Due to the presence of concentrated water geothermal heat exchanger must be resistive to corrosion [12, 30, 32 - 34] and no toxic. All pipes exposed to solar radiation must be fitted with UV resistant layers and furthermore must withstand the designing pressure and temperature conditions. Polymer based pipes (polyethylene, polypropylene and polybutylene) pipes are suitable for such installations. In order to avoid freezing problems (fracture due to expansion) glycol in various ratios is added to the heat transfer fluid to lower working temperature (down to -20 °C) [12, 30, 32 - 34]. When the working conditions are not extreme (temperate climate) the use of antifreeze agent is useless as the temperature of the heat transfer fluid does not drop below 0°C (the lack of glycol leads to improved performances in both cooling and heating modes).

OPEN LOOP GEOTHERMAL SYSTEMS

Introduction

Geothermal open loop systems are the oldest systems in this category, since the first of them occurred in the late 1940s [2]. Although they are simple in their nature they require careful planning, mainly regarding drilling and supply of ground water, the heat exchanger and the availability of water in order to obtain an efficient and reliable system. The Geothermal open loop exchangers (Figs. **14**

& **15**) use surface or ground water (from a lake, well, river, drilling or the sea) as a heat source or cooling and areas for disposal / of return water [48 - 50].

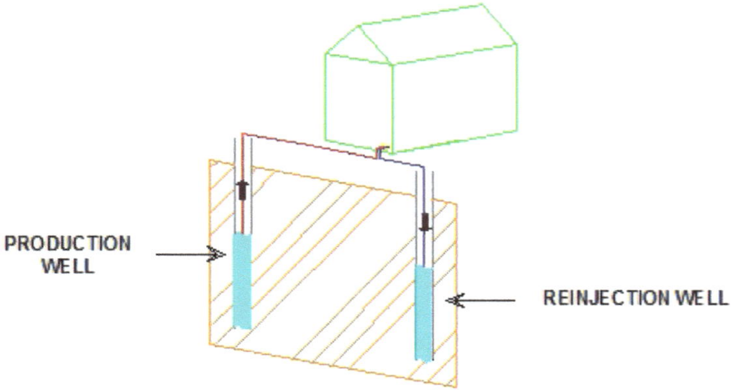

Fig. (14). Open geothermal heating and cooling system by discharging to underground water.

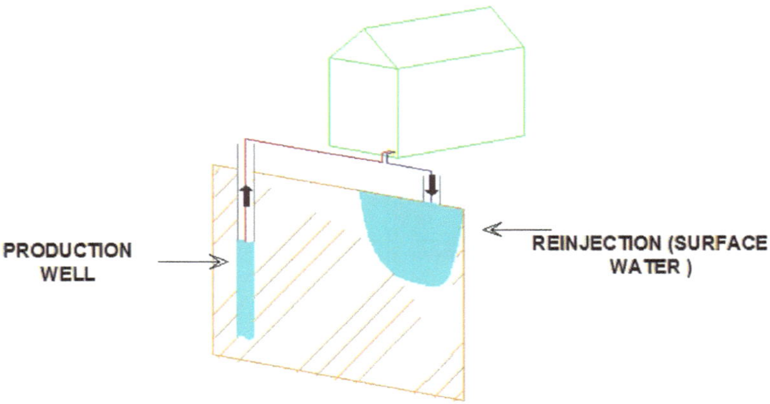

Fig. (15). Open geothermal heating and cooling system by discharging to surface water.

They draw water from the underground reservoir using drilling and usually intermediate water/water heat exchanger interposed between the geothermal heat pump and the open circuit, while they impart or absorb energy in the system before the water returns to the reservoir. This system is appropriate in areas with a shallow depth of aquifer. In this case, they exploit the property of the constant temperature that have underground reservoir waters during the whole year, regardless of the atmospheric conditions and the high specific heat of water (underground fluid). During the heating cycle in the winter the system works by pumping heat from the subsoil and it transfers it to the building while during the cooling cycle in the summer, heat is transferred from the building into the ground water. The heat transfer is accompanied by mass transfer of subsurface fluid and of course for the environmental balance of the system this fluid should return to the original water table.

The underground fluid is pumped from the ground to the tubes of the heat exchanger (evaporator). Depending on the type of heat exchanger, the cooling fluid comes in contact with the subsoil fluid (water with different elements) or surrounded by the water mains. The subsoil fluid exits from the heat exchanger and it advances to the underground area or surface disposal (Fig. **15**). The selection of the underground fluid pump is a critical parameter, since it must be able to exceed the friction in the tubes and transfer the necessary amount of water for the heat pump.

The advantages of open loop geothermal systems include among others the following:

- they have less heat transfer loss due to mass transfer;
- the cost of energy for the heat pump is less, since it is designed properly and it is selected with care.

Disadvantages are the following:

- Problems related to the availability of water, which can be costly.
- The production of the borehole may change with time and cause problems in the future.
- Permitting procedure and related costs normally required.
- They function *via* the fluid transfer (by water) through the heat exchanger, which may, depending on the fluid quality and potential fouling, create a long-term problem in the movement of the fluid and therefore the thermal efficiency of the system.
- The water having a suitable quality, could alter the quality of the time and use that causes various corrosion problems.
- The costs associated with the drilling of the water can be high and unknown at first.
- They require sufficient and stable supply of subsurface fluid to be efficient. This creates the need for special care and precision in manufacturing and their use in coastal areas, and in areas adjacent to lakes, rivers, streams, *etc.*
- Requirement for an acceptable method for return used water to the environment. This may be limited not only by environmental factors, but also by local and national regulations. Only in very shallow aquifers in whom reinjections impossible, used water could be returned to surface waters under the permission of the relevant national body.

Construction Issues

General Guidelines

Following the design and light construction of the system it should be decided [19, 20, 49]: the number of drilling, the drilling method, the depth of drilling, the diameter and position of the boreholes. For their construction, it will be required special permission from the competent authorities.

In production wells, submersible pumps should be used, made from materials that withstand corrosion. Also, in front of the heat pump system gravel traps must be installed. Regarding the injection wells the lower end of the intake pipe must extend below the water level at rest. The underground water must be cooled as little as possible during the course from the production bore hole to the heat pump, so as to ensure the proper temperature of the heat source. Also, the underground water that it is cooled in the heat pump, must not be further cooled on the way to the injection wells, so that it does not freeze within the tubes or injection wells. Ground water also should not come into contact with the air inside the circuit. Before connecting with the heat pump, it must be checked that the pipes do not leak and they must also be flushed. For small plants (5-10kW) one drilling well can be used. In those cases, the water at the outlet will be typically available in surface water bodies. For larger systems, a pair or more pairs of drilling wells are commonly used for production and approximately equal number of underground water injection wells for returning water.

Common Types of Open Loop Geothermal Systems

The different variants of open systems are based on the arrangement of boreholes and water disposal method. Usually, in residential and commercial applications a system of two drilling wells is used comprising one for pumping underground water (production drilling) and one for recharging the aquifer (reintroduction/reinjection drilling). In some cases, one drilling can be used for pumping and one for recharging. Open systems may also require different size drilling, based on the required amount of groundwater [19, 20, 49, 51]. The direct use of groundwater in heat pump units is sometimes used in very small commercial applications. The method is very susceptible to problems caused by the quality of water thus it is used only of very small applications, when the quality of the groundwater is excellent, and which requires extensive testing. The system of the standing column, is particularly preferred in areas with hard rock geology where boreholes do not produce a satisfactory water for conventional open type systems, and the water quality is excellent. The drilling depth typically extends to 300-450m and systems operate at temperatures between those for open and closed systems. Open indirect type systems use a heat exchanger between the building network and groundwater so as to eliminate the influence of the components of the underground water and allows the building network and underground water systems to operate at different flows for optimum operation. Water can be allocated either to a reinjection borehole or into surface waters if available.

Water Disposal

The water disposal can be either surface or underground methods [19, 20, 42, 52, 53]. Surface methods are easier to dispose used water. Disposal sites include lakes or streams near or far from the application site. For disposal of water in lake special permission are required. The transfer method should ensure avoidance of erosion and subsidence problems that may arise. Additional problems may arise in winter in case of ice conditions development. Long term effects on the level of underground water is possible if the pumping exceeds recharge. Also, water can be fed into channels where it is passed to the soil and will penetrate therein. This method is appropriate when the soil consists of highly permeable sand and gravel. Alternatively, the penetration will be slow and there must be provision for periodic maintenance. Another possibility is to discharge water in the rainwater network but which requires permission from the municipality. This option is usually excluded, mainly because of limited access to the rainwater network and that this method does not recharge the aquifer. Most underground methods return the water to the aquifer by reintroducing in a vertical reinjection well. This method keeps the groundwater and tends to reduce environmental problems. Another underground disposal method includes horizontal drainage or underground drainage field. The drainage must be able to accommodate the flow volume and must be sufficiently deep to avoid problems from winter ice.

Possible Problems

The use of open systems can ensure safe and environmentally friendly operation. Possible problems include [49, 51]:

• *Changes in Underground Water Temperature*

The water returns either hotter or colder and whether or not the discharge will affect the water depends on various factors such as the volume of water discharged, temperature, water flow is obtained, and other factors. If the water for cooling and heating is returned to the same aquifer at all times what the effect of temperature will tend to minimize. When disposing water underground large systems may incur a thermal plume (plume) which could interfere in another borehole.

• *Pollutants Leaking from Equipment*

The oils from mechanical equipment and refrigeration are usually sources of pollutants in heat pump systems. At residential systems since the heat pump is installed indoors, the risk of such pollution of the subsoil, is minimal. In larger commercial systems, this risk could be greater. However, most systems are equipped with automatic shutdown devices activated by pressure drop.

• *Leakage of Chemical Additives in the Water Leaving the Plant*

Chemical additives must not enter the returning water after as they change the classification of the borehole. Before using chemical additives, they have to update by the competent authorities.

• *Change in the Chemical Balance of the Water (Chemical Imbalance)*

As the underground water flows through the system, it can change in quality due to temperature, pressure and dissolved oxygen. Such changes can cause precipitation of insoluble materials such as oxides of iron, calcium carbonate, *etc*., and this consists the biggest problem. Salt formation may occur in the pipes, valves, heat exchangers and reduce system performance. The scaling may also block the reinjection drilling. Generally, injection wells require extensive maintenance in relation to the pumping bores. The control of water quality and the collaboration with the designer of the plant and suppliers of equipment can help avoid these problems. Also, the use of filters can help improve the problem by removing the fine particles from the water.

• *Continuous Pumping and / or Insertion of Boreholes*

Uncontrolled use of groundwater in an area may cause problems such as the decrease of the water table and interpolation between adjacent wells. The first occurs when the groundwater pumping rate is greater than the rate of re-introducing to the production well. The second may occur where the pumping wells are near each another, or the use of groundwater is high or the returns are small. The result is lower water levels in wells and lower yields. In some cases, the level of water levels may drop below the pump intake levels which results in dry boreholes. Whether it will appear interference between wells depends on: the transmissibility of the local groundwater, the rate of pumping of drilling and the distance between the boreholes. Aquifer with good transmissibility generally provides enough water for each user. The demand for underground water varies with the size and design of the pumping system. The over pumping can broaden the cone level drop to create interference with the extraction region from another borehole. A borehole may generally be positioned at a minimum 30m from an existing borehole.

ANALYSIS AND DEVELOPMENT OF NORMAL GEOTHERMY SYSTEMS

General Guidelines

For the analysis and design of normal geothermal installations, certain steps are followed and discussed in the following sections:

i. Definition of the installation purpose
ii. Imprinting of the region and the project land
iii. Finding and analyzing geological data area
iv. Compliance with the design criteria and investigation of possible restrictions
v. Imprinting of the air conditioning places - Calculation of maximum and minimum energy requirements
vi. Determining the type of installation, the required wells and their positions
vii. Specifications of the required drilling wells and geothermal heat exchangers, and
viii. Determination of the electrical and mechanical system equipment.

The purpose of the normal geothermal installation is the use of the subsoil heat capacity *via* the construction of a closed or open circuit to cover cooling or heating water needs. If the normal geothermal system is not required to fully meet the needs for summer or winter air conditioning, but it will operate as a complementary installation to cover partial loads, then operational conditions will change, thus changing the design and its size. For proper size assessment, the description of the project should include data for the area in which the project will be developed including data from a planning point of view, and the regulations governing it, as well as environmental conditions and constraints prevailing in the area. Under current legislation in Greece [55], in case of drilling wells or construction of trenches under the auspices of the development of space heating / cooling systems, those must be within the property limits.

For the rational assessment of the geothermal heat exchanger behavior and thermal efficiency but also for the evaluation of potential risks to be taken into account in the design and initial cost estimate, it is necessary to know the geological data of both the broader area and the specific position. From the geological morphology of the area, the following elements necessary for proper planning can be estimated [2]:

- It is necessary to provide data for the subsoil stability, as possible precipitation and the existence of subsurface caves, that may significantly alter the quantity and quality of the final drilling – geothermal heat exchanger.

- The characteristics of the materials that consist the subsoil and their quantities will be used to assess the thermo- physical characteristics of the subsoil. Such characteristics are thermal conductivity (average), specific heat (average), density and thermal diffusivity. All of the above are formed as a result of the quantitative ratio of the materials and the existence of water and porosity. These elements are essential for estimating average yields that will give the final performance of the system and the final size of normal geothermal installation.

The knowledge of the geological characteristics, beyond the specific point of the test drilling enables verification and more accurate forecasting. The prediction accuracy for the case of vertical geothermal heat exchanger systems is particularly important in regard to the cost of the initial installation, design and expected results.

The most important design criteria are [1, 19, 20] high performance, high reliability, cost-effectiveness and simplicity of installation. The systems should be as simple as possible, since this helps to minimize errors in the operation of facilities. The fact that there is a significant difference between the life of the installation elements in the subsoil and the facilities in the building should also be considered. The heat pump, the circulation pumps and most of the parts of the system, may change two or three times during the lifetime of the installations underground. Therefore, heat exchangers should continue to operate after about three generations of heat pumps. High performance in the installation may be achieved if the temperature difference between the heat source and the heat load is kept as small as possible, while minimizing potential losses (energy, pressure, temperature). Achieving high reliability over the life of the overall system is not only a matter of proper planning but it is also important work on the development of the plant, the total quality control and that the installation materials and spare - parts must be used which are certified, tested and approved for the respective requirements. To achieve high security of the system, both during installation and during operation, the specifications and requirements of manufacturers and local authorities should be strictly followed.

As concerns geology issues, it is necessary to use the appropriate construction method for the particular geological conditions of the site where it is installed. The geological requirements differ depending on the type of system to be installed. The knowledge of hydrogeological conditions is necessary for the design of open loop systems. The water type, the geometry, the groundwater level, soil texture, hydraulic properties and limits are among the factors that are important for the design. The ground water used for open systems may have limited performance, namely drilling capacity, and undesirable chemical composition. Also, the water table may be used for drinking water consumption or other uses which constitute a limiting factor. In this case, a possible alternative may be the closed loop system design. In closed systems, the above parameters are of less importance, but in some cases, may be the limiting conditions. Closed systems may be affected by the groundwater flow. For systems that heat is pumped, this is an advantage, whereas for heat and cold storage systems and cold it can be a disadvantage. A low level of underground water will limit the extraction of heat and cold, if no fill of drilling is used. The closed loop systems apply to all types of geology. However, the thermal properties and problems during the drilling of the borehole may constitute a limiting factor.

The climate plays an important role in the implementation of such systems as the ground temperature is influenced in the first meters of depth by the air temperature of the external environment. The type of climate can limit the use of some plant type. In hot climates with high humidity, there is a requirement for cooling temperature which allows the condensation, which in practice means that it is not possible to directly cool a building from the ground. In tropical climates with warm weather conditions and without seasons, a geothermal heat pump system has very low yield and there is need for large network while concerning UTES is not feasible. In hot dry climate with cold nights because of small conductivity a geothermal heat pump system has very low yield and there is requirement for a larger network. In that case, UTES can be used for cold storage. For the cases of Mediterranean climate with hot summer and mild winter, seaside areas with hot summer seaside and quite chilly winter and continental areas a geothermal heat pump system is feasible. UTES can be used for seasonal heat storage and cold. Finally, in Northern areas of the northern hemisphere or South areas with mild summer and cold winter due to discharge phenomena the system has very little performance and there is need for a large network while there is danger for ice formation in wells.

The design criteria that must be known by the side of the building is [2, 19, 20] are:

- The energy peak loads (kW) at design conditions.
- Annual and seasonal energy loads (MWh).
- The temperature requirements for heating and cooling distribution systems.

It is necessary to know the annual change of the load, while for larger projects monthly energy loads and peak loads need to be known. The building data also include information that is not directly linked to the demand for heating and cooling, but with the requirements for the development of the plant, such as available surface for drilling and any limitations on or under the ground.

The determination of the required drilling wells for geothermal heat exchanger construction is completed on the basis of thermal and cooling building requirements of the building and the ground performance in such area. Thermal characteristics of the underground are recorded, showing the possible potential thermal average transaction in W/m, both for the period of heat pump (winter), and for the period of heat rejection (summer). On that basis, the overall length of geothermal heat exchangers and the number of wells can be computed. The distribution of these quantities in a number of geothermal heat exchangers is determined by the following:

- The land extent and the area for the development of geothermal heat exchangers field. This is determined by the relevant legislation, the natural and urban boundaries of the land on which the facility will be constructed.
- The length of geothermal heat exchangers. The length of each heat exchanger is determined from the above, and it leads to the selection of the material of the heat exchanger and the final drilling depth.
- The required minimum distances between geothermal heat exchangers to ensure continued performance without the possibility of thermal fault. The safety distance depends on the thermal characteristics of the subsoil and the operation of the installation within the meaning of load development.

The final selection will be combined with the size of the pump closed network as the friction level that will be required is proportional to the heat exchanger length and inversely proportional to the pipe diameter. The heat exchanger material does not affect the final thermal characteristics of its operation and performance but only at a very limited extent. However, the material affects significantly its strength at operating pressures and the life of the heat exchanger. For this reason, the expected or budgeted drilling depth leads uniquely to material groups. For example, for a heat exchanger to be installed in drilling depth of 100 m pressure drop reaches under conditions 14 bar. In that depth and under this pressure levels a geothermal heat exchanger of similar strength can be used in the PEXa that these conditions are within the limits of strength. In the case of geothermal heat exchangers that will be placed at a depth of drilling 200m, the growing pressure touches under conditions 24 bar and is prohibitive for the materials of the two above geothermal heat exchangers.

The drilling positions are chosen so that there is no interaction between them, to avoid thermal short circuit, and to be able to visit and control network. In the study, the drilling coordinates are given according to attached area layout. Their exact locations are shown in the ground plan of the mechanical part of the study. In various European countries, there are restrictions for the selection of the drilling positions. For example, in Greece drilling positions should be at least: two meters from the property limits, five meters from the existing neighboring buildings of different ownership, five meters from the railway expropriated zone limit, ten meters from the main gas pipeline, five meters from the central underground pipelines for water supply, irrigation, and drainage, ten meters from high voltage electricity distribution lines unless between line and drilling mediates building and five meters of electricity distribution lines MV unless between line and drilling mediates building. The time sequence of operations is as follows: borehole drilling, mounting geothermal exchanger in each borehole, filling of boreholes with grout.

Cost Elements

The normal geothermal energy facilities depend on many parameters that mainly have to do with the earth's environment and also depend on the type of installation to be used [54]. The economical approach for the evaluation of the installations is related to the ratio of installation and operating costs. This relationship may distinguish the difference over time, focusing on the maximum operating time and the final actual user cost, to meet the required needs. A normal geothermal installation may cover the needs for heating and summer air conditioning of a building or part of it or of a building block and simultaneously domestic hot water needs. Thus, it can be compared with a group of installations or a conventional plant which also cover the same needs (*e.g.* a boiler, an air conditioning installation and a facility that produces hot water for use.

The cost in an energy facility includes various costs as industrial, environmental and energy costs, which are continuously valued at units of materials and energy and for each time period affected by many other factors. A technical study should take account both of these figures and in particular should characterize the energy and environmental footprint. The type of the system, open or closed, will create a completely different approach and therefore a different configuration of the facilities cost. In an open loop installation, the cost depends on the quantity, quality and the position of the water table, creating for each other installation, operation and maintenance costs. In closed loop systems, the cost depends on the thermophysical characteristics of the subsoil, that will define the size of the geothermal heat exchanger and the scope of the installation. In that case, the installation cost depends on the total area, the minimum maintenance costs and the need for a functional recycling water quantity. The key factors of the drilling costs related to soil are: depth, diameter, type of formation, hardness of the formation, supply of underground water, *etc*.

The determination of the costs of a normal geothermal installation depends on:

- The size and type of fluid transfer system to the heat pump (suction system and reinjection network, for the case of open loop geothermal systems). In the case of closed loop systems, the corresponding recycle network of the heat transfer fluid in the subsoil.

- The size and type of heat pump (power, weight, materials) and the water pump used to thermally support the heat pump.
- The power supply panel installation to cover the operation of the individual devices the various safety and control systems of the plant operation.
- The size and type of heat exchanger which is used for heat exchange between the fluid and the water of the heat pump. In the case of closed systems between fluid and refrigerant of the heat pump.

By comparison with the respective cost of a conventional unit used for heating, to determine the cost of a boiler it should be taken into account the size and type of the boiler - burner and type of fuel, the size and type of the fuel tank or the connection to the fuel system and the size and type of fire protection installation. Also, the power supply panel installation to cover the operation of the individual devices and the various safety and control systems of the plant operation should be considered. Respectively, to determine the cost of a summer air conditioning installation the following are taken into account:

- The size and type of water cooler or air-cooled heat pump central or local type.
- Power supply panel installation to cover the operation of the individual devices. In case of partial coverage, the building panel may be sufficient.
- The various safety and control systems often embodied in the unit.

Finally, to determine the cost of installation of domestic hot water (DHW), the size, nature and type of heater should be taken into account. The cost estimate of each installation must be described in detailed table quantities, types and their values as well as the list was supplemented with the corresponding installation costs.

In open loop systems, water is pumped from an underground aquifer from the production drilling and is reinjected in the injection wells. For such systems, the information necessary for the installation costing is the water pumping depth, its supply and the level and extent of reinjection network. Depending on the physical characteristics of the subsoil the depth, the range and drilling provision and plant size are determined. Underground water quality will influence the selection of the installation. In case of brackish or sea water the use of special heat exchangers with a high protection factor against corrosion is required, both in terms of the service life of the system and its credibility. Indeed, during the initial installation dual systems are often developed to switch between them and their parallel maintenance. This practice increases the initial cost of installation and operating costs in the amount of maintenance.

In Table **5**, indicative cost estimates for geothermal open loop and closed loop type geothermal facilities are presented. The cost to install a geothermal closed loop system with installed power 14 kW is shown, with respect to the normal geothermal system and not inside the building. The cost estimates concern the initial installation costs which absorbs all energy cost estimates for the production of equipment and materials used in a plant, the annual costs arising from the operation of the installation in 24-hour basis, throughout the year and the annual maintenance costs.

The simple payback method is applied to calculate the payback period, the annual saving, the annual primary energy saving and the annual CO_2 decrease in comparison with a conventional installation. Similar results are shown for open loop system.

Table 5. Economic estimations for Open loop and Closed loop systems.

	Open Loop System	Vertical Closed Loop System	Horizontal Closed Loop System
Installation cost (€)	22.600	27.985	20.420
Annual Operation Cost (€)	920	852	
Annual Maintenance Cost (€)	400	200	
Simple payback period (in years)	2,22	3,4	1,9
Annual saving (%)	80	78	83
Annual primary energy saving (%)	33	33	
Annual decrease CO_2 (tons)	3,34	3,34	

The conventional system in which the GSHP system is compared consists of boiler - oil burner for heating and split units (split-units) for summer air conditioning. The GSHP system installation cost does not include the low-temperature heating system in the building, because it is common to GSHP systems and conventional systems. For the conversion of electrical energy to primary energy the rate of 2.9 is used [55]. To calculate the annual CO_2 emissions reduction equation 1 toe = 10465,12 kWh_{th} = 3,2 TCO_2 is used.

The GSHP systems because of practical subsoil constant temperature have increased performance (high COP/EER), thus they have very low power consumption. The low operating costs is very important factor in choosing an energy system as continuous change in fuel prices depend on numerous economic and political factors. Furthermore, low operating cost leads to very large annual primary energy savings (35%) and money (60%), and the GSHP system repayment, both in closed and in open systems takes place in a short time (2.5-5.5 years).

Normal geothermy is a renewable energy source based on Article 2 of the European Directive 2009/28 / EC [1] and leads to a significant reduction of greenhouse gas emissions and to achieving the objective 20-20-20 of European legislation. The GSHP system applications lead to practically zero environmental impact, both during construction and during operation. Also, the above-ground space of geothermal facilities may be used in both private and public spaces for the creation of green areas, parks, squares, recreational areas, *etc.* contributing to the improvement of the microclimate, especially in urban areas.

Apart from the economic and environmental benefits of GSHP systems, they can be widely applied in many applications due to the following important advantages:

- Installation in any place because of non-exploitation of the geothermal field.
- Quiet and completely safe operation without combustion existence and very high degree COP / EER efficiency than other heat pump systems.
- Contribution to improving the energy efficiency of buildings [55].
- Uninterrupted operation with a continuous supply of heating / cooling / hot water (DHW) without the need for a conventional backup system.

CONCLUSIONS

The purpose of a normal geothermal installation is the use of the subsoil heat capacity *via* the construction of a closed or open circuit to cover cooling or heating water needs. If the system will operate as a complementary installation to cover partial loads, then operational conditions will change, thus changing the design and its size. Vertical closed loop geothermal heat exchangers are used when there is lack of surrounding space regardless soil composition. Horizontal closed loop geothermal heat exchangers are applied to sites where there is enough surrounding area. For small vertical closed loop systems up to 30 kW, the design is based on assessments. If the system is used to cover only heating loads the calculations may be based on the values of specific heat pumping rate (W/m). When it comes to the design of large vertical systems for working hours more than 2400 hrs, covering of cooling loads and demanding load more than 30 kW the geothermal loops have to be calculated in detail. For horizontal closed loop geothermal heat exchanger the design approach to calculate the total length is the use of tables that contain values of available specific heat pumping rate (W/m^2) from the soil depending on its composition. Finally, geothermal open loop systems although they are simple in their nature they require careful planning, mainly regarding drilling and supply of ground water, the heat exchanger and the availability of water in order to obtain an efficient and reliable system.

CONSENT FOR PUBLICATION

Not applicable.

CONFLICT OF INTERESTS

The authors declare no conflict of interests.

ACKNOWLEDGEMENT

Declared none.

REFERENCES

[1] European Directive 2009/28 / EC on the promotion of the use of energy from renewable sources and amending Directives 2001/77 / EC and 2003/30 / EC.

[2] M. Gr, *Vrachopoulos, "Notes for the graduate program Energy Production and Management.* NTUA Editions: Athens, 2004.

[3] B. Sigfússon, and A. Uihlein, *JRC Geothermal Energy Status Report Technology, market and economic aspects of geothermal energy in Europe.* JRC European Commission, 2015.

[4] N. Andritsos, "Geothermal energy Use, Country Update for Greece", *European Geothermal Congress,* 2013 Pisa, Italy..

[5] M. Fytikas, and N. Andritsos, *Geothermy.* Tziola Publications: Athens, 2004.

[6] M. Gr, "Definition of horizontal geothermal heat exchanger behavior", *9thNational Conference on Renewable Energy Sources,* 2009 Paphos - Geroskipou.

[7] Exploitation plan of normal geothermy and evaluation of application possibilities in the cross area - Municipality Moudros and Geroskipou, deliverable of the project "Cross-border plan of exploitation and evaluation of normal geothermy in the Municipalities of Moudros and Geroskipou" of the Community Initiative INTERREG IIIA Greece – Cyprus INTERREG IIIA.

[8] Developing an Integrated Methodology for Research and Structures Organisation for the Exploitation of low enthalpy geothermy in Insular Greece -Thermopolis, PEP North Aegean.

[9] European Geothermal Energy Council, http://www.egec.org.

[10] J. Gudmundsson, "The elements of direct uses", *Geothermics,* vol. 17, no. 1, pp. 119-136, 1988. [http://dx.doi.org/10.1016/0375-6505(88)90009-0]

[11] B. Lindal, "Industrial and other applications of geothermal energy", *UNESCOGeothermal Energy,* 1973.

[12] VDI 4640 Part 1: 2010-06, Thermal use of the underground Fundamentals, approvals, environmental aspects, 2010.

[13] VDI 4640 Part 2: 2001-09. Thermal use of the underground - Ground source heat pump systems. 2001. [http://dx.doi.org/10.1016/0375-6505(88)90009-0]

[14] General Report of the current situation of the regulatory framework for normal geothermy, Overview of the legislation of shallow geothermal energy in Europe, D2.2, Regeocities.

[15] J.W. Lund, and T.L. Boyd, "Direct Utilization of Geothermal Energy 2015 Worldwide Review", *Proceedings World Geothermal Congress 2015,* 2015 Melbourne, Australia.

[16] M. Antics, R. Bertani, and B. Sanner, "Summary of EGC 2013 Country Update Reports on Geothermal Energy in Europe", *Proceedings, European Geothermal Congress 2013,* 2013 Pisa, Italy.

[17] M.Gr. Vrachopoulos and I. Papageorgakis, "Exploitable heat derived from the sun and earth in Shallow Subsurface of Attica", Technical Abstracts TEE, 1, 1998.

[18] Technical Chamber of Greece, "Training material for Training Energy Inspectors", In: *module DE4 Technologies Renewable Energy*Athens, 2011.

[19] B. Sanner, *"Swallow Geothermal Systems, Ground Source Heat Pumps", International Geothermal Days Slovakia 2009.* Conference and Summer School, 2009.

[20] B. Sanner, Overview of shallow geothermal systems.*Geotrainet", Training Manual for Designers of Shallow Geothermal Systems.* Geotrainet, European Federation of Geologists: Brussels, 2011.

[21] VDI 4640 Part 3: 2001-09,Utilization of the subsurface for thermal purposes. Underground thermal energy storage, 2001.

[22] VDI 4640 Part 4: 2004-09, "Thermal use of the underground - Direct uses", 2004. 23. K. Terzaghi& R. Brazelton Peck, Soil mechanics in engineering practice, New York, Wiley, 1967.

[23] K. Terzaghi, and R. Brazelton Peck, *Soil mechanics in engineering practice.* Wiley: New York, 1967. The text is available online at: http://agrifs.ir/sites/default/files/Soils%2C%20Principles%2C%20Properties%20and%20Management%20%7BKhan%20Towhid%20Osman%7D%20%5B9789400756625%5D%20%28Springer%20-%202012%29_0.pdf

[24] Khan Towhid Osman, *Soils Principles, Properties and Management.* Springer: Netherlands, 2013.

[25] A. Papaioannou, *Thermodynamics: Principles and Laws- Pure Substances.* vol. Vol. I. Coral Publications: Athens, 2007.

[26] A. Benou, J. Choropanitis, E. Kontoleontos, and C. Karytsas, "GSHP systems in the built environment", *International Forum, Geothermal Energy in the Spotlight, Thessaloniki,* 2009.

[27] A. Benou, K. Karras, E. Kontoleontos, and C. Karytsas, *System of Geothermal Heat Pump for heating - cooling at Athens International Airport and Life Cycle Assessment, 4th NationalConference,Application of Renewable Energy Sources to an ambitious and credible national action program*Athens, 2010.

[28] A. Benou, D. Mendrinos, and C. Karytsas, *Ground Source Heat Pump Systems: Technology and Best Practicesk 6th National conference on Energy Efficiency and Renewable energy sources with international participation,* 2007

[29] T. Kasuda, and P.R. Archenbach, *Earth Temperature and Thermal Diffusivity at Selected Stations in the United States.* National Bureau of Standard Reports, US Department of Commerce, 1965. [http://dx.doi.org/10.21236/AD0472916]

[30] B. Sanner, BHE design examples.*Geotrainet Training Manual for Designers of Shallow Geothermal Systems. Brussels.* Geotrainet, European Federation of Geologists: Brussels, 2011.

[31] B. Sanner, "Guidelines, standards, certification and legal permits for ground source heat pumps in the European Union", *9th International IEA Heat Pump Conference,* 2008 Zürich, Switzerland.

[32] E. Sailer, D. Taborda, and J. Keir, "Assessment of Design Procedures for Vertical Borehole Heat Exchangers", *14th Workshop on Geothermal Reservoir Engineering,* 2015.

[33] *Geothermal Energy in 2011 ASHRAE Handbook - HVAC Applications.* American Society of Heating, Refrigerating and Air-Conditioning Engineers: Atlanta, 2011, pp. 34.1-34.20.

[34] S.E. Dehkordi, and R.A. Schincariol, "Guidelines and the design approach for vertical geothermal heat pump systems: current status and perspective", *Can. Geotech. J.,* vol. 51, pp. 647-652, 2014. [http://dx.doi.org/10.1139/cgj-2012-0205]

[35] A. Capozza, A. Galgaro, and A. Zarrella, *Linee Guida per la progettazione dei campi geotermici per*

pompe di calore. Ricerca Sistema Energetico: Milan, 2012.

[36] S.P. Kavanaugh, and K.D. Rafferty, Geothermal Heating and Cooling: Design of Ground-Source Heat Pump system, American Society of Heating, Refrigerating and Air-Conditioning Engineers.

[37] H. Zeng, N. Diao, and Z. Fang, "Heat transfer analysis of boreholes in vertical ground heat exchangers", *Int. J. Heat Mass Transfer,* vol. 46, no. 23, pp. 4467-4481, 2003. [http://dx.doi.org/10.1016/S0017-9310(03)00270-9]

[38] J. Claesson, and P. Eskilson, Conductive heat extraction by a deep borehole: analytical studies.*Thermal Analysis of Heat Extraction Boreholes.,* P. Eskilson, Ed., Department of Mathematical Physics, University of Lund, 1987.

[39] E.N. Guernsey, P.L. Betz, and N.H. Skan, "Earth as a heat source and storage medium for the heat pump", *Trans. Ashrae,* vol. 55, pp. 321-344, 1949.

[40] L.R. Ingersoll, O.J. Zobel, and A.C. Ingersoll, *Heat conduction with engineering and geological application.* McGraw Hill: New York, 1948.

[41] S.P. Kavanaugh, and K.D. Rafferty, Ground-Source Heat Pumps: Design of geothermal systems for commercial and institutional buildings, American Society of Heating, Refrigerating and Air-Conditioning Engineers 1997.

[42] S.A.M. Said, M.A. Habib, E.M. Mokheimer, N.A. Al-Shayea, and M.H. Sharqawy, "Horizontal Ground Heat Exchanger Design for Ground-Coupled Heat Pumps", *The International Conference and Exhibition of Ecological Vehicles and Renewable Energies,* 2009 Monaco.

[43] "Geotrainet Training Manual for Drillers Swallow Geothermal Systems, Project: IEE/07/581/S12.499061",

[44] M. McCorry, and G. Jones, *G.,Geotrainet Training Manual for Designers of Shallow Geothermal Systems.* Geotrainet, European Federation of Geologists: Brussels, 2011.

[45] P. Neuberger, R. Adamovský, and M. Šeďová, "Temperatures and Heat Flows in a Soil Enclosing a Slinky Horizontal Heat Exchanger", *Energies,* vol. 7, pp. 972-987, 2014. [http://dx.doi.org/10.3390/en7020972]

[46] P.M. Congedo, G. Colangelo, and G. Starace, "CFD Simulations of Horizontal Ground Heat Exchangers: A Comparison among Different Configurations", *Appl. Therm. Eng.,* vol. 33-34, pp. 24-32, 2012. [http://dx.doi.org/10.1016/j.applthermaleng.2011.09.005]

[47] C.S.A. Chong, G. Gan, A. Verhoef, R.G. Garcia, and P.L. Vidalec, "Simulation of Thermal Performance of Horizontal Slinky-Loop Heat Exchangers for Ground Source Heat Pumps", *Appl. Energy,* vol. 104, pp. 603-610, 2013. [http://dx.doi.org/10.1016/j.apenergy.2012.11.069]

[48] K. Rafferty, "Commercial Open Loop Heat Pump Systems", *ASHRAE J.,* 2009.

[49] C. Abesser, Open-loop ground source heat pumps and the groundwater systems: A literature review of current applications, regulations and problems, Energy Geoscience Program Open Report OR/10/045, British Geological Survey.

[50] K. Rafferty, "Water chemistry issues in groundwater heat pump systems", *ASHRAE Trans.,* vol. 110, no. 1, pp. 550-555, 2004.

[51] A. Morris, and A. Sheets, Geothermal Systems: System Types, Applicability and Environmental Impacts, University of Texas at Austin, UTSoA - Seminar in Sustainable Architecture.

[52] T. Vangkilde-Pedersen, C. Ditlefsen, and A.L. Højberg, "Shallow geothermal energy in Denmark, GEUS", *Geol. Surv. Denmark Greenl. Bull.,* vol. 26, pp. 37-40, 2012. [Open access: www.geus.dk/publications/bull.].

[53] D. Mendrinos, C. Karytsas, and B. Sanner, Project GROUND-REACH Reaching the Kyoto targets by

means of a wide introduction of ground coupled heat pumps (GCHP) in the built environment, Proceedings European Geothermal Congress 2007, Unterhaching, Germany.

[54] Z. Ziskatas, N. Andritsos, K. Diamantidis, and M. Fytikas, "Techno - Economic Study horizontal geothermal heat pump system", *8th National Conference on Renewable Energy Sources,* 2006 Thessaloniki.

[55] Regulation for Energy Performance of Buildings (KENAK), Greece, FEKB/407/09.04.2010.

CHAPTER 8

Measurement Systems for Renewable Energy

A.A. Nikoglou, P.K. Rouni and **E.P. Hinis***

Engineering Measurements Laboratory, School of Mechanical Engineering, National Technical University of Athens – NTUA, 9 Iroon Polytechniou, Zografos Campus, Athens 15780, Greece

Abstract: Renewable Energy Systems are designed to take benefit of the outdoor natural conditions. Their efficiency and long life depend on them (air flow, air velocity, solar radiation, temperature *etc.*). Continuous monitoring of outdoor natural qualities is essential for optimizing operation, preventing small and large scale damages and adapting design according local conditions. Common requirement for such applications is saving data for a long period of time and/or the ability to send them on line. This is why most of the instrumentation presented in this chapter is based on transducers: Flow transducers, temperature transducers, solar radiation transducers, humidity transducers, *etc*.

Keywords: Hot Wire Anemometer, Hygrometer, Magnetic Flowmeter, Nozzle Meter, Orifice Meter, Pitot Tube, Pyranometer, Pyrheliometer, Psychrometer, Rotameter, Resistance Temperature Detector, Turbine Flowmeters, Thermistor, Thermocouple, Venturi Meter, Vortex Shedding Flowmeter.

INTRODUCTION

The design and the efficient operation of Renewable Energy Systems are based on the exploitation of the conditions created by natural phenomena: The Wind, the Sun and the Water. Renewable Energy Systems are designed to take benefit of the natural conditions (air flow, air velocity, outdoor temperature, solar radiation, relative humidity *etc.*) and operate in field conditions: high/low temperatures (*e.g.* solar panels), sometimes tough ones like high wind speed (*e.g.* wind mills on high altitudes) or in other cases safety depends on water flow and pressure (hydroelectric power plants). In some cases, Renewable Energy Systems are expensive investments. The efficient operation and the long life of Renewable Energy Systems depend on the monitoring of the outdoor conditions. In order to continuously monitor the outdoor conditions, the field's measurement systems use transducers. In most cases, these measurement systems have to register, save and

* **Corresponding Author E.P. Hinis:** Engineering Measurements Laboratory, School of Mechanical Engineering, National Technical University of Athens – NTUA, 9 Iroon Polytechniou, Zografos Campus, Athens 15780, Greece; Tel: +302107722911, Fax: +302107722947; Email: ephinis@mail.ntua.gr

Emmanuel D. Rogdakis & Irene P. Koronaki (Eds.)

send all results of each measurement station to the operational control centre. There is also a need of portable measurement systems for exploring/check natural conditions before the installation of Renewable Energy Systems. There is a large variety of transducers for the accurate measurement of the above mentioned physical qualities. The ones most often used will be presented in this chapter along with the principle/method they are based on.

Fluid Flow Measurements

Flowrate measurement instrumentation provides either local measurements at a certain point of the flow or averaged values over an area of the flow. In any case, the measurements are expressed in terms of either volume flow rate or mass flow rate.

In industrial applications, the fluid may be liquid, gas, vapor or a two-phase mixture within a wide variety of temperature and flow rate. The flow may be confined inside pipes, semi- confined in open channels or flow in the form of thin liquid films. In each and every case, applying the ideal measuring technique is essential for optimizing the accuracy and the efficiency of the application.

Pitot

The Pitot tube is one of the most popular measurement instruments for measuring the velocity of fluids. The operating principle is based on the measurements of static and total pressure (static and dynamic pressure). A typical Pitot set up is illustrated in Fig. (**1**).

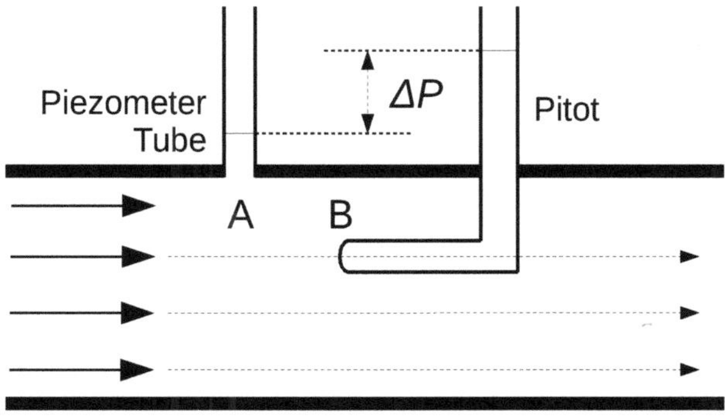

Fig. (1). Typical pitot set up.

The Pitot tube penetrates the fluid with the open end facing the flow while the piezometer tube is positioned upstream or downstream from the Pitot tube and vertically on the streamlines of the flow, measuring the total and static pressure, respectively [1]. By treating the fluid as incompressible and applying Bernoulli equation for the streamline passing from the center of the pitot and between points A and B (Fig. **1**), one can measure the pressure difference between these two points using a differential pressure transducer, and then estimate the fluid velocity as follows:

$$V_A = \sqrt{\frac{2(P_B - P_A)}{\rho}}$$

where:

- p_A is the static pressure (Pa)
- p_B is the total pressure (Pa)
- ρ is the fluid density (kg/m^3)
- V_A the velocity at point A (m/s) and
- V_B the velocity at point B which is equal to 0 (m/s)

Instruments that combine static pressure and total pressure measurements are known as pitot static tubes. A typical pitot static tube is illustrated in Fig. (**2**).

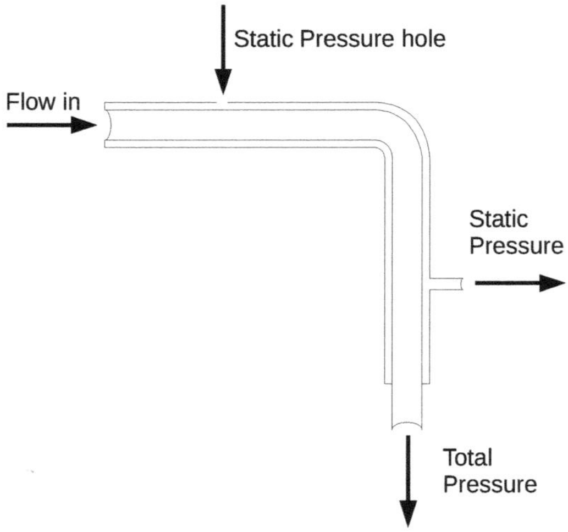

Fig. (2). Pitot static tube.

When applying Pitot tube, proper attention must be given to the alignment of the probe axis and the flow streamline. The angle between these lines is called yaw angle and it should be zero. Yaw sensitivity to velocity measurements depends on the design of the Pitot probe.

Variable Area Flow Meters

There is a class of fluid flow rate measuring devices, the design of which deliberately obstructs the flow, causing a pressure drop at a certain cross-section. By treating the fluid as incompressible and using Bernoulli and continuity equation before and after that cross-section, one can estimate the fluid flow rate indirectly by the measure of the pressure drop.

The most popular devices of the above class, which is also known as flow obstruction meters or differential pressure meters, are the venturi meter, the flow nozzle and the orifice. Typical designs of those are illustrated in Fig. (**3**).

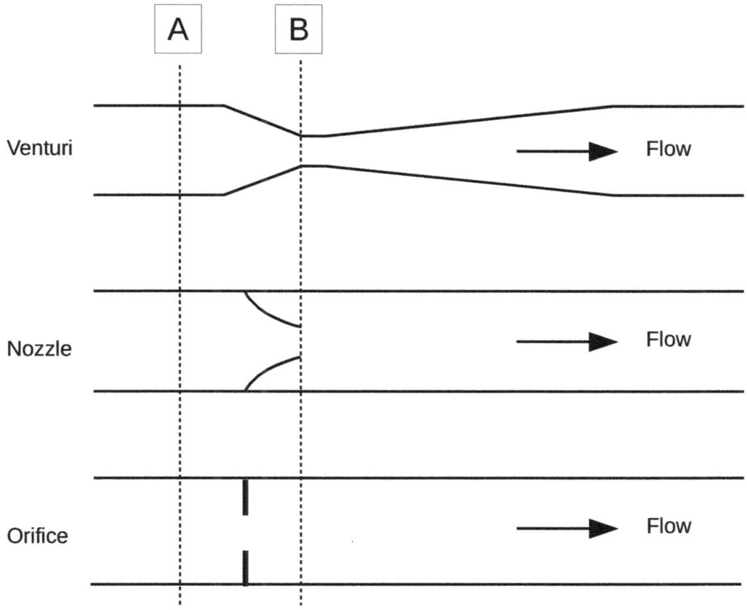

Fig. (3). Typical designs of a venturi, a nozzle and an orifice meter

More detailed information on how to perform measurements with these devices can be found in the American Society of Mechanical Engineers (ASME) standard MFC-3M-2004 [1].

Venturi Meter

Every Venturi meter consists of three main parts: a convergent nozzle, a cylindrical cone and a diffuser nozzle as can be seen in Fig. (**3**). The flow accelerates through the converging cone then flows with a constant speed through the cylindrical part and finally decelerates through the diffuser cone. Measuring the pressure drop between cross-sections A and B as shown in Fig. (**3**), one is given the volume flow rate for an ideal frictionless fluid by the following equation:

$$Q_i = A_A \sqrt{\frac{2(P_A - P_B)}{\rho\left(\frac{A_A^2}{A_B^2} - 1\right)}}$$

A_A and A_B are the cross-section areas (m^2) where the pressure drop is measured.

$$Q_r = C \cdot Q_i$$

In case of a real fluid, the volume flow rate is estimated by the following equation

where the constant $C < 1$, in called discharge coefficient and represents the pressure drop caused by the friction forces between the fluid and the inner part of the venturi.

Nozzle Meter

The Nozzle meter is similar to the venturi meter as can be seen in Fig. (**3**) with the essential difference that the nozzle meter does not include a diffuser cone and thus, larger pressure drops occur. Moreover, less space is needed and it can be installed in almost any piping system.

Orifice Meter

The Orifice meter is even simpler compared with the previous ones and consists of a thin plate with a circular hole in its center. It is vertically to the flow installed and the flowrate is indirectly estimated by the pressure drop across the orifice. A typical orifice meter can be seen in Fig. (**3**).

A common disadvantage of all these meters, compared to some other flow measuring techniques, stems from the square root relationship between the volume flow rate and the differential pressure. Assuming 3% measurement error in differential pressure at 50% of full scale, leads to 1.4% full-scale error in flow rate. However, the same error at 5% of full scale leads to 6.7% full-scale error in flow rate.

Hot Wire Anemometer

The hot wire anemometer is usually used for measurements in a turbulent flow where irregular fluctuations of the fluid speed exist. The major advantage of the hot wire anemometer is that it has very low response time and thus, high speed fluid celerity variations can be accurately measured.

The hot wire anemometer is a thermal transducer where the heat transfers from the wire that is exposed to a flow current is proportional to the velocity of the flow. The wire is made from materials that exhibit a change in electrical resistance concerning temperature changes such as platinum and platinum iridium alloy. The resistance – temperature relation can be represented with sufficient accuracy by the following equation:

$$R_a = R_r(1 + \gamma \Delta T)$$

where:

- R_a is the resistance at temperature T_a
- R_r is the resistance at reference temperature T_r and
- γ is the temperature coefficient of resistance of the wire

From the above equation, it can be clearly stated that the fluid instant velocity can be indirectly estimated by measuring the resistance R_a or the electric current through R_a.

Hot wire anemometers (Fig. **4**) can be classified goes here in two categories: as constant current anemometers (CCA) and constant temperature anemometers (CTA). In the first one, fixed electric current flows through the wire while the temperature of the wire varies. Wire temperature variations can be measured in terms of electrical resistance and thus, estimate the fluid velocity. In the second one, the electric current through the wire is constantly adjusted to maintain the wire temperature constant. The flow velocity is estimated by the electric power required to maintain the wire at a constant temperature.

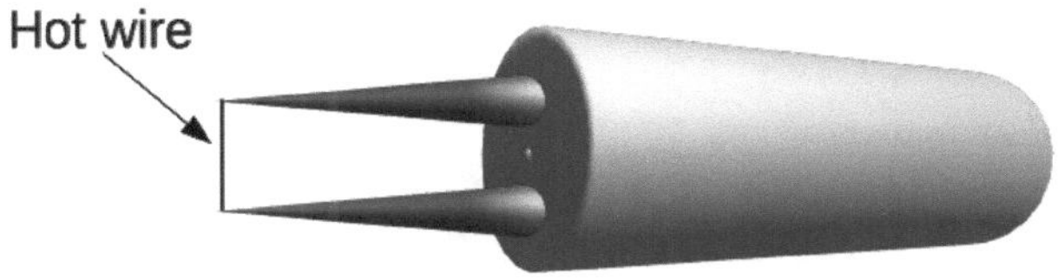

Fig. (4). A schematic illustration of a hot wire anemometer

Even though the word anemometer defines an instrument that measures wind speed, hot wire anemometers are widely used for measurements in various gases and liquids as well. The wire sensor is very thin and extremely fragile and thus, it is strongly recommended that it should be used in clean fluids free from floating particles.

The primary advantage of these transducers is the high frequency response that can reach up to 10 kHz, without significant spatial resolution loss.

Vortex Shedding Flowmeters

When a solid object obstructs the path of a fluid flow then vortices are developed on both sides of the object for long distances downstream, as shown in Fig. (**5**). That phenomenon is known as vortex Karman street and the frequency of the developed vortices is proportional to the fluid flow rate. The frequency of the vortices, the flow velocity and the characteristic length of the obstacle are all related by the dimensionless Strouhal number S_N, which is relatively constant over a wide range of Reynolds number. The Strouhal number is defined by the following equation:

$$S_N = \frac{f_s L}{V}$$

where:

- f_s the frequency of the vortices
- L the characteristic length of the obstacle, for example for circular or cylindrical objects it is the diameter of the circle or the cylinder.
- V: the velocity of the flow

Sensing the vortices vibrations using piezo-electric, ultrasonic sensors or strain gages, flowmeters provide one with the estimation of the fluid velocity by the measured frequency of vortices.

Fig. (5). Schematic illustration of a vortex flowmeter operation principle

In contrast with the hot wire transducers, the vortex shedding transducers are very robust and often used in harsh conditions, however the meter can be sensitive to vibrations. The meter turns down ratio is up to 40:1 and the accuracy approximately 1% of the flow rate.

Rotameter

The rotameter is a very popular flow rate measuring device and is often classified as variable area flowmeter or drag force flowmeter. It consists of a transparent tapered tube and a solid float, usually with twice the density of the measuring fluid, which is free to move along the axis of the tube (Fig. **6**). The rotameter is vertically installed and the flow enters from the bottom and exits from the top. At any flow rate within the measuring range of the rotameter, the fluid lifts the float up to a certain point where the viscosity and buoyancy forces are in balance with the gravitational force of the float, as expressed by the following equation:

$$F_D + \rho_f U_b g = W_b$$

where:

- F_D is the drag force of the fluid on the float, $F_D = \frac{1}{2} C_D \rho_f V_f^2 A_b$
- C_D is the drag coefficient of the float
- A_b the frontal area of the float
- ρ_f is the density of the fluid
- U_b is the volume of the float
- V_f is the fluid velocity in the annular area between the float and the tube
- W_b is the weight of the float, $W_b = U_b\ \rho_b$
- ρ_b is the density of the float

Due to the drag force F_D complex dependencies on the density, viscosity and the velocity of the fluid as well as on the size and shape of the body (float), drag coefficient C_D is used, allowing to collect all the effects into a single equation. From the equilibrium equation, and the drag force equation, the fluid velocity V_f is estimated as follows:

$$V_f = \sqrt{\frac{2U_b(\rho_b - \rho_f)}{A_b \rho_f C_d}}$$

The annular orifice surface area A_w at any given position y of the float above zero position is estimated by the following equation:

$$A_W = \frac{\pi}{4}\left[(D + ay)^2 - d_b^2\right]$$

where D is the internal diameter of the bottom of the tube (y=0), d_b is the maximum diameter of the float and α is a constant that describes the temper (the change in cross-section diameter per unit change in height) of the tube.

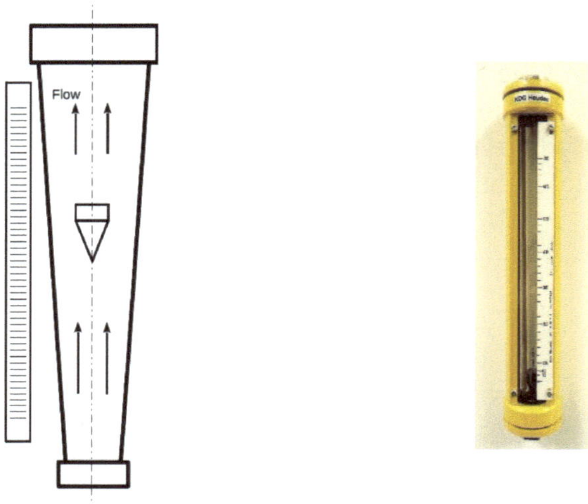

Fig. (6). Schematic illustration and a picture of rotameter.

The rotameter provides direct measurement of the flow rate by the position of the float inside the tube. The position is detected visually by using a graduate scale, or magnetically in cases of an opaque tube.

Turbine Flowmeters

A turbine flowmeter consists of a small propeller inside a tube. Using the mechanical energy of the fluid, the propeller rotates. The rotor speed is proportional to the flow rate of the fluid. By using a permanent magnet and a coil or a proximity sensor, one can determine the rotating frequency as well as the flow rate of the liquid, gas or vapor.

Turbine flowmeters possess very high levels of accuracy and a wide temperature and flow rate working range. However, due to moving parts (rotor) they are prone to failure compared with other measuring devices *i.e.* vortex shedding meters.

Magnetic Flowmeter

The magnetic flowmeter, also commonly known as mag meter, can be applied in case of a conducting fluid, *e.g.* tap water that contains ions. The operating principle is based on Faraday's law of electromagnetic induction, which states that the voltage generated across any conductor that moves proportional to the magnetic lines of a magnetic field with constant speed is proportional to the length and the velocity of the conductor and of the magnetic field strength as well. It is understood that the magnetic flowmeter has a linear response with the flow rate as can be seen from the following formula describing the Faraday's law:

$$e = Bul$$

where:

- B is the magnetic field strength
- u is the speed of the conductor
- l is the length of the conductor and
- e is the developed voltage across the conductor

In the magnetic flowmeter, the conductor is the measured fluid itself and the length is defined by two electrodes that are positioned diametrically on the inner surface of the flow tube without penetrating the flow, as can be seen in Fig. (**7**). Magnetic flow meters have accuracy of 0.5 to 2.0% approximately.

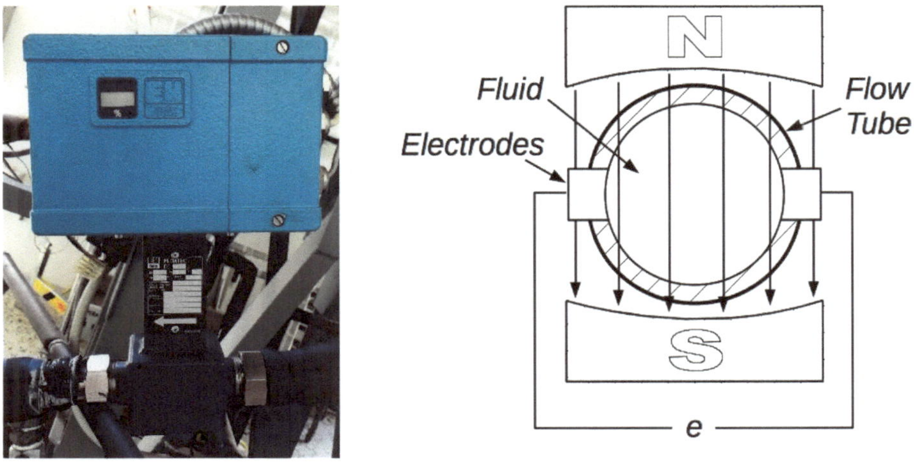

Fig. (7). Magnetic flow meter connected in a pipe system and schematic of the operation principle.

Air Velocity and Direction

For the assessment of the air velocity the three-cup anemometer, in Fig. (**8**), can be used. It consists of an axle with three cups on top mounted vertically to the wind. The cups catch the wind; they are moving around making the generator spin to producing an electric palm. Meteorological stations have also a wind vane to point the wind direction. Their operation is quite simple based on a magnetic switch. When the magnet passes over the switch, the switch closes.

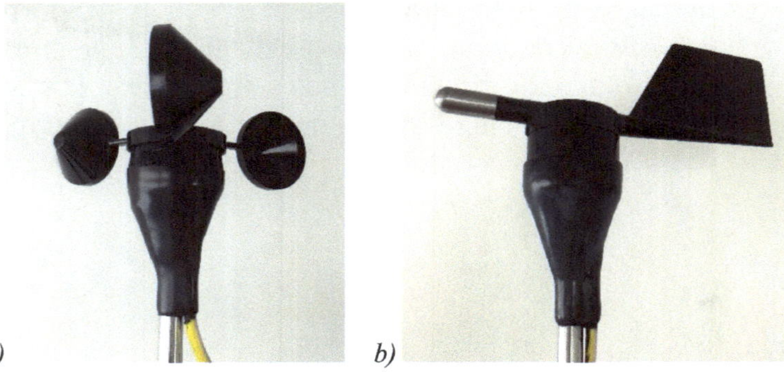

a) *b)*

Fig. (8). a) Three cup Anemometer, b) Wind Vane.

TEMPERATURE MEASUREMENTS

Temperature is the degree of hotness or coldness of an object or a substance measured on a definite scale. The first temperature scale defined by the freezing point (32 °F) and boiling point (212 °F) of water with 180 divisions, was introduced by Fahrenheit in 1715. Later, the centigrade scale defined by the same points and 100 divisions, followed (freezing point 0 °C, and boiling point 100 °C). Since 1948 this scale has been known as Celsius and it is the most popular temperature scale. Two other scales are used for absolute temperature measurements named Kelvin and Rankine.

Temperature is an abstract quantity compared with other quantities such as mass or length and it is defined by the impact of temperature changes on certain materials.

Thermistor

The term *thermistor* stems from the more descriptive term "thermally sensitive resistor" and refers to elements with electrical resistance that varies significantly with respect to the temperature compared with typical resistors. Thermistors are classified as semiconductors meaning that their electrical resistance lies between common conducting materials and common insulators. Oxides of nickel, cobalt, manganese and sulfides of materials such as iron, aluminum or copper are used for the fabrication of thermistors.

These temperature sensors are classified in two main categories depending on the type of correlation between temperature and resistance variations as follows:

- N-type, where a temperature increment implies a resistance decrement (negative correlation)
- P-type, where a temperature increment implies a resistance increment (positive correlation)

For N-type thermistor the resistance-temperature relationship is defined by the following equation:

$$ln\left(\frac{R}{R_0}\right) = \beta\left(\frac{1}{T} - \frac{1}{T_0}\right)$$

where:

- R is the thermistor electrical resistance at temperature T
- R_0 is the thermistor electrical resistance at reference temperature T_0
- β is a constant of the thermistor

Thermistors are commercially available in a large variety of resistances from a few ohms to several megaohms, (two examples in Fig. **9**). Choosing the optimum resistance for an application depends on the temperature measuring range of the application. Thermistors have an accuracy of ±0.1 ^{0}C approximately and they are used for measuring temperatures within the range of 0 to 100 ^{0}C without compromising the sensor accuracy and sensitivity. The 0-100 ^{0}C range is limited compared to the other available transducers analyzed below. Other disadvantages of thermistor are that their operation is not linear and they are fragile.

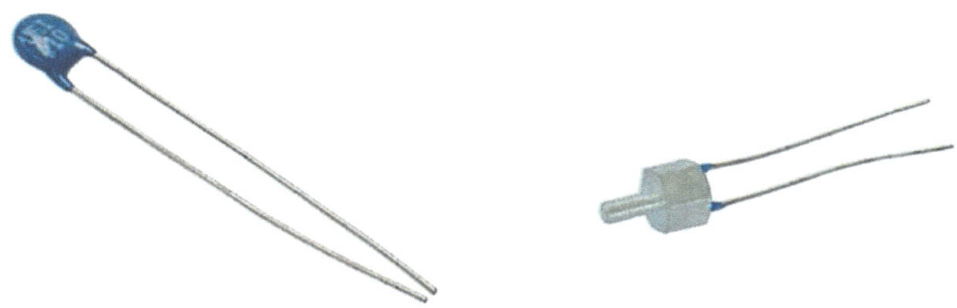

Fig. (9). Two examples of commercially available Thermistors.

Thermocouple

The thermocouple is the most widely used temperature sensor in every application in which temperature measurements are needed. It is very simple to construct as it merely consists of two dissimilar wires coupled at a junction, as shown in Fig. (**10**). A thermocouple directly produces electric voltage that stems from Seebeck effect, also known as electromotive force (*emf*) or thermoelectric voltage, *E*, and is used for estimating the temperature at the junction of the thermocouple.

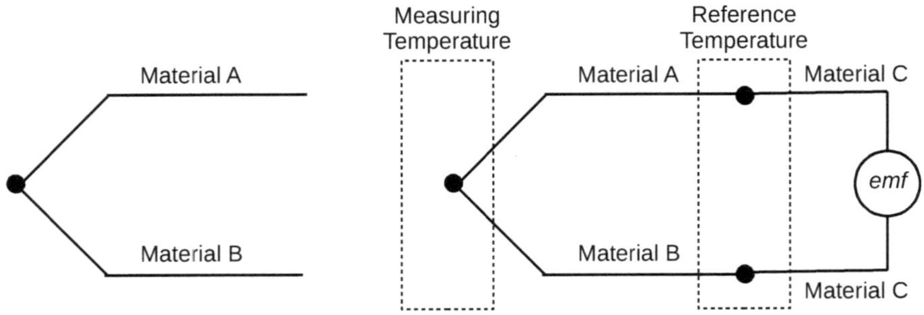

Fig. (10). Thermocouple sensor and thermocouple circuit [2].

The thermoelectric voltage, *E*, depends on the temperature difference between the temperature of the junction of two metal wires at the measuring point and the temperature of the free ends of the metal wires. We assume that there is no temperature difference between these two free ends. The voltage E is measured using an AC special voltmeter because the voltage for temperature difference of 1K is in the microvolt range.

The reference temperature must remain constant and accurately known. Two common methods are utilized for setting and controlling the reference temperature. The first one consists of an adiabatic bath with distilled water and ice at 0°C, which accommodates the reference joints. This method is not practical and is used in lab conditions to check the calibration of instrumentation.

The second method consists of an electric circuit equipped with a thermistor or a resistance thermometer that provides automatic voltage compensation for reference junctions following ambient temperature conditions. The compensation is achieved by adding algebraically an electric voltage which varies with the ambient temperature. This electric circuit is known as 'Electric Ice Point' or 'cold junction compensation'. No voltage is produce by the thermocouple circuit if the measured temperature is equal to the cold junction (reference) temperature.

If a thermocouple is short-circuited, a new measuring point is produced at that point. In case there is a break in the measuring circuit, the instrument will show the cold junction temperature.

The implementation of 'Electric Ice Point' in the thermocouple circuits facilitate their use and now there are applied in almost every industrial, scientific or routine installation because of their wide temperature ranges (approximately from -270 up to 2800 °C), their accuracy and durability.

There are a lot of types of thermocouples and many ways of applications, each one with its own unique characteristics. The Types of thermocouples used are presented in Table (**1**) according EN 60584 [3]. The symbol of Type *e.g.,* type T, identifies a specific temperature-voltage relationship, not a particular chemical composition. Thermocouples of a given type may have variations in composition as long as their resultant temperature-voltage relationships remain within specified tolerances. All materials manufactured in compliance with the established thermoelectric voltage standards are equally acceptable.

Table 1. Thermocouple types [3].

Thermocouple Types		
Type	**Composition**	**Temperature Range °C**
B	Pt-30% Rh versus Pt-6% Rh	0 to 1820
E	Ni-Cr alloy versus Cu-Ni alloy	-270 to 1000
J	Fe versus Cu-Ni alloy	-210 to 1200
K	Ni-Cr alloy versus Ni-Al alloy	-270 to 1372
N	Ni-Cr-Si alloy versus Ni-Si-Mg alloy	-270 to 1300
R	Pt-13% Rh versus Pt	-50 to 1768
S	Pt-10% Rh versus Pt	-50 to 1768
T	Cu versus Cu-Ni alloy	-270 to 400

Commonly used thermocouples for environmental applications are Type 'J' and for industrial applications Type 'K'.

Mineral-insulated thermocouples, (Fig. **11**), are designed for industrial use (chemical plants, pipelines, shock resistance *etc.*). They consist of a stainless steel sheath, in which thermocouple wires are embedded in compressed magnesium oxide.

For all (except type K thermocouples for the case of temperatures above 0 °C), the general form for the thermoelectric voltage, *E*, as a function of temperature in measurement point t_{90} is [4]:

$$E = \sum_i^n c_i \, (t_{90})^i, \text{ where } E \text{ is in microV and } t_{90} \text{ in °C.}$$

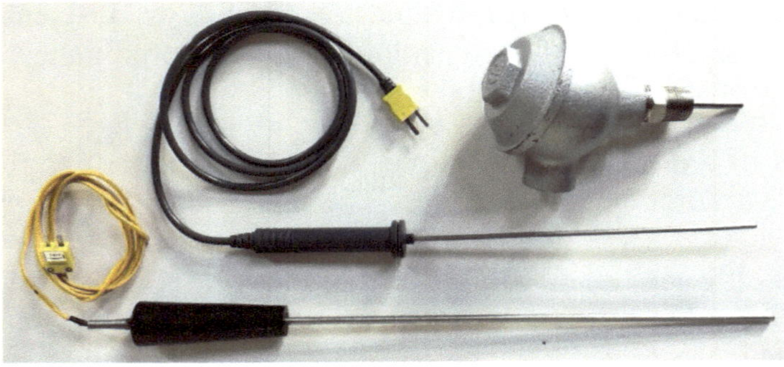

Fig. (11). Thermocouples for different applications.

For type K thermocouples and temperatures above 0 °C, there is an additional term to account for a magnetic ordering effect [4]:

$$E = \sum_{i}^{n} c_i \,(t_{90})^i + a_0 e^{a_1 (t_{90} - a_2)^2}$$

where E is in microV, and t_{90} is in °C and e is the natural logarithm constant. The constants α_0, α_1, and α_2 have the values [4]:

α_0=0.118597600000x10^0, α_1=0.118343200000x10^{-3}, α_2=0.126968600000x10^3,

Metrology standards Institutions (*e.g.* NIST) and manufactures provide tables or diagrams [4], of the thermoelectric voltage, E, which have been measured and interpolated over a range of temperatures, for each thermocouple type. This way one can estimate the temperature (°C) from a known thermoelectric voltage (μV) or the thermoelectric voltage (μV) from a known temperature (°C).

The main advantage of temperature measurement using thermocouples is the ability to save measurement data in a digital registry and the ability for continuous remote monitoring. To take benefit of these advantages, one has to connect the thermocouple free ends with an 'Electric Ice point' and input voltage to an Analog to Digital Converter (ADC). The voltage data can be converted to temperature values using the available from the manufacturer, functions of the relevant type of thermocouple. A thermocouple used for surface temperature measurement is presented in Fig. (**12a**).

There are also portable Digital Thermometers, which are equipped with Cold Junction Compensation, Analog to Digital Converter (ADC), Digital to Analog Converter (DAC) and are calibrated for a specific type of thermocouple. When the

Digital Thermometer is connected to the specific type of thermocouple can provide temperature measurements. Digital Thermometers can also take Temperature measurements according a preset period of time (from seconds to hours), keep registry or send them on line. A Digital Thermometer is presented in Fig. (**12b**).

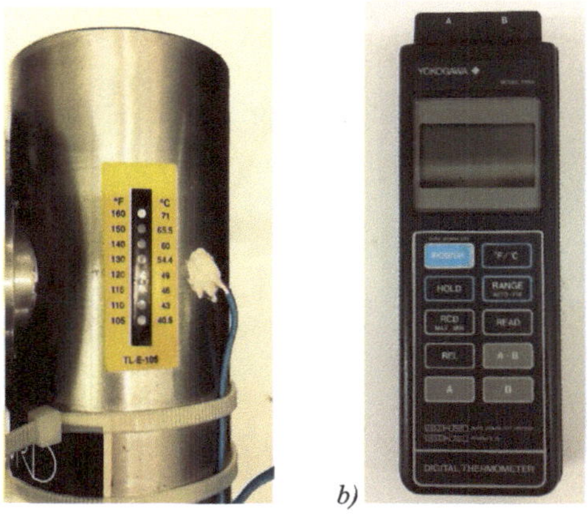

a) *b)*

Fig. (12). a) Thermocouple used for surface temperature measurement b) Digital Thermometer.

The accuracy of temperature measurement using thermocouples depends on their type, the height of temperature and the resolution of the ADC used. In general, the accuracy of Digital Thermometers waves from ±0.5 °C to ±2.5 °C [5, 6].

Electrical Resistance Temperature Detector

The electrical Resistance Temperature Detectors (RTD) are also used for outdoor temperature monitoring but are more expensive than thermocouples. The electrical resistance of the metal wire used (platinum, Nickel) changes as a function of the temperature. Their main advantage is their stability. Platinum has the higher resistivity compared with other noble metals (Au, Ag, Cu) and as a noble metal avoids contamination. The RTD is a more linear device than the thermocouple, but it still requires curve-fitting [7].

SOLAR RADIATION MEASUREMENTS

As far as it concerns solar energy systems ground-based measurements of solar radiation are needed. The Global Horizontal Irradiance falling on to the earth's surface G (from sky + solar disk) consists of the Direct Normal Irradiance D from the sun projected on the surface (multiplied by the cosine of the angle –i- between the solar disk and the normal to the horizontal surface) and the horizontal irradiance D from the sky excluding the sun [8].

$$G = B \cdot \text{Cos}(i) + D$$

For such measurements the following instrumentation can be used:

Thermopile Detectors

Thermopile detectors covered by black coating are used for ground-based measurements of solar radiation. Thermopile detectors consist of a large number of thermocouples serially connected on a silicon substrate [9], (photosensitive area). The black coating absorbs the incoming radiation and generates an output voltage proportional to a local temperature difference or temperature gradient.

Bolometer

In bolometers, the photosensitive area uses a bolometer resistance made up of thermoelectric conversion materials, so the resistance temperature coefficient is the primary cause in determining bolometer sensitivity [9]. Since bolometer sensitivity does not depend on the size of the photosensitive area, detectors can be fabricated that have a small photosensitive area yet no drop in sensitivity.

Pyranometer

Pyranometer is used for the measurement of the Global Horizontal Irradiance. It measures the solar radiation flux density (W/m^2) from the hemisphere above, within a wavelength range 0.3 μm to 3 μm. It consists of a thermopile detector which is protected by one or two glass domes and responds to radiation with a 2-pi hemispherical field of view (Fig. **13**). The domes protect the sensing element and prevent long-wave radiation to enter [10].

Fig. (13). a) Pyranometer [11], b) Pyrheliometer [11].

Pyrheliometer

Pyrheliometer is used for the measurement of the Direct Normal Irradiance from the sun in (W/m^2), (Fig. **13**). It consists of a thermopile type radiometer with a narrow (~5°) field of view and a flat window [8].

RELATIVE HUMIDITY MEASUREMENT

In areas where average relative humidity ranges 40%-78% a minimal layer of water vapor is established at the front solar cell. This way the solar energy is subjected to loss in absorption/reflection causing a decrease of the produced energy by the solar panels [12]. This is why humidity conditions have to be taken into consideration in Solar Energy systems design and monitoring of solar facilities. Humidity measurement is among the most difficult problems in basic meteorology. One of the first devices for relative humidity measurement was Hydroscope, which used a human or animal hair under tension. Hair tends toward retaining moisture and its length changes with humidity.

Nowadays, the following instruments are commonly used:

Psychrometer (Wet-and-Dry-Bulb Thermometer)

The Psychrometer's operation is still based on the construction proposed by R. Assman since 1825. Two thermometers are used: The first one is a classical thermometer (with a dry bulb) while the bulb of the second one is covered by a wet wick (wet bulb). Psychrometer comes with a mechanism for ventilating both thermometers. The evaporation caused by ventilation lowers the temperature of the wet bulb. The temperature difference of the dry bulb thermometer compared with the temperature shown on the wet bulb thermometer, is used as a measure of the relative moisture in the air according a range produced by experimental calibration. Obviously, Psychrometer cannot be used in monitoring systems but it can be used for field measurements and for checking the operation of Hygrometers (see below) [10].

Hygrometer

Hygrometer used for routine monitoring systems is based in the logarithmic increase of electrical resistance of salts, such as Lithium Chloride, as relative humidity in the air decreases. Hygrometer consists of a metal wire impregnated with lithium chloride solution. Relative humidity changes result in changes of the electrical resistance of the metal wire, because of the change of the electrical resistance of Lithium Chloride. Another type of Hygrometer, with a faster response, is based in detecting the change in capacitance of a thin polymer film due to humidity change [9].

TECHNICAL CHARACTERISTICS OF MEASUREMENT INSTRUMENTATION

Essential criteria for the selection of the appropriate instrumentation (for a given application and available budget) are the technical characteristics of the instruments, such as [13 - 16]:

- Range: the maximum and minimum values of the measurand that can be measured.
- Sensitivity: the ratio of incremental change in the output to its incremental change of the measurand in input.
- Resolution: the smallest incremental change in the measurand that will result in a detectable increment in the output signal.
- Accuracy: The maximum difference that will exist between the actual value (measured by a primary or good secondary standard) and the indicated value at the output of the sensor.
- Response Time: The time required for a sensor output to change from its previous state to a final settled value.

A large number (more than 40) of measurements should be provided in order to achieve better accuracy and check repeatability and reproducibility of results. Important digits in the reported results depend on the total uncertainty of the measurement procedure.

Taking into account that most measurements concerning Renewable Energy Sources take place on the field, the instrumentation to be used should respond additionally to the following qualifications:

- The transducer circuit should have overload protection to withstand overloads.
- The transducers output signal should not get affected by environmental disturbances, like extreme temperature, extreme relative humidity *etc.*
- The transducer should withstand most possible mechanical strains.

CONSENT FOR PUBLICATION

Not applicable.

CONFLICT OF INTERESTS

The authors declare no conflict of interests.

ACKNOWLEDGEMENT

Declared none.

REFERENCES

[1] ASME, *Measurement of Fluid Flow in Pipes Using Orifice, Nozzle and Venturi.* ASME: USA, 2004.

[2] S.E. Simopoulos, *Mechanical Measurements.* National Technical University of Athens: Athens, 1989.

(in Greek)

[3] IEC, *EN 60584:Thermocouples - Part 1: EMF specifications and tolerances.* BSI: UK, 2013.

[4] G.W. Burns, M.G. Scroger, G.F. Strouse, M.C. Croarkin, and W.F. Guthrie, *Temperature-Electromotive Force Reference Functions and Tables for the Letter-Designated Thermocouple Types Based on the ITS-90.* NIST: USA, 1993. [http://dx.doi.org/10.6028/NIST.MONO.175]

[5] http://www.omega.com/prodinfo/thermocouples.html

[6] JUMO, *Data Sheet 90.1000.* JUMO GmbH & Co. KG: Germany, 2009.

[7] http://www.omega.com/techref/rtd-measurement-and-theory.html

[8] NREL (US), U.S. D.O.E., D.R. Myers, *Solar Radiation Modelling and Measurements for Renewable Energy Applications: Data and Model (nrel/cp-560-3360)*, International Expert Conference on Mathematical Modelling of Solar Radiation and Daylight: Challenges for the 21st Century, Edinburgh, Scotland, 2003.

[9] http://www.hamamatsu.com

[10] E.P.A. U.S.A., *Meteorological Monitoring Guidance for Regulatory Modelling Applications - EPA-454/R-99-005*, U.S.A. ENVIRONMENTAL PROTECTION AGENCY, February 2000.

[11] http://www.hukseflux.com

[12] Manoj Kumar Panjwani, Dr. Ghous Bukshsh Narejo., Efficiency of Solar Cell (photovoltaic), *International Journal of Engineering Research and General Science* - iJERGS, Vol. 2, no. 4, June-July 2014

[13] K. Kalantar-zadeh, B. Fry, and N-E. Sensors, *Nanotechnology-Enabled Sensors*, Springer Science + Business Media LLC, New York, 2008

[14] ASTM, *Manual on the use of thermocouples in temperature measurements.* ASTM: Baltimore, 1990.

[15] J.E. Hardy, J.O. Hylton, T.E. McKnight, C.J. Remenyik, and F.R. Ruppel, *Flow Measurement Methods and Applications.* John Wiley and Sons, Inc.: USA, 1999.

[16] O. Ernest, *Doebelin, Measurement Systems: Application and Design.* McGraw-Hill, 2004.

SUBJECT INDEX

Emmanuel D. Rogdakis & Irene P. Koronaki (Eds.)

C

H

R

S

www.ingramcontent.com/pod-product-compliance
Lightning Source LLC
LaVergne TN
LVHW070945210926
848482LV00002B/96
* 9 7 8 1 6 8 1 0 8 7 2 0 7 *